BIOTECHNOLOGY OF MICROBES AND SUSTAINABLE UTILIZATION

R.C. Rajak
Professor
Department of Biological Science
Rani Durgavati University Jabalpur (M.P.)

SCIENTIFIC PUBLISHERS (INDIA)
P.O. Box 91 JODHPUR

Published by:
Pawan Kumar
Scientific Publishers (India)
5-A, New Pali Road, P.O. Box 91
Jodhpur - 342 001
Tel.: +91-291-433323 & 624154
Fax.: +91-291-512580
E-mail: scienti@sancharnet.in
artrans@satyam.net.in

ISBN: 81-7233-314-5

Lasertype set: Rajesh Ojha
Printed in India

PREFACE

The diversity of microorganisms is the most precious biological resource on the planet earth today, which plays a pivotal role in the function and maintenance of earth's ecosystem of the biosphere. Microbes perform unique and indispensable activities in the circulation of matter in the world on which all other organisms depend. Microorganisms provide the foundation for biotechnology i.e. the basis for new product discovery, bioremediation and genetic manipulation for new products and processes for food, health-care, environment and industry. During the last few years, microbial biotechnology has made tremendous progress and contributed significantly in the field of Agriculture, Industry, Environment and Health.

Intensive farming system during the last century with the application of chemical fertilizers and pesticides coupled with heavy irrigation has resulted in increased production but also caused depletion of soil organic matter and micronutrients associated with non-recycling of organic manures. With ever-increasing population, it is necessary to continue towards increase in productivity per unit area without further deterioration in soil health with maintenance in micronutrient balance. For this it is peremptory to have an integrated nutrient management approach with inclusion of organic manures, appropriate legumes in cropping system, use of biofertilziers, biopesticides, along with discrete use of chemical fertilizers. Kehri and Chandra in their article on "Organic farming and sustainable agriculture" highlighted that organic farming is a system of agriculture that encourage healthy soils and crops through such practices as nutrient-cycling of organic matter such as compost and crop-residues, crop rotations, proper tillage and the avoidance of synthetic fertilizers and pesticides. Microbial inoculants of live and latent cells of efficient strains of nitrogen fixing and phosphate solubilizing microorganisms augment the extent of availability of nutrients in a form, which can be easily assimilated by plants and play an important role in sustainable agriculture. Similarly biopesticides based on microbes and plants alongwith their application in the management of insect, pests and pathogens of plants, has been an area of intensive research during the last few years. All these important area of research along with other areas such as : Allelopathy, Tissue culture, Aflatoxins etc. have been very well covered by the articles included under Agriculture Biotechnology.

Industrial Biotechnology, the technology that involves commercial exploitation of the biosynthetic capabilities of microorganisms, has been very well dealt in an article by Dr. Abhay Despandey, who presented a very succinct account of strengths, weaknesses, opportunities and threats of commercial biotechnology. An article by Dr. Ganguli highlighted the significance of biodiversity in the discovery of novel secondary metabolites of therapeutic importance. In other articles included under Industrial Biotechnology, Microbes have been shown to be the source of several bioactive compounds which may be of pharmaceutical importance or biosurfactants associated mainly with the problem of substrate transport or enzymes involved in biodegradation of organic substances. An article on *in vitro* production of alkaloid through tissue culture is also included.

In the face of the increasing severity of environmental damage evidenced during this century, there has been a growing awareness among all, that actions are needed to maintain and restore environmental quality. Microbial biotechnology is one of the several competing technologies for the maintenance of environment quality. Articles dealing with environmental biotechnology in this book cover significant areas of research such as Biomining, Bioremediations, Microbial degradation of Petrochemicals and other organic wastes etc. Singh, *et al.*, in an article in Medical Biotechnology highlighted the potential of *Mycobacterium habana*, a Vaccine Candidate for Leprosy.

The articles included in this book have been contributed and presented during the International Symposium, by the scientists of India and abroad, who are actively engaged in the area of research dealing with various aspects of biotechnology. I express my gratitude to all the contributors for their keen interest, support and co-operation. I am also aware that inspite of all care, some of the errors might have appeared in the book for which readers may send their suggestion and comments. I hope the book will be of great help to researchers, teachers and post-graduate students interested in biotechnology in general and microbial biotechnology in particular.

I am extremely grateful to my teachers, Prof. G.P. Agarwal, founder Head, Department of Biological Science, R.D. University Jabalpur and Prof. S.K. Hasija, former Head, Department of Biological Science, R.D. University, Jabalpur, two eminent mycologists of the country, who always inspired and encouraged me. I also thankfully acknowledge the help extended to me by my colleagues of Department : Prof. S.P. Gautam, Dr. P.K. Singhal, Dr. Y.K. Bansal, Dr. Anjana Sharma, Dr. S.N. Bagchi, Dr. A.K. Pandey, Dr. S.S. Sandhu, and Dr. Kamal Prasad etc. I am thankful to M/s. Scientific Publishers, Jodhpur for excellent publication of the book.

R.C. Rajak
Professor
Department of Biological Science,
Rani Durgavati University Jabalpur (M.P.)

CONTENTS

SECTION – 2
INDUSTRIAL BIOTECHNOLOGY

SECTION – 3
ENVIRONMENTAL BIOTECHNOLOGY

SECTION – 4
MEDICAL BIOTECHNOLOGY

SECTION – 1

AGRICULTURE BIOTECHNOLOGY

CHAPTER – 1

ORGANIC FARMING *vs.* SUSTAINABLE PRODUCTION

Harbans Kaur Kehri and Sudhir Chandra

Department of Botany, University of Allahabad, ALLAHABAD - 211 002

1. ORGANIC FARMING

According to IASA (International Alliance for Sustainable Agriculture), 1990, organic farming is a system of agriculture that encourage healthy soils and crops through such practices as nutrient cycling of organic matter such as compost and crop residues, crop rotations, proper tillage and the avoidance of synthetic fertilizers and pesticides.

2. WHY ORGANIC FARMING ?

As a result of the "green revolution" - the successful introduction of new crop varieties and associated fertilizer and pesticide packages - per capita food production in Asia has risen steadily over the past so many years. However, in large parts of the tropics, with less fertile soils, the per capita food production has actually been declining during the past 20 years. Even the environmental problems were created due to over exploitation of land, extension of cropping and deforestation. Because of a very heavy irrigation, some of the large irrigated areas are seriously affected by salinization. Increased use of pesticide and artificial fertilizers is also causing environmental problems. Particu- larly the degradation of soil fertility and the scarcity of fuelwood indicate the graveness of the situation. This has led to the increasing removal of forest and natural grassland cover, loss of biodiversity and degradation of soil through nutrient depletion and erosion.

In several populous countries, including China, India, Indonesia and Mexico, agricultural production stagnated, whereas populations continued to grow. According to a data, growth in world production in 1973 was achieved partly by ploughing highly erodible land and partly by drawing down water tables through over pumping for irrigation.

Farmers are overpumping and overploughing with impressive results in the short run, but the short run is running to a close. The marginal lands brought under the plough during 1970s become exhausted and as irrigated areas shrink because of falling water tables in key food-producing areas, the growth in world food production is slowing down.

Moreover, most of the agrochemicals have clearly shown the adverse effects on soil structure, soil microflora, quality of water, food, fodder and food materials. Traces of banned DDT and HCH isomers have been found in soil, cereals, pulses, vegetable oils, human milk, butter, fish, meat eggs, vegetables, fruits, animal feed and drinking water. W.H.O. has estimated globaly atleast three million people are poisoned by pesticides every year out of which 20,000 die. A majority of pesticides deaths are reported to the occurring in developing countries, which use only 25% of the global pesticides. The past two decades have therefore seen great interest in the development of farming systems characterized by a relatively inexpensive level of input, a high efficiency of internal resource use and hence more sustainable production in both economic and ecological terms.

Under the circumstances, organic farming seems to be an answer for making safe food and environment. It is an alternative to an intensive agriculture where more emphasis is on the use of biological and natural inputs which do not confront with the nature e.g. organic manures, plant origin pesticides, biopesticides, biofertilizers, bioherbicides, resistant varieties and cultural practices.

3. COMPONENTS OF ORGANIC FARMING

Soil is an important component of all the agricultural systems. It is often described as consisting of solid particles, water, gaseous elements, humus and raw organic matter. An extremely important aspect which is often forgotten is that the soil is also the dwelling place for a large variety of living creatures. This soil life includes soil flora (microflora such as bacteria, actinomycetes, fungi and algae) and fauna (microfauna, such as protozoa; mesofauna such as nematodes and collemboles; microfauna, such as beetles, centipedes, millipedes, ants and termites; megafauna, such as earthworms, rodents and moles). These organisms play a major role in many soil processes and soil plant interactions, such as soil formation, creating soil structure, mineralization to free nutrients for plant growth, building up humus, nitrogen fixation, phosphate solubilization and uptake of nutrients by plant roots. There are strong interdependencies between roots and soil life, as roots excrete substances that stimulate soil life, which again release nutrients that are taken up by the roots.

Soil microbial life provides free labour for the farmers, if the right conditions have been created and organic matter with the right composition (C/N balance) has been fed at the right moment. Then soil life can do its work effectively in synchrony with crop production.

4. ORGANIC MATTER

Soil organic matter is an important regulatory of numerous environmental constraints to crop productivity. Mineralization of decomposing residues is a major source of plant nutrients in highly weathered soils with little inherent mineral fertility. The activities of microorganisms and soil fauna serve to promote soil aggregation (16), leading to reduced erosion (12) and greater moisture infilteration (13). Labile carbon compounds, complex toxic aluminium and manganese species, characteristic of many highly weathered tropical soils, resulting in a more productive rooting environment (8,9). Other important benefits resulting from the maintainance of soil organic matter in organic farming include nutrient retention and storage (24), increased buffering capacity in low-activity clay soils (23) and an increase in their otherwise poor water-holding capacity (12). However, a decrease in total soil organic matter as a result of land management and increased soil aeration is an almost universal event in both the temperate regions and the tropics (2). Degradation of the soil system through soil organic matter loss results from soil tillage (7). Nutrients are replaced primarily by chemicals and the farmers no longer attached to high value to manuring, making the soil poor in organic matter and buffered nutrients, and more susceptible to drought and pests. In another words, the productivity and stability of the farm system is decreasing.

Overcoming soil organic matter decline is an important component in the development of more sustainable agroecosystems (19). No doubt, the loss of soil organic matter reduces the nutrient storage and retention capabilities and an inevitable consequence of land management, no cropping system be regarded as sustainable. The need for renewable, higher crop yields because of increasing human population pressures, the expected rise in living standards and the reduction in per capita land holding requires a greater understanding of the cropping systems themselves and a balance between human needs and environmental concerns (22).

Yet in many tropical cropping systems, little or no agricultural residues are returned to the soil. This leads to a decline in soil organic matter (3-4, 12, 17, 24) which frequently results in lower crop yields (12) or plant biomass productivity (24). Short of the often uneconomical and massive application of organic residues and the avoidance of burning during land preparation and few other management options e.g. inclusion of deep-rooted companion crops, trees and symbiotic legumes into traditional cropping systems, coupled with effective residue management, crop rotations, agroforestry and mulches may improve the organic matter in soils, which is central to the issue of "sustainable agriculture".

5. MANAGING ORGANIC MATTER

(a) Manure handling and improvements : Although farmers generally do not dispute the importance of manure to improve soils, many farmers use it primarily for other purposes or handle it inefficiently. Thus a valuable, locally available resource is not used to its fullest advantage for agriculture. There is a need for improved collection, composting, storage and transport of dung and urine to reduce the nutrient losses.

(b) Farmyard Manure : Good quality farmyard manure (FYM) is perhaps the most valuable organic matter applied to a soil. It consists of a mixture of cattle dung, the bedding used in the stable and of any remnants of straw and plant stalks fed to cattle. It improves soil due to its content of principal nutritive elements and its ability to (i) improve the soil tilth and aeration, (ii) increase the water holding capacity, (iii) stimulate the activity of microorganisms that make the plant-food elements in the soil readily available to crops. The supply of organic matter, which is later converted into humus is a property of FYM.

But more than 50% of the cattle dung produced in the country today is burnt as fuel and thus lost to agriculture. Sometimes there is thremedous waste because of faulty methods used in preparing and storing FYM. There is a need to give considerable attention to improve the quality of manure and use of preservatives on manure. The quality of manure may be improved by the concentrated feeds given to cattle. The practice of penning cattle, sheep and goats in the summer may also be encouraged. Folding of 7,000 sheep for one night is said to add the equivalent of 149.3 quintals of cattle dung.

(c) Composting : Another method of augmenting the supplies of organic matter is the preparation of compost from farmhouse and cattle- shed wastes of all types. It is the breakdown of organic materials by microorganisms and soil fauna to give a humus and product called compost. It is an important technique for the recycling organic waste (weeds, crop residues, waste from post-harvest processing, dung, nightsoil, urine etc.) and for improving the quality and quantity of organic fertilizer. Compost is a slow-release organic fertilizer which stimulates soil life and improves soil structure. It also has beneficial effects on the resistance of plants to pests and diseases. Good quality raw material and proper handling are decisive for the quality of the compost obtained. By mixing in mineral additives e.g. rock dust, rock phosphate, urea fertilizer or chalk, the nutrient content of the compost can be improved.

(d) Sewage and sludge : The liquid waste, like sullage and sewage contain large quantities of plant nutrients and may be used for growing vegetables, fodder crops, sugarcane etc.

(e) Night-soil or poudrette : Human excrements are potential source of soil improvement. Public health authorities in many towns make the necessary arrangements for its conservation and conversion into a form in which it can be safely used as a mixture. Dehydration of night-soil as such or after admixture with absorbing materials e.g. soil, ash, charcoal and sawdust, produces a poudrette that can be easily used as a manure.

(f) Green manuring : Despite special efforts at increasing the supplies of FYM and compost, the supply of farm and other organic manures is scarce and ever becoming more costly. Green manuring, wherever feasible, is the principal supplementary means of adding organic matter to the soil. Trees, shrubs, cover crops, grain legumes, grasses, weeds, ferns and algae provide green manure, an inexpensive source of organic fertilizer to built up or maintain soil organic matter and fertility. The cumulative effects of continued used of green manures are important, not only in terms of nitrogen supply but also with regard to soil organic matter and other elements such as phosphate and microelements which are mobilized, concentrated in the top soil and made available for plant growth.

Deep-rooted green manure crops in a rotation can help to recover nutrients leached to the subsoil. Under high rainfall conditions, especially at the start of the wet season, permanent deep-rooted systems, as in some trees, are needed for recycling. Some leguminous cover crops such as *Crotalaria, Pueraria* and *Centrosema* also have deep-root systems.

Green manure crops can be planted in different combinations and configurations in time and space.

* improved fallow i.e. replacing natural fallow vegetation with green manure crops to speed up regeneration of soil fertility and permit permanent cultivation.
* alley cropping, a form of simultaneous fallow in which quickly growing trees, shrubs or grasses are planted in rows and are regularly cut back; the prunings are used as mulch or worked into the soil in the alleys in between the rows,
* integration of trees into cropland, as is found in several traditional farming systems.

* relay fallowing by sowing bush legumes among the food crops after these have established and, in the dry season, using the cut green biomass as mulch or working it into the soil.
* live mulching, in which rows of food crops are sown into a low but dense cover crop, of grasses or legumes e.g. *Centrosoma pubescence*, *Pueraria phaseoloides*, *Arachis prostrata*; strips of cover crop are removed by hand or killed by herbicides when the food crops are to be sown, thus reducing soil tillage operations to zero.
* shaded green manures
* azolla and blue-green algae

6. BIOFERTILIZERS

The crisis of non-renewable sources of energy is catching the attention of every one in the world as the deepening of the crisis is going to affect the life on the earth to a great extent. Efforts are being made, the world over, to tap renewable sources of energy to the maximum extent for the supplementation of energy resources.

Biofertilizers are one of them which are used in agriculture to combat the ill-effects of chemical fertilizers. These are the microbial inoculants of live and latent cells of efficient strains of nitrogen fixing, phosphate solubilizing or cellulolytic microorganisms used for application of seed, soil or composting areas with the objective of increasing the numbers of such microorganisms and accelerate certain microbial process to augment the extent of the availability of nutrients in a form which can be easily assimilated by plants. The term may also be used to include all organic resources for plant growth which are rendered in an available form for plant absorption through microorganisms or microorganisms plant associations or interactions.

Biofertilizers do not form toxic chemicals like N_2O and are available at cheaper rates than the chemical fertilizers.

(a) Nitrogen-fixing bacterial inoculants

For agricultural production there can not be a better technology than utilization of inexaustable source of atmospheric nitrogen which is present to the tune of 75,000 tons over an hactare of land and as such is of no use to the plant. This nitrogen if properly harnessed through biological means can go a long way in saving chemical fertilizers on one hand and to save soil from these chemicals on other hand.

Rhizobium is the most researched and well known among all the biofertilizers. Symbiotic nitrogen fixation by *Rhizobium* with legumes contributes substantially to total BNF. *Rhizobium* inoculation is a well known agronomic practice to ensure an adequate nitrogen nutrition of legumes in place of fertilizer N.

Non-symbiotic nitrogen fixation is mostly dominated by *Azotobacter, Beijerinkia, Azospirillum* and blue green algae. Plant response to inoculations with azotobacters and azospirilla in cereals and non-cereals are often reported in terms of increased grain yield, plant biomass yield, nutrient uptake, grain and tissue N contents, nitrogenase activity, early flowering, tiller numbers, greater plant height, leaf size, increased enzyme levels in plant parts, increased number of spikes and grains per spike, increased root length and volume, reduced insect and disease infestation. The worldwide crop responses to inoculation with azotobacters and azospirilla indicate that in most of the cases increase in yield is statically significant.

Certain algae are also used in field application to improve the nitrogen content in the soil. *Anabaena, Nostoc, Aulosira, Tolypothrix, Scytonema, Plactonema* etc. are used frequently.

Azolla, which is a tiny, floating, fresh water omnipresent fern and which has a *Anabaena azollae* as blue green algae symboint, has a potentiality to fix upto 900 kg N/ha in the soil.

(b) Phosphate solubilizing microorganisms

Phosphorus is one of the most important plant nutrients which is required in optimum amounts for proper growth of plants and soil microorganisms. Most of the tropical soil have inadequate supply of available phosphorus. Phosphorus is taken up by plants generally as primary orthophosphate ion H_3PO_4. Due to chemical fixation of phosphatic fertilizers by soils, the efficiency of utilization of phosphorus is of the order of 10 to 15% only. Therefore, the solubilization of plants is of practical importance. Different groups of bacteria and fungi have been reported to render inorganic phosphatic compounds soluble. The genera of bacteria such as *Pseudomonas*, *Bacillus* of fungi *Aspergillus* and *Panicillium* have been reported to be active in solubilization process of insoluble phosphates such as tricalsium, ferric, aluminium and rock phosphates and bone meal. In earlier work carried out in USSR 5-10% yield increase due to inoculation with *B. megatarium* var. phosphaticum (14). In India, 10 trials showed significant increases in yields in wheat, rice, maize, chickpea, pigeonpea, soybean, groundnut etc.

(c) Plant growth promoting rhizobacteria

A group of rhizosphere bacteria (rhizobacteria) that exert a beneficial effect on plant growth is referred as plant growth promoting rhizobacteria or PGPR (21). PGPR belong to several genera e.g. *Actinoplanes, Agrobacterium, Alcaligenes, Amorphosporangium, Arthrobacter, Azotobacter, Bacillus, Cellulomonas, Enterobacter, Erwinia, Flavobacterium, Pseudomonas, Rhizobium* etc. (25). Currently Pseudomonas spp. are receiving much attention as PGPR. Fluorescent Pseudomonas strains also suppress major plant pathogens. In field trials with wheat, potato, sugarbeet and zinnia conducted on experimental and commercial scale 40 out of 63 trials showed significant increases in yields.

(d) Vesicular-arbuscular mycorrhiza (VAM)

The symbiotic association between plant roots and fungal mycelia is termed as mycorrhiza which is formed by non-septate members of Zygomycotina (15). These fungi are found associated with majority of agricultural crops. They are ubiquitous in geographic distribution occurring with plants growing in arctic, temperate and tropical regions alike. These fungi infect the host and spread inside the root. They posses special structures known as vsicles and arbuscules. The arbuscules help in the transfer of nutrients from the fungus to the root system and the vesicles, which are saclike structures store P as phospholipids, There is little host specificity for VAm but the competitive ability of a given species with native strains may influence the dominance of certain endomycorrhizal fungus in a root system. VAM have been associated 3 with increased plant growth and with enhanced accumulation of plant nutrients mainly P, Zn, Cu and S mainly through greater soil exploration by mycorrhizal hyphae. It has also been suggested that VAM stimulate plant growth by physiological effects other than by enhancement of nutrient uptake or by reducing the severity of diseases caused by soil pathogens (5).

The results of field trials conducted indicated that VAM inoculations increased yields significantly in around 50% trials and the response varied with soil type, soil fertility and VAM cultures.

7. BIOPESTICIDES

In ecologically sound agriculture, ways of controlling a pest outbreak are limited. As natural pesticides are less effective than chemical ones, ecological pest management is based on understanding the life cycles of pests and preventing the build up of excessive pest populations.

A number of natural enemies of the pests, such as birds, spiders, mites, fungi, bacteria, viruses or plants, suppress the pest population and reduce the damages caused by them. Such biological control measures are cheap, efficient, selective and ecologically sound. A specific form of biological control involves the use of microbes mixed with water or another fluid and applied as chemical pesticides. But these have many advantages over chemicals; their effect is generally selective, they do not damage useful organisms and man, and resistance to them is not likely to develop. However, they must be applied more often, because they disintegrate more rapidly.

The best known and most widely used microorganism is the bacterium *Bacillus thuringiensis*. During sporulation it produces a protein which is toxic to most caterpillars. No susceptible species have become resistant to *Bacillus thuringiensis* (6, 11). *Metarhizium, Beauveria, Verticillium, Trichoderma, Agrobacterium radiobacter* etc. also effective in controlling the pest and diseases.

The microbial pesticides produced by Western companies with the aid of high-technology equipment are usually expensive but, in developing countries these can be produced using cheap materials and simple locally available equipments. To control the sapsucking bug *Mahanarva anisopilae*, a fungus produced locally in simple laboratories, using botlles of sterilized rice as a substrate. Likewise, to control the stemboring caterpillar *Ostrinia nubilalis* in China, 0.4 million hectares of maize are treated with locally prepared *Bauveria bassiana*, a fungus produced on steamed but not sterilized rice, bran or corn stalks (18).

8. PLANT-DERIVED PESTICIDES

Using plant-derived pesticides is as age-old control method. Unfortunately this knowledge is rapidly being lost, particularly where chemical pesticides have been introduced. Using natural pesticides is cheap and does not make farmers dependent on external inputs.

Plant products like neem, pongamia, tulsi, sorghum, marigold and bougainvillea have been found effective against a number of diseases and insect-pests. Many neem based formulations like neemark, neemguard, nimbcidin, welgro, agricel, neoconeem and limonol are available in great demand.

Scientific interest in neem is based on its traditional use for pest control. In many Indo-Pakistan households for example, neem leaves are placed between folds of clothing and pages of books to protect against termites, silverfish and other household pests. For the same reason, neem leaves are mixed with stored wheat, rice, maize, sorghum and other grains.

The use of neem cake for controlling nematodes and other soil-borne pests is increasing even among affluent growers of cardamum, citrus, vegetables and sugarcane in India.

In Cameroon, good results against, leaf eating catterpillars, aphids and garden bugs werer obtained with the sprays of *Datura stramonium, Ricinus communis, Lobilia columnaris, Carica papaya* etc. (20).

9. BIOHERBICIDES

Weeds are undesirable plants growing in the field competing with the crop for the nutrients, space and moisture. Biological agents like insects, fungi, bacteria etc. are effective against number of weeds. *Neochitinia eichorniae* and *N. bruchi*, are proving effective against water hyacinth and beetle *Zygogramma biocolorata* is effective against congress grass, *Parthenium*. There are number of fungal and bacterial pathogens parasitizing different weeds. Two mycoherbicides have been registered in USA for commercial use for weed control. One of the formulations of soil-borne fungus is *Phytophthora palmivora* against strangler vine Morrenia odorata and the other formulation of *Colletotrichum gloeosporoides* against selective control of Aeschynomene virginica in rice and soybean.

(a) Cultural Practices

Cultural practices modify and manifulate the environment to the disfavour of pests by disrupting their reproductive cycles, eliminating their food, destroying their weed hosts or making the environment more favourable for predators, parasitoids and antagonists and provide a favourable conditions for the better plant growth. Alley cropping, Relay cropping, intercropping, crop rotation etc. improve the organic matter in the soil while practices like removing crop debris and diseased plants, collection and destruction of egg masses, pruning infested parts, frequent shallow ploughing, flooding, crop rotation, adjusting time of sowing/ planting, avoiding monoculture and soil solarization reduce the incidence and severity of insect-pests diseases and weeds in the field.

10. EPILOGUE

Concept of sustainable agriculture is mainly related to the management and conservation of resource base and orientation of technologies in such a manner as to ensure the continued satisfaction of human needs for present and future generations, which should be non-degrading to the environment, technologically appropriate, economically viable and socially acceptable.

The organic manures, biofertilizers, biopesticides, bioherbicides etc. help in protecting the environment from the deleterious effect of fertilizers and pesticides on one hand and improve the performance of crops on the other hand. All these are helpful in maintaining the ecological functions, such as producing organic matter, nutrient pumping, creating a nutrient reservoir in the matter, nutrient pumping, creating a nutrient reservoir in the soil, natural crop protection and controlling erosion, properly. All these functions contribute to the continuity and stability of farming and they can be seen as producing internal inputs and the system is very much similar to mature natural ecosystems, in which nearly all biomass produced is reinvested to maintain fertility and biotic stability of the system.

REFERENCES

1. Ahmed S. *ILEIA* Newsletter 6 : (1990) 7.
2. Ayodele O J. *Biology and Fertility of Soils* 8 : (1986) 519-525.
3. Bouwman A F. Land use related sources of greenhouse gases : Land use policy, London, U.K., Butterworth (1990a).
4. Bouwman A F. In : Soils and the Greenhouse effect. Chichester, UK, John Wiley (1990b).
5. Dehne H W. *Phytopathology*, 726 : (1982) 1115-1119.

6. Gips T. Breaking the peasticide habit : alternative to 12 hazardous pesticides, Mineeapolis, IASA (1987).

7. Follet R F & Schimel D S. Soil Science Society America J. 53 (1989) 1091-1096.

8. Hargrove W L & Thomas G W. In : Chemistry in the soil Environment, (M. stelly, ed.) Madison, Wisconsin, USA, American Society of Agronomy (1981).

9. Hue N V, Craddock G & Adams F. Soil Science Society of America J. 50 (1986) 28-34.

10. TASA, Planting the future : a resource guide to sustainable agriculture in the Third world. Minneapolis : IASA (1990).

11. Kumar R. Insect pest control with special reference to African agriculture, London, Arnold (1984).

12. Lal R. In : Advances in Soil Science (Volume 5), New York, USA (1986).

13. Lavelle P. Biology and Fertility of Soils 6 (1988) 237-251.

14. Mishustin E N & Naumova A N. Microbiologia 31 : (1962) 543- 578.

15. Morton J B & Benny G L. *Mycotaxon*, 37 (1990) 471-491.

16. Oades J M. Plant and Soil 76 (1984) 319-337.

17. Post W M & Mann L K. In : *Soils and the Greenhouse effect.* Chichester, UK, John Wiley (1990).

18. Prior C. Biological pesticides for low-external-input agriculture, Wallingford, CAB International (1989).

19. Sanchez P A. In : Soils and the Greenhouse effect, New York, USA, John Wiley (1990).

20. Schrimpf B & Dziekan I. ILEIA Newsletter, 6 (1989) 12-13.

21. Schroth M N & Hancock J G. Ann. Rev. Microbiology, 35 (1981) 453-476.

22. Swindale L D. Sustainability the impact of agricultural development on the environment : An IARC point of view. *Proccedings of CGIAR Mid-Term Meeting, Berlin, Germany, Washington DC, USA (1988).*

23. Swift M J & Sanchez P A. Nature and Resources 20 (1984) 1-10.

24. Woomer P L & Ingram J S I. The Biology and Fertility of Soils : TSBF Report, 1990, Nairobi, Kenya, TSBF (1990).

25. Weller D M. *Ann. Rev. Phytopathology*, 26 (1988) 379-407.

CHAPTER – 2

BIOLOGICAL ACTION OF HERBICIDES ON CYANOBACTERIA — AN OVERVIEW

Surendra Singh, Pallavi Datta and Rekha Patel*

Algal Biotechnology Laboratory, Department of Biological Science Rani Durgavati Vishwavidyalaya, Jabalpur - 482 001, India

Cyanobacteria or blue green algae are a remarkable group of prokaryotic, photosynthetic micro-organisms. Their principal mode of nutrition, oxygen-evolving photosynthesis, however is similar to that which operates in all other nucleate or eukaryotic algae and in green plants (7). The vegetative organization in the group ranges from unicellular as in Chroococcales to filamentous as in Oscillatoriales and Nostocales to multiseriates filamentous branched forms as in Stigonematales. The members of Nostocales, exhibit three cell types namely, vegetative cells, heterocysts and spores (akinetes). Heterocysts are the well established sites of aerobic nitrogen fixation while spores, in general, are perennating reproducing units formed under unfavourable growth conditions. All cyanobacteria, although basically photoautotrophic, do not exhibit photoheterotrophic or chemotrophic potential. A large number of cyanobacteria exhibit diazotrophic mode of nutrition and property of diazotrophy is found to be present in unicellular, non heterocystous filamentous and heterocystous filamentous forms. The dual capacity of fixing atmospheric CO_2 and N_2 by the same cell or different cells of same filament of cyanobacteria makes them attractive as source of nitrogenous biofertilizer in wet land rice agriculture independent of fossil fuel (23, 25).

Ecological factors like high temperature, high humidity, alkaline pH and water logging conditions are the reasons for nitrogen- fixing cyanobacteria occurring abundantly in wet land rice-agriculture (20). Sustained productivity of rice field year after year in absence of nitrogenous fertilizer was the emperical reason for assigning a role of diazotrophic cyanobacteria in regulating the nitrogen budget of paddy soil. In fact, this emperical observation has become the guiding principle for present day attempts in large

* Corresponding Author

scale use of N_2-fixing cyanobacteria as biofertilizer in India and some its neighbouring countries (25, 12). *Azolla*, the water fern, play a leading role as biofertilizer in rice field because of its diazotrophic endosymbiont *Anabaena azollae* (19).

In modern agriculture weed control by extensive use of herbicides is a common practice to facilitate tremendous increase in productivity. It appears that use of herbicides in modern agriculture can not be renounced. It is just like necessary evil. The continuous application of herbicides not only adversely affect undesired weeds (targetted species) but also important microbial population of the soil including cyanobacteria (non target species). More than half of the herbicides like Triazine, DCMU, CMU, Paraquat, Propanil, Chloro-fluoron etc. used in agriculture are designed to inhibit photosynthesis of weed plants and others are designed to inhibit various other metabolic biosynthetic pathways.

Cyanobacteria which are very similar to the higher plants in terms of their photosynthetic ability, growth and multiplication invites similar growth inhibitory actions of herbicides. Extreme sensitivity of diazotrophic cyanobacteria to the common rice field herbicides is a major cause of concern for successful exploitation of cyanobacteria as source of nitrogenous biofertilizers. The observed reduction of cyanobacterial strains (native or transient) from such agricultural land which recieves regular heavy treatment of herbicides may lead to complete elimination of cyanobacterial flora over a period of time.

There are various groups of herbicides known to affect different metabolic processes of weed plants. These groups are as follows :

(i) Photosynthesis inhibiting herbicides:

Exp-DCMU, Atrazine, Propanil, Diquat and Paraquat.

(ii) Inhibitors of Aromatic amino acid biosynthesis :

Exp - Glyphosate (Round up, Glycel etc), Sulfonyl urea.

(iii) Inhibitors of glutamine synthetase (GS) enzyme :

Exp - Phosphinothricin (PPT).

(iv) Inhibitors of lipid biosynthesis :

Exp - 1, 3, cyclohexanedione (CHD), Aryloxy phenoxy propionate (ADPP).

(v) Inhibitors of growth promoter harmone metabolism (Auxin type) :

Exp-2,4-D, MCPA.

(vi) Mitotic Inhibitors :

Exp - Dinitroanilines, Trifluralin, Ben-fluralin and Carbamates like Propham and Barban.

(vii) Inhibitors of protein Synthesis :

Exp - Alachlor and Butachlor.

There are reports of cyanobacteria showing natural tolerance to a variety of herbicides (4,11) affecting several different biochemical pathways.

Glyphosate is a potent inhibitor of enzyme-3-enol-pyruvylshikimate- 5-phos-phate synthetase, which is involved in the biosynthesis of aromatic amino acids, phenyl alanine, tyrosine and tryptophan (6, 22). *Synechocystis 6803* and *Anabaena variabilis* showed tolerance towards the free acid of glyphosate, with little or no inhibition of growth even at 20 mM compared with total inhibition of growth by 5 mM free acid found for other bacteria (13, 3, 10). These two species showed some tolerance to the isopropyl amine salt of slyphosate although this compound was more toxic than the free acid. Round up was most toxic formulation though it was far less toxic towards *Synechocystis 6803* than towards *A. variabilis*. The metabolic pathway inhibited by glyphosate is common to bacteria, cyanobacteria and higher plants, therefore it would be interesting to examine impact of glyphosate on different cyanobacterial species to understand the biology of glyphosate actions.

Phenoxyacetic herbicides have auxin like activity and are used to control the growth of weeds with many crops including rice. The exact mode of action of phenoxyacetic herbicides on biological system is not known. The fast response of photosynthetic oxygen evolution to toxic levels of 2,4-D suggest that the primary action is on photosynthesis. Most studies concerning the effect of auxinic herbicide on cyanobacteria has been done with 2, 4-D (2, 4-Dichlorophenoxy acetic acid). Most reports indicate that cyanobacteria can readily tolerate upto 1mM of 2, 4-D. The lethal level generally appears with herbicide concentration ranging between 4-5 mM (8).

Sodium salt of 2, 4-D (8% active ingredient) commonly applied for the control of aquatic weeds, was used to observe its effect on the growth and nitrogen fixation of a heterocystous bloom forming blue-green alga *Anabaenopsis raciborskii* (1). The alga could tolerate upto 800 ugml- in liquid culture media with and without nitrate and upto 900 ugml- on to agar plates. Nitrogen fixation was inhibited in presence of its higher concentration.

In contrast with the extensive literature regarding the interaction of 2, 4-D with cyanobacteria little work has been done on the effect of MCPA which is the most commonly used phenoxy acetic herbicide in Spanish rice fields.

The herbicide monuron is an inhibitor of growth and heterocysts differentiation. The biological effect of the herbicide was seen on *Nostoc muscorum* under different carbon and nitrogen supplemented conditions (24). The monuron supplemented cultures of *N. muscorum* exhibited a very meagre growth in both N_2 and NO_3 media, and failed to grow with 3.4×10^{-1} mM or higher concentrations. In addition, monuron significantly inhibited the heterocysts frequency of the organisms in N_2-medium. The effects of monuron are similar to those of DCMU on *N. muscorum* (14). DCMU (Diuron) has widely been used in photosynthetic studies. There are great variations in the toxicity of different photosynthesis inhibiting herbicides towards some cyanobacteria (16, 17).

DCMU inhibit electron transport in the region of PS-II. Its application to oxygenic photosynthetic organisms is known to result in abolition of photochemically generated reducing power ($NADPH_2$) without causing any adverse effect on the generation of ATP through cyclic photophosphorylation. It has been reported that DCMU-inhibition of both growth and heterocyst differentiation required a photosynthetically fixed carbon supply and that glucose can effectively substitute for photosynthetically generated organic carbon in growth and differentiation (14).

Machete (Butachlor) and Lasso (Alachlor) are both pre-emergence amide herbicide extensively used in rice-agriculture for prevention of emergence and growth of weeds. The mutagenic action of the above herbicides have been examined on various strains of *Nostoc muscorum* and found to be highly mutagenic (15).

There are very little information of naturally occuring cyanobacterial strains showing resistance to double or multiple rice-field herbicides. For the first time a natural isolate of aerobic dizotrophic *Gloeocapsa* sp. from the wet land rice field conditions in the presence of rice field herbicides Machete and Basalin was found highly resistant than *N. muscorum* under photoautotrophic nitrogen fixing conditions (18). Successful biotechnology involving the use of cyanobacteria as biofertilizer in modern rice agriculture would require them to have the additional property of resistance to various rice-field herbicides.

The effects of four pre-emergence herbicides Basalin, Butachlor, Oxyfluorfen and Pendimethalin on *N. commune* and *N. muscorum* were studied (9). Except oxyfluorfen, below normal (50%) concentrations of herbicide had a significant effect on biological activity of these cyanobacteria. The order of herbicide toxicity was oxyfluorfen Pendimethalin Hiltaklor Basalin. *N. muscorum* was more tolerant to herbicides than *N. commune.*

Arozin is a totally new rice field weedicide which is based on Anilofos and belongs to organophosphorus group of compounds. Its action on cyanobacterial flora is little known (5, 21) which needs further investigation.

It is clear from on going discussion that most of the studies on the interaction between herbicides and cyanobacteria were mainly focussed on growth, nitrogen fixation, photosynthesis or mutation (26, 2). However, little information is available on the effect of herbicides on other biochemical processes in cyanobacteria. Such a knowledge may be valuable (i) in elucidating the mode of action of particular herbicides and (ii) for protecting the ecologically important group of cyanobacteria from toxic action of these chemicals by inducing tolerance/ resistance in them. Unfortunately, no sincere effort has been made so far to see the herbicide degrading properties of cyanobacteria. These aspect invites attention of cyanobacteriologist for multidirectional exploitation of cyanobacteria under biotechnological programmes.

ACKNOWLEDGEMENTS

We are thankful to the Head, Department of Biological Science for facilities and U.G.C. New Delhi for financial assistance.

REFERENCES

1. Das B, Singh PK. *Arch. Environ. Contain. Toxicol.* 5 : (1977) 437-45.
2. Das B, Singh PK. *Microbios. Lett.*, 5 : (1978) 103-7.
3. Fischer RS, Berry A, Gaines, CG Jensen, RA. *J. Bacteriol.* 168 (1986) : 1147-54.
4. Friedberg D, Seijffers J. *Arch. Microbiol.* 150 (1988) : 278-81.
5. Goyal D, Roy Choudhary P, Kaushik BD. Acta Bot. Ind. 19 (1991) : 25-8.
6. Jaworskii EG, *J. Agric. Food Chem.*, 20 : (1972) : 1195- 98.
7. Kratz WA, Myers J. *Am. J. Bot.*, 41 : (1955) 282-87.
8. Leganes F, Fernandez E. *Arch. Environ. Contain. Toxicol.* 22 : (1992) 130-34.
9. Likhitkar VS, Tarar JL. *Ann. Pl. Physiol.* 10 : (1996) 74-77.
10. Pipke R, Amerin N, Jacob GS, Schaffer J, Kishore GM. *Eu.J.Biochem* 165 : (1987) 267-73.
11. Powell HA, Kerby NW, Powell P. *New Phytol.* 119 : (1991) 421-26.
12. Roger PA, Kusassooriya SA. IN *International Rice Research Institute*, pp. 112. (1982) Los Bonas : Philippines.
13. Schulz A, Sost D, Amrhein N. *Arch. Microbiol.* 139 : (1984) 121-23.
14. Singh HN, Vaishampayan A, *Environ. Exp. Bot.*, 18 : (1978) 87.

15. Singh HN, Vaishampayan A. *Environ. Exp. Bot.* 19 : (1979) 6-12.
16. Singh J, Tiwari DN. *J. App. Bacteriol.* 64 : (1988) 365- 76.
17. Singh J, Tiwari DN, *Pes. Biochem. Physiol.* 31 : (1988) 120-28.
18. Singh LJ, Tiwari DN, Singh HN. *J. Gen. App. Microbiol.* 32 : (1986) 81-8.
19. Singh PK, *In National Symposium on increasing rice field in Kharif.* CRRI. Cuttack (1978).
20. Singh RN. *In Indian Council of Agriculture Research,* PP. 17-9. New Delhi (1961).
21. Singh S, Thakur IPS, Kohli K. *All India Botanical conference,* University of Jammu, Jammu (1995).
22. Steinrucken HC, Arnhein N. *Biochem. Biophys. Res. Commun.,* 94 : (1980) 1207-12.
23. Stewart WDP, Rowell P, Kerby NW, Reed RH, Mechray GC. *Phil Trans. R. Soc. Lond.* 317 : (1987) 245-58.
24. Vaishampayan A. *Microbios Lett.* 28 : (1985) 105-111.
25. Venkataraman GS. *In Today and Tomorrow's Printers and Publishers,* PP. 75. (1972) Faridabad, Haryana
26. Venkataraman GS. *In Nitrogen fixation by Free living Microorgranisms.* PP. 207-48. (1975) Cambridge University Press.

CHAPTER – 3

RESEARCH PRIORITIES IN RELATION TO PATHOGENS POSING PROBLEM IN FORESTRY PRACTICES

Jamaluddin

Tropical Forest Research Institute, Jabalpur (M.P.), India

1. INTRODUCTION

National Forestry Policy, 1998 emphasizes on increasing forestry productivity for restoring ecological balance and conservation of country's natural heritage and biodiversity on one hand and to meet the demands for timber, fuel, fibre, paper and pulp, and fodder on the other. To increase the forest productivity, losses due to various abiotic and biotic factors must not cross the acceptable limits. Therefore, development and protection of forests, one of the world's greatest renewable naturel resources, for maximum productive and protective functions, are the prime objectives of forestry. Protection of forests and forest produce from diseases constitutes one such objectives.

Diseases are known to be the most destructive agents of planting stock in forest nurseries and consequently the plantation yield directly affect the forest productivity. However, unlike agricultural crops, the looses due to diseases are not so apparent in plantations and natural forests till partial or total death of trees occurs or the expected growth rate is not achieved. Therefore, evaluation of the importance of tree diseases is often based on appraisal of the lethal effects of diseases.

The out break of an epidemic disease, depends on the presence of a large number of susceptible hosts, an aggressive pathogen and favourable environmental conditions. In forest nurseries; a continuous crop of planting material either in beds or containers aided with inappropriate management practices such as over- watering, over crowding, shifting injuries to root system, use of same bed area year after year for same plant species, etc. create such prerequisites for development of epidemics.

In plantation, monocultures provide a uniform crop of hosts and in case a virulent pathogen is present, epidemics occur during conducive environmental conditions. In case of exotics either the indigenous pathogens or the pathogens produced with the hosts may cause epidemics under congenial climatic conditions.

2. DISEASES OF FOREST SEEDS

High quality seeds of different promising tree species are required for raising quality planting materials. Quality of seeds is deteriorated due to diseases which not only account for seed abortion but also affect the germinability of seeds and survival of seedlings. In India, unlike agriculture much attention has not been paid to adopt suitable and scientific seed technology to improve the quality of seeds and seedling production. Seeds carry numerous potential microbes as contaminants which incite diseases in forest nurseries or serve as primary inocula for various seedling diseases. Such potentially dangerous pathogens are transmitted to the field along with infected seedlings used as planting material and pose several practical problems. Failure of seedlings in forest nurseries at the pre-and post-emergence stage, are mainly attributed to the use of poor quality seeds which carry harmful spermoplane microflora. Recently Jamaluddin (12) has reviewed the work on diseases of tropical tree seeds. Besides reducing the germinability and viability, microflora also damage the seed contents during storage. The problem becomes acute particularly for those seeds which are exploited in seed industries for oil and other edible purposes.

Among the various microorganisms, fungi are the most important, affecting the seeds even at their early developmental stages. Fungi infect the radical, plumule, embryo, cotyledons and seed coat during the development and maturation stage, and reduce seed yield both qualitatively and quantitatively. The quality is lowered due to discolouration, seed abortion, production of shrunken seeds, rotting of seeds, necrosis and reduced germination. Seed rot is most common in deteriorating the seed contents and some toxins like aflatoxins are also produced. *Aspergillus flavus, A. terreus, Fusarium* spp. *Penicillium* spp. are well known seed borne fungi producing toxins.

Most of the tropical forest seeds become infected by pathogens in natural stands, plantations and seed orchards. Jamaluddin *et al.*, (15) have reported infection in pods of *Pongamia pinnata* which is transmitted to the seeds. The seeds of *Albizia lebbek* were also found infected by *Fusarium pallidoroseum* in the pods during maturation, the fungus affected the seed production by 20-30 per cent. A large number of tropical tree seeds are found infested with several fungi belonging to diverse taxon.

Majority of fungal pathogens associated with seeds cause seed- rot, pre-or post-emergence damping-off, root-rot, wilt and some foliage diseases in the nurseries. Seed-rot is often associated with *Aspergillus* infection, while root diseases such as damping-off, root-rot and vascular wilts are caused by seed borne infections of *Fusarium, Macrophomina, Phytophthora, Pythium, Rhizoctonia, Sclerotium* and *Verticillium* spp.

Viral diseases damage the seeds by developing discolouration, causing shrinkage and reduction in seed size. Seedling of such infected seeds develop yellowing, whitening, little leaves, stunting and mosaic symptoms, Bacterial pathogens, particularly species of *Pseudomonas, Xanthomonas*, etc. cause seed rot and foor-rot and pose problems in germination and viability.

3. DISEASES OF FOREST NURSERIES

The forest nurseries are the producer agencies of planting stock required for various plantation programmes. Several biotic and abiotic factors create problems in developing healthy planting stock. Fungal diseases, namely foliar, stem and root diseases, play important role in the forest nurseries by affecting the overall health of the planting material. Some important contributions on nursery diseases in the region have been made (14, 20.

4. ROOT AND COLLAR ROT DISEASES

The health of the root system is perhaps the most important factor in the total health of plant. A large part of the root system is generally killed before any obvious symptoms appear. The root of seedlings especially soon after germination of seeds are very delicate and underdeveloped and prone to infections. Most of the soil-borne pathogens are responsible for causing wilt, root-rot and damping-off in nurseries (26). The important root diseases are being mentioned below :

Pre-and post-emergence damping-off diseases are quite common in forest nurseries in moist as well as dry climates which sometimes have been reported to the extent of 70 per cent mortality due to post-emergence damping-off. The pathogens commonly associated with pre-emergence damping-off are species of *Aspergillus, Fusarium, Penicillium, Rhizopus* etc.

The fungi known to cause post-emergence damping-off are species of *Fusarium, Phytophthora, Pythium, Rhizoctonia* and *Verticillium*, all of which have wide host range. While all tree species are susceptible, *Eucalyptus* and *Pinus* spp. suffer more from these diseases than others. Among other species in which damping-off has been reported are *Acacia* spp., *Albizia* spp., *Azadirachta indica, Cassia* spp., *Dalbergia* spp., *Delonix regia, Dendrocalamus strictus, Holoptelia integrifolia, Lagerstroemia* spp., *Leucaena leucocephala, Oroxylum indicum, Pongamia pinnata, Tectona grandis*, etc.

Collar rot is another common disease in forest nurseries under moist climatic conditions in which young seedlings are attacked at collar region resulting into toppling over of seedlings. *Rhizoctonia solani* and *Sclerotium rolfsii* are common collar rot causing fungi in forest nurseries. The most affected hosts are *Ailanthus triphysa, Bombax ceiba, Eucalyptus* spp.

Wilt disease, like damping-off, is also very common in forest nurseries particularly in dry climates in central India. *Fusarium* spp., *Macrophomina phaseolina, Rhizoctonia solani* state of *Thanetephorus cucumaris* and *Verticillium* spp. are mostly responsible for these diseases while tree species are susceptible when congenial conditions for infection exists.

Set rot of poplar cuttings is a serious problem in north Indian states of U.P. Haryana, Himachal Pradesh and Punjab (22, 23) which adversely affects the raising of poplars in nurseries. The disease is caused by *Botryodiplodia palmarum* and damage has been reported more than 50 per cent under ideal disease conditions. This disease is also observed under shade house condition at Jabalpur affecting as much as 40 per cent cuttings.

5. FOLIAR DISEASES

Foliar diseases are considered significant in affecting the productivity of quality planting stock but the information is lacking about the actual damage they cause. However, in forest nurseries the foliar diseases may pose serious problems to seedling health by causing premature defoliation apart from checking or reducing photosynthetic activity of the leaves thus ultimately affecting the nutrition of the plant and also increasing the respiration activity and altering overall physiology of the plants. The infection of foliar parts may results in weak vigour and stunted growth of planting material. The common diseases found in nurseries are leaf spots, leaf blight, blotch, shot holes, leaf rust, powdery mildews, etc.

6. DISEASES OF PLANTATIONS

Plantation are being intensively managed to increases productivity by reducing genetic variation, eliminating competing vegetation, maintaining optimum stand density, and other silviculture operations. In general the disease risks are higher in pure stands than that of mixed stands. Some of the following diseases of plantations need priority attention.

7. ROOT DISEASES

Root rots and vascular wilts are two important root diseases encountered in most tree plantations.

The pathogens decay root tissue by attacking both lignin and cellulose and affected trees loose their foothold in the soil which are generally uprooted during high wind velocity. Disease spread is through root to root contact between diseased and healthy plants and rhizomorphs.

Root rot caused by *Ganoderma lucidum* is most common in tree plantation of broad leaves species and a wide host range is affected in tropical dry climate (3). The root rot mortality is severe in reforested stands not cleared from residual roots and stumps. The affected hosts are *Acacia* spp., *Albizia* spp., *Azadirachta indica, Boswellia serrata, Cassia* spp., *Casuarina equisetifolia, Cocos nucifera, Dalbergia* spp., *Delonix regia, Eucalyptus* spp., *Jacaranda mimosaefo, Leucaena leucocephala, Mangifera indica, Melia azadirachta, Peltophorum ferruginium, Pinus roxburghii, Pongamia pinnata, Populus* spp. *Pterocarpum marsupium, Tectona grandis, Terminalia* spp. etc.

Brown root rot occurs in *Populus* spp. having injuries to root system and tree trunks particularly through agroforestry operations (29). Up to 12 percent mortality has been reported due to this disease in poplars.

Root rot occurs in *Tectona grandis* plantations through basal injuries (3). Root shoot cuttings also get infected at nursery stage which in turn transfer the inoculum to the plantation sites and root and butt rot is established in *T. grandis* plantations.

Jamaluddin *et al.*, (16) studied charcoal root-rot in tropical pines in India and also in hardwood species (26). High intensity of charcoal root-rot in Casuarina by *Macrophomina phaseolina* has been recorded in nursery and plantations in Orissa by Jamaluddin *et al.*, (19). The infection occurred during dry period. Moisture stress in the site is one of the most favourable predisposing factors for this disease. The plant start dying from top to downwards.

Amylosporus campbelli is the potential pathogen causing serious root rot to plantations of different species. It has already been reported on *Dendrocalamus strictus* (31), *Casuarina equisetifolia* and *Parkia javanica* (personal observations). In moist and warm climates of peninsular India the Pathogen may pose serious threat to plantations.

Wilt disease of *Dalbergia sissoo* caused by soil inhabiting vascular parasite *Fusarium solani* is the only vascular wilt disease which has been reported from northern parts of India (2). Recently Mehrotra and Pandey (24) reported Verticillium wilt in *Delonix regia* assuming serious proportions in Punjab.

8. STEM DISEASES

Several canker diseases occur in tree plantations.

Gummosis and canker of *Leucaena leucocephala* are caused by *Fusarium pallidoroseum* throughout the country. Besides causing gummosis and stem canker, the pathogen also attack seeds, seedlings and leaves.

Botryodiploida

Cankers occur in basal regions of species of *Eucalyptus* and *Populus* Bark bursts, sun scorch and heat injuries are main centres for establishment of pathogen. Botryodiplodia canker usually remain localized but in severe infection, girdling occurs at infected basal part resulting into mortality.

Cryphonectria cubensis and *C. gyrosa* cause stem cankers in *Eucalyptus* spp. in Kerala which have been reported up to 50 per cent. Mortality of about 10 percent plants due to cankers has been witnessed in high rainfall areas.

Valsa eucalypti and *V. eucalypticola* cause extensive juvenile twig and branch cankers in *Eucalyptus* spp. in Kerala resulting in die-back of about 30 percent plants.

Hendersonula toruloidea cause serious stem and twig canker in *Anacardium occid-entale, Cassia siamea, Gmelina arborea, Leucaena leucocephala, Eucalyptus camal-dulensis*, etc.

The only stem wilt recorded in the country is that of *Casuarina equisetifolia* caused by *Trichosporium vesiculosum*, which establishes through lopping injuries to stem. The cambium is attacked and killed by the pathogen which progresses the sapwood, blocking vessels and causing wilt. In advanced stages of the disease the pathogen attacks roots and further spreads through root contact. The disease has been reported to cause up to 90 per cent mortality in Tamil Nadu and Orissa (13, 17, 30).

Heart rots have been attributed as the greatest single cause of damage to forest crops, which destroy the most valuable heart wood in living trees and render it unfit for use. Some common heart rot are of *Acacia catechu* by *Phellinus badius*, of *Azadirachta indica* by *Phellinus caryophylli*, of *Pongamia pinnata* and poplars by *Perenniporia tephopora* of *Shorea robusta* by *Hymenochaete rubiginosa* and *Phellinus caryophylli* in trees of old age.

Bamboos have emerged as the most important non-wood forest produce because of their fast growth, adaptability, and myriads of end uses. Heavy mortality in *Bambusa nutans* plantations in coastal areas of Orissa due to blight disease of culms by *Sarcocladium oryzae* has been reported (18). The disease cause heavy losses to the farmers in coastal belt of Orissa. The disease can be easily controlled by spraying fungicides during July or after onset of monsoon. Rot of emerging culms is caused by *Fusarium moniliforme*.

9. DISEASES OF HIGH FORESTS

In high forests, diseases generally remain in an endemic state. However, due to continuous disturbances in natural ecosystem, diseases may erupt in epidemic proportions.

Wide spread mortality of dry sal forest in eastern Uttar Pradesh, Bihar and Madhya Pradesh have been reported (6, 27). The cause have been identified as deteriorating site conditions the most important single factor being lessened amount of soil moisture. Pathogens are found to cause secondary invasion.

Root rot caused by *Aurificaria shoreae* occurs in severe form in wet sal forests where annual rainfal is 2000 mm and above. The disease caused extensive damage to the crop in West Bengal and Assam as compared to dry sal forests in M.P. (11).

Among the different causes responsible for loss to the standing crop in forests, decay is the most important as it causes heavy loss of timber. Sal has the reputation of being one of the most naturally durable timbers in service and is not easily subjected to decay. However, some heart rot fungi cause decay of economic significance such as, *Hymenochaete rubiginosa, Phellinus caryophylli, P. fastuosus, Deadalea cubensis, D. sulcata* and *Coriolopsis sprucei.* An appraisal of loss of timber due to decay carried out by Bakshi and Singh (5) revealed that out of 704 trees examined 70 percent showed heart rot and total loss of timber was about 10%. However, extensive surveys of sal forest in different states are needed to estimate the quantum of loss of timber due to heart rot in the country.

In high sal forest, the incidence and volume of decay may be high. In derelict coppice sal in parts of eastern India, a high incidence of decay in the stool wilt limit any attempt to raise rotation age beyond 15-20 years when decay in the stool progress in the trees (4).

Heart rot in dry teak forests in western part of Gujrat state is a serious problem incurring heavy loss of timber. An assessment of heart rot revealed about 50 per cent incidence resulting in considerable loss of valuable timber. The fungi associated with heart rot are *Perenniporia tephropora* and *Rigidoporus lineatus.* Large scale dying in coppice teak has been noticed from many places of M.P. (28). The dry coppice teak forests in M.P. are worked under coppice system on a rotation of 30 to 50 years. At the end of rotation or before, tree commonly exhibit unsoundness (heart rot) at the base and this often extends up to varying heights (2-4 meters) in the stem.

10. MANAGEMENT OF DISEASES

In order to minimize the losses the management of diseases can be attempted through chemical, biological, cultural and silvicultural measures. To restore the ecological balance, being gradually disturbed by continuous unmanaged pesticidal control operations of diseases, the need of biological control is being greatly felt. Certain innocuous fungi, bacteria, amoebae and mycorrhizae can be used to manage some destructive pathogens. There is a need to gather information based on understanding and consideration of biological, economical, social and environmental processes which provide the basis for integrated disease management as part of over all forest management.

(a) Chemical control

The diseases of seed-borne nature and in forest nurseries are effectively controlled through use of fungicides. The chemicals are also effective in controlling the diseases in plantations and seed orchards.

In a disease control operation, correct recognition of pathogens and use of environmentally safe fungicides with minimum effective doses are essential. Many diseases such as damping-off caused by species of Cylindrocladium, Fusarium, Pythium, Rhizoctonia, etc. are controlled through fungicides. A large number of fungicides are used as seed treatment (seed dressing), seed bed treatment, soil drenching or foliar application. Chemical seed treatment is probably the cheapest and most effective control measure of seed borne pathogens. Some of the commonly used fungicides in nurseries are Bavistin, Captan, Dithane M-45, Dithane Z-78, Fytolan and Sulfex, etc. Harsh and Gupta (9) have dealt in detail with the control of root-diseases in forest nurseries.

The number and frequency of application of fungicides should be related to biology of the pathogen being controlled, prevailing weather conditions and management practices followed in nurseries. Generally a second follow up application is safe.

The chemical treatment for control of diseases in plantation can be adopted if it is economically feasible and the losses due to pathogens have crossed the tolerance level. In seed orchards and small plantations systematic and non-systemic fungicides can be used for control of diseases. Some fungicides used as frequent sprays are adapted by the pathogens and the diseases remain uncontrolled, in such cases two chemically different group of fungicides may be used as alternate sprays.

(b) Biological control

Certain antagonists like *Trichoderma* spp., *Gliocladium* spp. and actinomycetes are applied for controlling the pre-and post-emergence damping-off and root-rot diseases of seedlings. Important diseases are caused by species of *Corticium*, *Rhizoctonia* and *Gliocladium virens*. Use of Rhizobium and VAM fungi is effective in boosting up the growth of seedlings and protecting the seedlings from the infection by soil pathogens.

Unlike agriculture very little information is available on the biological control of diseases in forest nurseries and plantations. A review by Sharma and Shankaran (1988) on biocontrol of rust and leaf spot diseases indicated that microbial control has considerable potential against plantation pathogens. More recently Agarwal (1) reviewed the information on biocontrol of pathogens. Some mycorrhizal fungi are of practical importance to nursery management as they can protect root against pathogens like *Fusarium*, *Phytophthora*, *Pythium*, *Rhizoctonia* and *Sclerotium* spp. (25). Considerable attention has also been paid to the role of vesicular-arbuscular mycorrhizae in disease control. Jalali (10) found that mycorrhizal inoculation results in significant reduction in host infection by *Rhizoctonia solani*. Recently Harsh & Gupta (9) have successfully controlled the post-emergence damping-off disease of *Moringa pterygosperma* caused by *Fusarium acuminatum* by using *Trichoderma harzianum* treated wheat straw and also by VAM culture. Application of antagonists and hyperparasites like *Cladosporium*, *Darluca*, *Trichothecium*, *Tuberculina* and *Verticillium*, can control the soil or foliar pathogens.

(c) Culture and Silvicultural Measures

There are some indirect measures which can be attempted to minimize the losses from diseases.

I. The healthy seeds should be selected for raising the seedlings in nurseries.

II. The seeds produced in dry areas have comparatively less diseases than those produced in warm humid areas.

III. The control of diseases of nursery seedlings is an integral part of nursery management practices. Harsh (7) has mentioned that low sowing density, avoidance of over-crowding of seedlings and excess watering in seed beds and polypots can minimize the incidence of several root diseases in nurseries.

IV. The quality seedlings free from infection should be used to raise plantations free from diseases.

V. The pathogens develop serious hazards in close density plantations. Mixed planting with resistant species is effective in minimizing the losses.

VI. Injury to the stand crops through silviculture operations should be avoided and lopping and thinning wounds should be treated with dressing paints.

VII. Seedlings fortified with mycorrhizae or Rhizobia in nurseries should be used for planting as they encourage the growth and check the pathogens.

VIII. Proper understanding of epidemiological parameters of diseases in nurseries and plantations has considerable implications on control strategies of diseases. Therefore, proper and timely surveillance by technical personnel will help in monitoring the disease situation and use of control measures.

(d) Integrated management of diseases

Losses caused by many forest diseases can be effectively minimized through management practices that promote tree and stand vigour under different site and stand conditions.

The major objectives in disease management are prevention of losses and reduction of cost of control operations. The strategies for integrated disease management include avoiding, excluding and eradication the pathogen, protecting the host from infection, developing resistant hosts and killing the pathogens in infected hosts. The disease causing organisms interfere with the management practices and involve risks or uncertainty in decision making. Integration and manipulation of all the available control measures to get a healthy crop of plants by checking the disease below the economic threshold level at an effective cost is, therefore, required for over all management of tree species.

REFERENCES

1. Agarwal, G.P. 1990. Biological plant protection recent development. Presidential address, 78th Indian Science Congress, Bot. Soc. Indore.
2. Bakshi, B.K. 1954. Nature, 174, 278.
3. Bakshi, B.K. 1976. Forest pathology Principles and practices in forestry FRI & Colleges, Dehradun.
4. Bakshi, B.K. 1979. Some concepts in forest disease management. Paper presented at Seminar on forest Production at FRI, Dehradun.
5. Bakshi, B.K. and Singh, Sujan 1970. Int. Review for. Res. 3 : 197-251.
6. Boyce, J.S. and Bakshi, B.K. 1959. Indian forester, 85 : 585- 588.
7. Harsh, N.S.K. 1991. Factors involved in disease incidence and development in forest nursery. Reg. Meeting silvi., Bhubaneshwar 4 pp.
8. Harsh, N.S.K., Rai B.K. and Rai, Harbinder 1994. Jour. Temp. For. 10 : 319-321.
9. Harsh, N.S.K. and Gupta B.N. 1993. TFRI Inf. Series No. 1. pp 31.
10. Jalali B.K. 1989. Mycorrhiza News. TERI New Delhi.
11. Jamaluddin 1991. Jour. of Trop. For 7 : 342-344.
12. Jamaluddin 1995. *Proc. Seed Technology workshop.* Institute of forest Genetic & true breeding coimbatore, 19-22 July, pp 22.
13. Jamaluddin 1998. Indian forester 124, 1 : 59-61.
14. Jamaluddin, Dadual, V.S. and Soni, K.K. 1981. Indian forester 107 : 500-504.
15. Jamaluddin, Dadual, V.S. and Soni, K.K. 1983. Ind. Jour of for. 6 : 287-288.
16. Jamaluddin, Dadual, V.S. and Soni K.K. 1984. Indian forester, 110 6 : 552-57.
17. Jamaluddin, Dadual, V.S. Namdeo, R.K. and Soni, K.K. 1989. *Proc. in National Seminar Casuarinas*, Neyrchi. 18-19 Dec.

18. Jamaluddin, Gupta B.N., Bohindar, S.C. and Dadual, V.S. 1992. *Jour. Trop. For.* 8 : 252-61.

19. Jamaluddin, Gupta, B.N. and Bohindar, S.C. 1993. Regional Meeting of Silviculture and Research Workes of Central Region, Bhubaneshwar.

20. Jamaluddin, Verma, R.K. and Nath, V. 1994. Indian J. Trop. Biol. 2 : 466-468.

21. Jamaluddin and Dadual, V.S. 1989. Charcoal root rot in forest Tree species seminar on forest Prot. FRI, Dehradun June 29-30.

22. Khara, H.S. 1989. Trees disease problem in punjab. Seminar on forest Production at FRI, Dehradun.

23. Malhotra, M.D. 1993. Some destructive nursery diseases and their management IUFRO symp. On impact of diseases & insect pest in tropical Forest. KFRI, Peechi.

24. Mehrotra, M.D. and Pandey, P.C. 1992. In Tropical forestry in India by D.N. Trivedi, Int. Book. Distributor Dehradun.

25. Mosse, B. 1973. Ann. Rev. Phytopath. 11. 171-191.

26. Nath. V. and Jamaluddin. 1993. Indian Jour. Trop. Biod. 1 : 154-159.

27. Prasad, R. and Jamaluddin 1985. Indian Forester 111 : 250- 271.

28. Prasad, R. and Jamaluddin 1989. *J. Trop. For.* 5, 11 : 72-75.

29. Singh, Sujan and Pandey, P.C. 1989. Indian Forester 681-689.

30. Soni, K.K. Shrivastava, K.K. Kalyani, K.B. and Raja, R.R. 1989. Proc. National Seminar on Casuarinas Neyrchi 18-19 Dec.

31. Tahir, M. Soni, K.K. and Jamaluddin 1992. Myforest 28 : 88- 98.

CHAPTER - 4

Mycorrhizae IN ECOSYSTEMS : AN ECOFRIENDLY APPROACH FOR IMPROVED PLANT GROWTH

*S.K. Dwivedi and P. Dwivedi**

School of Environmental Sciences, B.B. Ambedkar University, Lucknow - 226 025, INDIA

ABSTRACT

Mycorrhizae enhance plant growth through increased nutrient uptake, stress tolerance, disease protection, nutrient cycling and biological diversity. VAM fungi result in protection of the root system against the root pathogen due to improved phosphorus uptake. Mycorrhizosphere soil influences the plant growth. The colonization of roots by VAM fungi favours nodulation by *Rhizobium*. The interaction between phosphate solubilizing bacteria and VAM increased the plant growth. VAM has significant effect on host physiology and biological interactions in the rhizosphere. Population of pathogenic nematodes is also affected by VAM fungi.

1. INTRODUCTION

In recent years the use of chemicals as fertilizers and pesticides is being practised to increase the agricultural production. But these chemicals have adverse effects on the environment and the whole biota including ground water and food stuffs too. Keeping this in view several countries have banned the use of such non-biodegradable chemicals including some pesticides, herbicides, insecticides etc. Under the circumstances the use of chemical fertilizers have also been minimized to some extent, and an alternative method is needed for the control of plant diseases, and to increase the crop yield. To save the ecosystem and to increase the crop productivity green manure is being encouraged. The present article aims to review the impact of mycorrhizal fungi on improvement of plant growth in relation to nutrient uptake and to control pathogenic forms associated with the crops which cause severe damage to the crop as well as on the crop yield.

*Department of Botany, Arunachal University, Rone Hills, Itanagar - 791 111

Mycorrhiza (fungus root) is symbiotic, non-pathogenic or weakly pathogenic association of fungi and the roots of plants. All higher plants are mycorrhizal including monocotyledons and dicotyledons, conifers, and some pteridophytes and bryophytes. The ectotrophic mycorrhizas are known as characteristic mycorrhizas of the forest trees represented by several basidiomycetous fungi viz. species of *Boletus, Cortinarius, Russula, Tricholoma* etc. in which the fungus is on the surface of roots, sometimes forming a Hartig net. The endotrophic forms are exemplified with orchids and plants of Ericaceae in which the fungus invades the roots where it is confired to distinct layers. In the ectendomycorrhiza there is a Hartig net but the fungus also may penetrate deep in the roots. The other types are the pseudo in which the fungus is often parasitic, the vesicular-arbuscular (VAM) in which penetrating hyphae reach the cortical cells and are coiled and branched therein. The arbuscular mycorrhiza is a common type of symbiosis in at least 80% of higher plant genera including the cultivated ones. The cultivation of fungi causing arbuscular mycorrhiza has been impossible so far in axenic culture, otherwise their application for mycorrhizal association with roots on large scale for improvement of plant growth by artificial inoculation would have been easier to achieve the great benefit.

The influence of certain mycorrhizal fungus is such that certain plants depend for growth on them e.g. Orchids, the seeds of some plants germinate only when right fungus is associated with them. In forest trees mycorrhizas are formed with a number of different fungi but in some cases the association is specific e.g. strains of the *Rhizoctonia* - like fungus with certain orchids, *Armillaria mellea* (vahl ex Fr.) Kummer with a Japanese orchid (*Gastrodia elata*) (1). However, it is difficult to present a brief definition of mycorrhiza because the surface and environment of roots harbour populations of a variety of microorganisms, most of which behave as commensal than symbiotic with the roots. The availability of nutrients in the soil environment around roots vary due to activities of mycorrhizal fungi and antibiosis prevailing on the root surface. The fungus and the host co-exist in a physiologically, ecologically and reproductively active state for long periods- the state to be known as mutualistic symbiosis.

The majority of higher green plant and large numbers of fungal species are involved in mycorrhizal associations. Harley and Harley (2) have reported that in the British flora, 80% of the angiosperms species, 100% of the gymnosperms and 70% of Pteridophytes were potentially mycorrhizal. The fungal species forming mycorrhizas belong to all the larger taxonomic divisions. About eight genera and more than a hundred species of Endogonaceae belonging to Zygomycetes are mycorrhizal. At least twenty genera comprising several hundred species of Ascomycetes, particularly belonging to Pezizales, some to Helotiales and Elaphomycetales, form ectomycorrhiza in angiospermous and gymnospermous trees; at least one species of *Hymenoscyphus* (Helotiales) forms mycorrhiza with plants of Ericaceae. Of the Basidiomycetes, 5000 or more species of many genera of Hymenomycetes and Gastromycetes are involved in ecto- and ectendo- mycorrhizas of angiosperms and gymnosperms. Among Deuteromycetous fungi, many sterile ones, which may be Ascomycetous and Basidiomycetous in perfect-states, form ectomycorrhizal association e.g. *Cemococcum geophilum* - very common and widespread on many hosts; and *Oidiodendron griseum* forming ericoid mycorrhiza (3, 4). On the basis of informations gathered through researches on mycorrhizas, a large number of angiosperms, gymnosperms, pteridophytes, many bryophytes and a few species of algae associate with about one-quarter of the known species of fungi in non-pathogenic or mutualistic symbioses. In addition, lichens show fungal associations of a similar kind with algae - the fungi growing in association with algae are about 300 genera and about 13,000 species, the genera of algae are very few-perhaps less than 10.

With regard to the concept of mutualism, De Bary (5) defined it as parasitism in which parasite and host mutually and permanently live and support one another. According to Harley (6), the definition of De Bary which indicates simply a give - and - take between symbionts is not as objectionable as the derived term 'mutual benefit' for the same relationship. The benefit can be judged to the host or to fungus by determining the selective advantage to them of any traits or characters that can be measured and their way of life, and their symbiosis seem to have given them a selective advantage i.e. a survival value. There appears to be ecological specificity in vesicular - arbuscular mycorrhizal association which can be demonstrated in glass house by inoculation of plants with mycorrhizal fungus/fungi.

2. NUTRIENT SIGNIFICANCE IN MYCORRHIZAL ASSOCIATIONS

According to Frank (7) mycorrhizal fungi are supplied with carbon compounds by their hosts and that their hyphae, which penetrate the soil, especially into the humus, help to uptake nutrients into both the partners of the symbionts. In forest, soils are deficient in nitrate although ammonium and organic nitrogen compounds are present for supply of nitrogen. The significance of mycorrhizas in the absorption of compounds of nitrogen, phosphorus and potassium into the symbiotic systems has been proved by many experiments of early work (8). The significance of hyphae as the extension of the absorbing surface has been emphasized in these experiments. The first direct demonstration of the significance of the fungal hyphae in absorption and translocation of nutrient material to the host was made by Melin and Nilsson (9) who reported by growing pine seedlings in axenic conditions in which isotopically labelled nutrients, phosphate, ammonium, nitrate, glutamate and cations (sodium and calcium) were fed to the hyphae and they were absorbed and traced through the hyphae to the host upto the leaves. In similar experiments $^{14}CO_2$ fed to the leaves was traced to the hyphae in photosynthetic product which were translocated to them. The absorption of different substances by mycorrhizas from the soil has been further studied and demonstrated using tracers and other modern techniques in low concentrations as expected to be used in soil. It was discovered that the environmental factors affecting absorption of nutrients are similar to those affecting the absorption mechanisms of plants viz., temperature, concentration, availability of carbon substrate, oxygen supply and metabolic inhibitor (c.f. (6)).

Mycorrhizas possess a large number of hyphal connections with the soil in form of branching hyphae and strands that ramify profusely in soil substrate. In some cases profuse branching and ramification is absent and hyphae are either smooth or provided with spine like structures on their surface. The quality of fungal tissue involved in the mycorrhizas for absorption of nutrients in different kinds of mycorrhiza has been reported in Ericaceae (10), vesicular arbuscular type (11-14), and in case of Ectomycorrhiza (15-17, 9). As to the plant nutrition, carbohydrates are transported from the host to the fungus and nitrogen, phosphate, potassium and calcium are absorbed by hyphae of fungi and transported to the host plant (18). Adenosine - Triphosphatase is present in both host and fungus membrane and is active in plasma membranes of the host (19). Acid phosphatase activity in the cytoplasm of senescent hyphae in the mantle and in the Hartig-net has been recorded (20). Ho (21) has reported the increased nitrate reductase activity in forest trees associated with mycorrhizal fungi.

The work of Melin on the carbon nutrition of mycorrhizal fungi, particularly movement of photosynthates to the fungus, was supported by Nelson (22) They reported a change of pattern of translocation in mycorrhizal as compared with non-mycorrhizal pine seedlings. In twelve hours after photosynthesis of $^{14}CO_2$ had started, 54% of the labelled products had been translocated to the roots of mycorrhizal plants. Further, using excised mycorrhizas of *Fagus*, it was possible to feed the host tissue with [^{14}C] sucrose and showed that it was translocated to the fungus tissue where it accumulated in quantity greater than the value remained in the host. It was converted into the fungal carbohydrates in the fungus in the form of mannitol, trehalose and glycogen that are not readily absorbed by the root tissues due to their insolubility in water. The movement of carbohydrate from host to fungus in vesicular - asbuscular mycorrhiza has been estimated and the process appears to be less active than into ecto-or ericoid mycorrhiza (23).

The phosphate is available to mycorrhizas from inorganic ortho-phosphate present in the soil. Sometimes phosphatases hydrolyze phytates which constitute a large fraction of the phosphate of humic soils which are present in ecto-and ericoid mycorrhizas. These enzymes are more active than those on non-mycorrhizal roots are reported by many workers (24-26). The absorption of phosphate in ectomycorrhizas has depicted that a large quantity of phosphate absorbed from low concentrations, some 80-90% accumulates in the fungus but 10-20% passes on into the host (c.f. (6)). The work on phosphate absorption by mycorrhizas has been done by many other workers (27-29).

The researches on absorption of nitrogen compounds by mycorrhizas have been done similarly by many workers (30-35). It has been reported that eutrophic soils where nitrate is present, mycorrhizal plants do not adsorb nitrogen readily compared to non-mycorrhizal plants - particularly in case of vesicular - arbuscular plants which are rich in such soils. They absorb ammonium and organic nitrogen readily than the other kind of nitrogenous compounds. Ericoid mycorrhizas and ectotrophic mycorrhizas hydrolyze the insoluble nitrogen compounds of humus. The nitrogenous compounds are absorbed in hyphae and then translocated to the host. Smith *et al.*, (33) have reported glutamate dehydrogenase and glutamine synthetase in mycorrhizas. The nitrogen uptake and phosphate metabolism in mycorrhizas go hand - in - hand. Further the nitrogen fixation in leguminous plants is favoured by mycorrhizae infection of roots (35, 36).

The different kinds of mycorrhizas are based on the particular characters of their own-hence they are classified in different categories. The ectomycorrhizas are well suited to the habitat where nitrates are deficient and they absorb phosphates and ammonium compounds from soil. They are provided with strong fungal sheath and produce large fruit bodies of Basidiomycetes which need high demand of carbon. Thus they are found with larger trees in forests which produce enough carbon compounds during photosynthesis.

The strong fungal sheath harbours good amount of nutrients and carbon compounds. The VAM are well suited in eutrophic soil and fungi associated absorb phosphate compounds. They need less carbon nutrients and are generally confined to the smaller plants of herbaceous nature. Ericoid mycorrhizas are suited to peat and humus soils which contain phenolic compound and are acidic in nature. In addition to carbon, phosphorus and nitrogen compounds the role of mycorrhizal fungi in plant nutrition by absorption of compounds of zinc, molybdenum, copper and potassium is much advantageous to forest trees and crop plants. Enhanced absorption of potassium sulphate, zinc and strontium by some plants has been emphasized (37); calcium, magnesium, iron and potassium compounds were absorbed in larger quantities by mycorrhizal plants than by non-mycorrhizal ones (38). All these kinds of mycorrhizas are beneficial in different ecosystems and can be used as bioinoculants for enhanced growth of plants. They can be utilized in waste lands to convert them in fertile land, particularly the uncultivable land. The soil erosion in India is a great problem which degrade the fertile top soil with certain level of depth due to run-off with rain water. It affects the crop productivity too. For reclamation of waste lands, bioinoculants need encouragement in the form of Blue-green algae, symbiotic and non-symbiotic microorganisms, Actinorrhiza, mycorrhizas etc. Improvement in tree seedlings due to mycorrhizas has been observed in forest trees, particularly pine trees in the countries where development of forestry is being practised viz., Canada, USA, Nigera, Congo, Brazil, Mexico, Phillipines etc (39). Similarly, the reclamation of wastelands has gained attention of workers in the relevant subject during seventies by vesicular - arbuscular fungi and ecto-mycorrhizal fungi in Pennsylvania and Scotland (40, 41). The practice of inoculation of mycorrhizal fungi with roots of plant seedlings of forest trees opens a wide scope in India for development of forestry on strong foot. There is, scope for extensive work on utilization of mycorrhizae and other microorganisms as bioinoculants for improvement of plants growth and conversion of barren lands into fertile land to combat deficiency of crop production. Their usage can be beneficial against chemical fertilizers that are costly to farmers and sometimes show serious harmful effects to crops. Since VAM have proved to be efficient biofertilizers and at the same time ecofriendly, several techniques have been developed in recent years for their mass production. Notable amongst them are pot culture, aeropoic culture, roto organ culture, etc. (42).

3. EARLY WORK IN INDIA ON *Mycorrhizae*

In India the work on mycorrhizas commenced in early sixties and the knowledge on symbiotic relationship between fungi and roots of higher plants was then scaty. During early stage of the work on mycorrhizae, ectomycorrhizas attracted attention of mycologists in this country with different forest trees. The research on mycorrhizas involves a number of disciplines in plant science viz., mycology, plant pathology, microbiology, physiology, ecology and silviculture among other related subjects. The earlier work in India started in forest soil and trees with emphasis on the role of mycorrhizas in forest development. The ecto-mycorrhiza forming fungi were isolated from roots and cultured in artificial medium. The work was later extended to the nurseries bed and experiments were carried out to study the role of different nursery management practices like shading of nursery beds, use of biocides and nutrients on the development of mycorrhiza in nursery seedling stock. The endomycorrhiza has been less studied compared to ectomycorrhiza. Many fungi forming ectomycorrhiza belong to Basidiomycetes and in India several of them are recognised as *Amanita, Boletinus, Boletus, Armillaria, Cantharellus, Clitopilus, Clitocybe, Cortinarius, Entolome, Lactarius, Lepiota, Paxilhus, Russula, Polyporus* etc. - all belonging to Hymenomycetes, while *Rhizopogon* and *Scleroderma* among Gasteromycetes. The identification of these fungi is based on sporophores.

Different techniques have been employed to isolate mycorrhizal fungi in pure culture but VAM fungi could not be grown for profuse growth on culture medium. A detailed account of mycorrhizal fungi and their role in forestry has been given bvy Bakshi (43) in this country. The VAM fungi isolated from the forest ecosystem in India include several genera and species such as Glomus (*G. macrocarpus, G. macrocarpus, var. geosporus, G. mosseae), Sclerocystis (S. Coremioides), Gigaspora (G. calospora), Acaulospora Endogone (E. gigantea, E. microcarpa)* etc. The detailed taxonomic descriptions of the fungi forming ecto-mycorrhiza in forest soils and trees, the seasonal distribution of the some of VAM genera and their distribution pattern in soil have also been given by Bakshi (43).

There is much scope of the work on mycorrhizal fungi in this country in different ecosystems particularly in forest and cropland ecosystems, and their role in improvement of plant growth.

4. BENEFICIAL ASPECTS OF Ecto-*Mycorrhizae*

(a) Enhanced fertility of soil

Uptake of nitrogen, phosphorus, potassium, aluminium, iron etc. is enhanced. The surface area of root system is increased due to mantle formation by hyphae around roots by ectomycorrhizal fungi; some other aspects viz., encouragement of other microfloras around and on roots for improved mineralization, enzyme like phosphatase activity is enhanced, production of siderophores which chelate nutrients around root system and making the iron available to the mycorrhizal fungus and to the host plants, production of certain acids which assist in weathering of minerals, the enhanced production of external hyphae in soil and thus widening of the absorption surface of roots, decomposition of litter due to mass accumulation of hyphae around roots and in soil which provide released energy to the plants and these are of much significance.

(b) Development of resistance in plants against diseases

These fungi develop resistance in plants against root diseases by protecting root system from invasion of pathogens or by production of toxic substances in the form of hyphal excretions. The mass of hyphae around feeding roots establish antibiosis in the soil environment due to which pathogen die and roots are protected from attack. There is possibility of development of competition amongst microorganisms in soil around roots for nutrients and well developed hyphae around roots compete and succeed in damaging other microbes adjacent to roots including pathogens. Krishna and Bagyaraj (44) reported the decline of severity of disease caused by *Sclerotium rolfsii* due to *Glomus fasciculatum.* Garrett (45) has emphasized that mycorrhizal association, in addition to symbiotic relationship, serves as defence of host against pathogens and so was the presumption of Schenck (46) who opined that mycorrhizal fungi offer an alternative method for biological control of diseases. Zak (47) proposed some mechanisms through which ecto- mycorrhizal fungi can exercise protection of plants from diseases viz, by utilizing surplus carbohydrates in the roots thereby declining the amount of chemicals stimulatory to pathogen(s) causing physical barrier by forming hyphal mantle, secreting chemicals that may be inhibitory to pathogens - supporting the microbial population in the rhizosphere. Toxic chemicals are secreted by mycorrhizal fungi infecting Orchid tubers viz., Orchinol, Coumarin, phenolic compounds in response to the infection of *Rhizoctonia repens* (48). VAM may limit nematode activity and improve plant growth free of diseases (49).

Many leguminous plants in nature establish both nitrogen fixing microbes and mycorrhizal symbiotic system. The nitrogen fixing ability is enhanced due to mycorrhizal association of roots in some cases. There are reports on VAM association as well as *Rhizobium* association with roots of leguminous plants, particularly with mycorrhizal fungi viz., *Gigaspora margarita, Scutellospora pessica, Glomus fasciculatum, Sclerocystis clavispora* etc. (50). The effect of insertion of protoplasts of nitrogen - fixing bacteria into mycorrhizal fungi for improved nitrogen fixation as pointed by Giles and Whitehead (51) need encouragement towards this kind of fruitful studies. Thus there is need for mass production of ectomycorrhizal fungi in culture for commercial formulation for availability to forest tree growers. The attempt in this direction is being made by Abbott and USDA laboratories for such purpose with certain ectomycorrhizal fungi (52). Similarly, protoplast fusion with soil culture microorganisms which are harmless to plant growth may be employed in an attempt to produce a large amount of VAM fungi, as evinced by a few findings (53, 54).

The fruitful attempt to widen the studies on arbuscular- mycorrhizae would be to study the formation of arbuscular-mycorrhizas by artificial root inoculum in case sufficient inocula of such fungi in axenic culture is prepared. The inclusion of the work on effect of bacteria on the mycorrhizal infection would amplify the significance of dual inocula in the improved growth of plants.

ACKNOWLEDGMENTS

The authors thank Prof. R.S. Dwivedi for his suggestions and criticism during the preparation of this article.

REFERENCES

1. Ainsworth GC, Ainsworth & Bisby's dictionary of the fungi (5th Edition), CMI, Kew, Surrey, (1963), pp. 519 + Plates XIV.
2. Harley JL & Harley EL, New Phytol (Supplement), 105 (1987).
3. Couture M, Fortin JA & Dalp Y, New Phytol, 95 (1983) 375.
4. Burgeff H, Mikrobiologie des Hochmoores Stuttgart, Gustav Fischer Verlag, (1961).
5. De Bary; Comparative morphology and biology of the fungi, Mycetozoa and Bacteria (Eng. Trans. by Garnsey HEF & Rev. Balfour JB), Clarendon Press, Oxford, (1887).
6. Harley JL, Mycol Res., 92 (2) (1989) 129.
7. Frank AB, Bericht der deutschen botanischen Gesellschaft, 3 (1885) 128.
8. Hatch AB, *Black Rock Forest Bulletin*, 6 (1937) 1.
9. Melin E & Nilsson H, *Botaniska Notiser*, 111 (1958) 251.

10. Read DJ & Stribley DP, In : *Endomycorrhizas, (Sanders FE, Mosse B & Tinker PB eds.)*, Acad, Press London & New York, (1975) 105.
11. Hepper CM, *Soil Biol. & Biochem*, 9 (1977) 15.
12. Tinker PB, *Physiologie vegetale*, 16 (1978) 743.
13. Bethenfalvay GJ, Brown MS & Pacovsky RS, *New Phytol*, 90 (1982) 537.
14. Tisdale JM & Oades JM, *J. Soil Res.*, 17 (1979) 429.
15. Harley JL & McCready CC, *New Phytol.*, 51 (1952a) 56.
16. Harley JL & McCready CC, *New Phytol.*, 51 (1952b) 342.
17. Brierley JK, *D. Phil thesis*, Oxford University, (1953).
18. Soderstrom B. Finlay RD & Read DJ, *New Phytol*, 109 (1988) 163.
19. Gianinazzi S, Dexheimer J, Gianinazzi - Pearson V & Marx C, *Plant and Soil*, 71 (1983) 211.
20. Dexheimer J, Aubert - Duffresne MP, Gerrd J, Tacon F Le & Mousain D, Letters Botaniques, 113 (1986) 343.
21. Ho I, Can J Forest Res., 17 (1986) 31.
22. Nelson CD, In : *Formation of wood in Forest trees, (Zimmerman MH ed.)*, New York, Maria Moors Cabot Foundation, (1964) 235.
23. Bevege DL, Bowen GD & Skinner MF, In : *Endomycorrhizas, (Sanders FE, Mosse B & Tinker PB eds.)*, Acad. Press, London & New York, (1975) 149.
24. Bartlett EM & Lewis DH, *Soil Biol. & Biochem*, 5 (1973) 249.
25. Williamson B & Alexander L, *Soil Biol & Biochem*, 7 (1975) 195.
26. Mitchell DT & Read DJ, *Trans. Brit. Mycol. Socl.*, 76 (1981) 255.
27. Ling Lee M, Chilvers GA & Ashford AE, *New Phytol*, 75 (1975) 551.
28. Strullu DG, Grellier B, Garree JP, McCready CC & Harley JL, *New Phytol*, 103 (1986) 403.
29. Cox GC, Moran KJ, Sanders FE, Nockolds C & Tinker PB, *New Phytol.*, 84 (1980) 649.
30. Read DJ & Stribley DP, Nature, London, 244 (1973) 81.
31. Stribley DP & Read DJ, *New Phytol.*, 77 (1976) 63.
32. Stribley DP & Read DJ, *New Phytol.*, 86 (1980) 365.
33. Smith SE, ST John BJ, Smith FA & Nicholas DJD, *New Phytol.*, 99 (1985) 211.
34. France RC & Reid CP, *Can, J. Bot.*, 61 (1983) 964.
35. Asai T, *Jap. J. Bot.*, 13 (1944) 463.
36. Subramanian L & Dwivedi RS, In : *Mycorrhiza Round Table, IDRC* Canada, (1987) 417.
37. Gilmore AE, *J. Amer. Soc. Hort. Sci.*, 96 (1971) 35.
38. Routien JB & Dawson RF, *Amer. J. Bot.*, 30 (1943) 440.
39. Marx DH & Ruehle JL, In : *Proc. 1st Asian Conf. on Mycorrhizae, Madras Univ.*, Madras (1988) 336.
40. Daft MJ & Hacskaylo E, *J. Appl. Ecol.*, 13 (1977) 523.

41. Daft MJ & Nicolson TH, *New Phytol.*, 73 (1974), 1129.

42. Sahay NS, Sudha, Singh A & Varma A, *Ind. J. Exptl. Biol.*, 36(1998) 1069.

43. Bakshi BK, *Mycorrhizae and its role in forestry. PL 480 Project report.* F.RI and College, Dehra Dun, (1974) 89.

44. Krishna KR & Bagyaraj DJ, *Can. J. Bot.*, 61 (1983) 2349.

45. Garrett SD, *Biology of root infecting fungi*, Cambridge Univ. Press, Cambridge, (1960).

46. Schenck NC, *Plant Dis.*, 65 (1981) 20.

47. Zak B, *Annual Rev. Phytopathol*, 2 (1964) 377.

48. Gaumann E, *C.R. Acad Sci.*, 250 (1960) 1944.

49. Hussey RA & Roncadori RW, *Plant, Dis.*, 66 (1982) 9.

50. Dela C, Manalo RE, Aggansen MQ & Tambalo NS, *Plant and Soil*, 108 (1988) 111.

51. Giles KL & Whitehead HCM, *Plant and soil*, 48 (1977) 143.

52. Sharma S, Madan M & Vasudevan P, In : *Proc. Nat. Acad. Sci. India*, LXV, Sec B (IV) (1995) 345.

53. Barret V, Lemke PA & Dixon RK, *Appl. Micro. Biotech.*, 30 (1989) 381.

54. Kropp BR & Fortin JA, *Can. J. Bot.*, 64 (1986) 1224.

CHAPTER – 5

EXPLOITING *Mycorrhiza* FOR OILSEED CROP PRODUCTION

Anil Prakash and Vandana Tandon

Institute of Microbiology and Biotechnology, Barkatullah University, Bhopal 462 026

1. INTRODUCTION

Mycorrhiza is the term commonly used to denote the symbiotic association between plant roots and fungal mycelia. By now it is well established that uptake of phosphorus in plants is improved by VAM fungal associations (7). Allen *et al.*, (1) reported the increase of cytokinins and chlorophyll by VAM fungal associations. Oilseeds are an important group of agricultural crops in terms of quantity and value. India occupies a prominent place among the oilseed produced countries owing to the large variety of oilseed crops grown. With the exception of palm, palm kernel, olive and tung oil, India produces all the other principal seeds which form the basis of the seed crushing industry. Considerable information on increase in yield of oilseed crops utilizing chemical fertilizers is available but the literature regarding use of alternative strategies for sustainable production of oilseed crops is very meagre and dispersed. The focus of this contribution is to appraise the role of mycorrhizal fungi in oilseed production.

2. OILSEED IN INDIA

Oilseed constitute an important group of agricultural crops and are grown in an area of about 23 mha in India. Fat and edible oils are mostly derived from oilseed crops such as groundnut, palm, soybean, sunflower, sesamum, coconut etc. (30). The main oilseed crops in order of importance are, groundnut, rapeseed and mustard, seasme, linseed and castor. Recently crops, like sunflower, safflower, soybean, coconut etc. are also gaining importance as oil yielding crops. "The major oilseed producing states in the country are Karnataka, Gujarat, Maharashtra, Andhra Pradesh, Rajasthan, Orissa, Uttar Pradesh and Tamil Nadu. Oilseed crops are generally grown under rainfed areas (80%) and in soils deficient in nutrients in contrast to cereals which are grown largely under irrigated conditions. Therefore, the yield of oilseed crops are relatively lower and depend much upon the rainfall for their successful cultivation. Despite the steady progress achieved in oilseed production in our country, oils have to be imported every year. Oil import reached a maximum of some 18.2 lakh tonnes at a cost of Rs. 1061 crores in 1988. The burden of importing oils was partly reduced by increasing the production of soybean and sunflower during the year 1989-90 and consequently, only about 6.1 lakh tonnes of oil valued at Rs. 328 crores had to be imported during the year 1989-90 (19). Madhya Pradesh popularly known as the soybean state also contributes to India's total produce of other oilseeds like Groundnut, Sesame, Niger, Sunflower, Rape/Mustard, Linseed and Safflower. An overview of the different district of M.P. engaged in oilseed cultivation indicate that majority of area is put under soybean cultivation and there exists a wide gap in production of soybean and other oilseed in this state. However M.P. contributes 20% and 22% of total area and production respectively of total oiseeds in India. In the past years the area put under oilseed cultivation in M.P. has been increasing but the production and yield has not increased in that proportion (Fig. 1). The reason for this stagnation in productivity might be the non-provision of plant nutrient inputs in proportion to the crop demands. With a few exceptions, nearly 70% of the samples collected from different districts of M.P. and tested in soil testing laboratory at JNKVV, Jabalpur, were low to medium in available P. Thus P is the most limiting plant nutrient threatening the sustainability of crop production in the soils (29).

3. NEED FOR SUSTAINABLE AGRICULTURAL PRACTICES

Field level studies giving different fertilizer treatments and combinations for increased yield and oil content have been conducted satisfactorily for most oilseed crops. Chemical fertilizers are used to boost

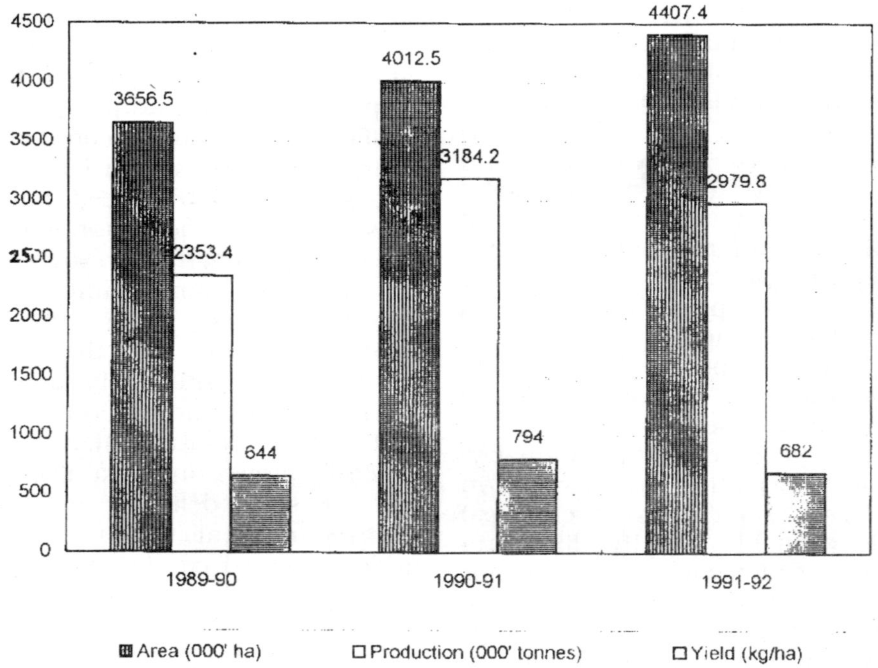

Fig. 1. Area, Production and Yield of Oilseed crops in Madhya Pradesh (1989-90 to 1991-92)

oilseed production, but these fertilizers usually supply only the major nutrients and some times even impede the uptake of other nutrients, thereby upsetting the entire mineral balance. Although the use of chemical fertilizers has immediate beneficiary effect on the plant and soil health, in the long term the chemical fertilizers have a negative impact on soil. These effects necessitate the development of sustainable management practices. One approach for sustainable plant production system is the decrease in chemical inputs to economic but non polluting levels through the development of alternative strategies to ensure increased productivity. Biological routes of improving soil fertility are operated by microorganisms who either synthesize plant usable forms of nutrients or increase the availability and root accessibility of nutrients already present in the soil as in case of phosphorus.

Utilization of various microorganisms, especially VAM fungi as biofertilizer has the potential to increase oilseed production. Recent researches on plant nutrition and vesicular arbuscular mycorrhizal (VAM) associations indicate that these fungi play a major role in nutrient acquisition from soils deficient in phosphorus and other elements as well as better water uptake under stress (30). Fungi forming vesicular arbuscular mycorrhiza are of increased interest due to the widespread distribution of these fungi in the soils of various climatic zones and their ability to colonize major agricultural crops thereby enhancing plant productivity and improving their mineral nutrition especially phosphorus (14, 18, 22). The feasibility of using VAM fungi to reduce the need for fertilizer application, particularly phosphorus has been examined by Menge (20).

Mycorrhizal infection enhances plant growth by increasing nutrient uptake *via* increase in the absorbing surface area, by mobilizing sparingly available nutrient sources or by excretion of chelating compounds or ectoenzymes leading to active absorption of soil phosphates (18). VA mycorrhizal hyphae are also known to transport nitrogen towards the host plant roots over distances of several centimeters in quantities that are sufficient to improve plant growth significantly (11). There are many results on effects of infection on concentration and amount of K in shoots, these results are however inconsistent and difficult to interpret. Experiments involving soybean inoculation with *Glomus mosseae* revealed a relation to improved K rather than P nutrition. External hyphae of VAM also show the capacity to uptake and transport Ca to host plant though the rates of this transport are very low compared to P uptake. Numerous reports indicate enhancement of Cu and Zn uptake by VAM infection (18).

Preliminary studies have been made on the association of VAM fungi with oilseed crops such as soybean (2), groundnut (25), Sunflower (32), Safflower (10), coconut (24). Exhaustive studies on the potential use of VAM in increasing growth and yield of soybean (3, 4, 26, 23, 33) and groundnut (8, 15, 16) have been made.

4. RESPONSE OF OILSEED CROPS TO Mycorrhizal INOCULATION

An insight into the previous work done in relation to oilseed crops reveals both positive and negative growth responses to VAM infection at pot and field level. Positive responses resulting in improved plant growth and nutrient uptake has been shown in groundnut (2), Mustard (6, 9) and sunflower (13, 19). Inoculation of *Glomus fasciculatum* in groundnut resulted in increased uptake of nitrogen, phosphorus, potassium, zinc, manganese and iron (16). Combined inoculation of *Acaulospora laevis* and *Glomus elunicatum* with suflower has been beneficial in promoting the growth of the plant. Mustard plants in association with *Glomus mosseae* have shown enhanced P uptake (6). Young *et al.*, (33) have reported an increase in seed yield in soybean by 7-45% in mycorrhizal plants. Manoharachary *et al.*, (17) reported greater number of VAM fungi and propagule association with sesame than sunflower, safflower, castor or groundnut (32) have shown abundance of *Glomus* sp. with sesame in red laterite soil having a pH of 7.2. Sulochana (30) conducted a range of studies on VAM in relation to sesame and reported two fold increase both in macro and micro elemental uptake in pot experiments on Gowri cultivar inoculated with *Gigaspora margarita* and *G. fasciculatum*. Field experiments resulted in an increase in plant growth and biomass, N.P.K. level, number of leaves, total sugars and yield, associated with greater alkaline/ acid phosphatase activity in early growth period of crop after VAM inoculation. Kharif sesamum was reported to support greater VAM association than the rabi crop. Gowri cultivar of sesamum, due to its high level of colonization can be used as a host for multiplication of mycorrhizal fungi.

Our studies involving the inoculation of indigenous mycorrhizal culture in Sesamum indicate an increased biomass nutrient uptake and yield (Table 1). Also, rock phosphate fertilizer at appreciably lower application rate in association with mycorrhizal fungi produces maximum biomass and nutrient uptake. Mycorrhizal inoculation also leads to an increased P uptake with both rock phosphate and single super phosphate fertilizer.

Besides increased growth, increased seed production has also been reported in mycorrhizal oilseed crops. Young *et al.*, (33) showed VAM inoculation can increase soybean yields by 7-45% over the non-inoculated treatment depending on the soil type and fungal species. In contrast growth depression following VAM infection has been observed in soybean (3) and sunflower (12). Such negative responses to VAM fungi have also been reported in other crop plants and this has been attributed to various factors such as introduction of pathogens along with mycorrhiza inoculum (8), competition between the plant and mycorrhizal fungus for photosynthate (Smith, 1980) or phosphorus (3), low P concentration in plants (31) or high P status of Plants particularly in roots (12, 20). However, the exact mechanism involved in growth depression in mycorrhizal plant has not been fully understood. Yield reductions following (28) much more detailed and exhaustive studies on VAM symbiosis, particularly with respect to oilseed crops are required before we can realistically utilize VAM fungi for commercial use in field.

Table 1. Response of *Sesamum indicum* to mycorrhizal inoculation at 45 kg/ha Rock Phosphate application

Treatment	Percent colonization	Plant Height (cm)	Plant dry wt (g/plant)	No. of capsules produced	Capsule dry wt. (g)	Dry weight of seeds produced/plant	% Increase over control
Control (sterile soil)	—	6.23	3.74 ± 0.17	13.66 ± 0.57	3.21 ± 0.17	1.76	
Mycorrhizal (45 kg/ha R.P. added)	72.33 ± 3.21	24.06	6.5 ± 0.81	19.66* ± 0.57	5.63* ± 0.70	3.09	75

REFERENCES

1. Allen MF, Moore JS, Christensen M. Canad. J. Bot. 58 : (1980) 371.
2. Bagyaraj DJ, Manjunath A, Patil RB. New Phytol. 82 : (1979) 141.
3. Bethlenfalvay GJ, Brown MS, Pacovsky RS. 72 : (1982) 889.
4. Carling DE, Brown MF. Soil Sci Soc Ann. J. 44 : (1980) 528.
5. Daft MJ, AA El-Giahni, F.E. Sanders. B. Mosse, P.B. Tinker, Academic Press New York, (1975) 581-592.
6. Esther JD, Dube JN, Acta. Bot. Indica. 16 (1988) 210-12.
7. Gray LE, Gerdemann JW. Plant and Soil. 30 : (1969) 415-22.
8. Hall IR, Kelson A. N.Z.J. Agric. Res. 24 : (1981) : 221-22.
9. Jalali BL. In Rapeseed Mustard Annual Report ICAR (1984) 323.
10. Janaki Rani S, Manoharachary C. In Mycorrhizal symbiosis and plant growth. (DJ Bagyaraj, A Manjunath, eds.) Proc. of 11 National Conference on Mycorrhizae, Bangalore, India (1990) 152.
11. Johnson A, Jokobson I, Jensen ES. Plant and Soil 160 : (1994) 1-9.
12. Koide RT. New Phytol. 99 : (1985) 449-62.
13. Koide RT, Li M. New Phytol. 111 : (1989) 35-44.
14. Krikun J. In Plant Roots : The Hidden Half. (Yoav Waisel ed.), Amran Eshel Vzi Kafkasfi pp. 767-786. New York : marcel Dekker, (1991).
15. Krishna KR. Ph.D. Thesis, University of Agricultural Sciences, Bangalore University, Bangalore, India (1981).
16. Krishna KR, Bagyaraj DJ. Plant and Soil 77 : (1984) 405-08.
17. Manoharachary C, Sulochana T Ramarao P. Mycorrhiza News 2 (1) : (1990) pp 2-4.
18. Marschner H, Dell B. Plant and Soil 159 : (1994) 89-02.
19. Mehrotra VS, D.Phil. Thesis, Allahabad Univ., Allahabad (1993).
20. Menge JA, Munnecke DE, Johnson ELV, Leonard RJ. New Phytol. 80 (1978) : 575-78.
21. Menge JA. Canad. J. Bot. 61 : (1983) 1015-24.
22. Pate JS. Plant and Soil. 159 : (1994) 1-10.
23. Paula MA, Siqueira JO, Hoshika E. Revista Brasileire de Ciencia do solo 14 : (1990) 151-56.
24. Ramesh CR. Proc. 5th Ann Symp. Plantation Crops. (1984) 548- 54.
25. Rao AS, Parvathi K. Plant and Soil. 66 : (1982) 133-37.
26. Ross JP, Harper JA. Phytopath. 60 : (1970) 1552-56.
27. Smith SE. Biol. Rev. 55 : (1980) 475-10.
28. Stribley DP Tinker, PB, Snell Grove RC. J. Soil Sci. 31 : (1980) 655-72.
29. Subba Rao A, Sammi Reddy K, Takkar PN. Fertilizer News. 40 (12) : (1995) 87-95.
30. Sulochana T, Manoharachary C, RamaRao P. Curr. Sci. 58 (9) : (1989) 519-20.

31. Thompson JP. Can J. Bot. 64 : (1986) 2282-94.

32. Vijayalakshmi M, Rao AS. Proc. Indian National Sc. Acad Part B (Biol-Sci) 53 : (1987) 279-81.

33. Young CC, Juang TC, Guo HY. Plant and Soil 95 : (1986) 2455- 45.

CHAPTER - 6

HETEROGENICITY IN THE DISTRIBUTION OF Endo-*Mycorrhizal* SPECIES IN SWEET POTATO SOILS OF INDIA

V.S. Harikumar and V.P. Potty[1]

Department of Post Graduate Studies and Research in Botany, Sanatana Dharma College, Alleppey - 688 003, INDIA.
[1]Central Tuber Crops Research Institute, Sreekariyam - 695 017, INDIA.

ABSTRACT

Arbuscular mycorrhizal isolates belonging to the species of *Acaulospora, Gigaspora, Glomus, Sclerocystis* and *Scutellospora* were obtained from sweet potato soils of India. The distribution and composition of species varied with locations. Heterogenicity in the distribution of AM species was more apparent in the sweet potato soils of Kerala.

1. INTRODUCTION

Arbuscular mycorrhizal (AM) fungi, a predominant component of soil microflora, form symbiotic association with crop plants and are known to be helpful in ameliorating plant growth (1). Though most of the agricultural soils possess an indigenous AM population, the effective manipulation of these fungi in crop production systems require a proper understanding of their biodiversity in natural conditions and the competitive abilities of species in promoting plant growth. The present paper reports the mixing of AM species in the sweet potato soils of India.

2. MATERIALS AND METHODS

Root region samples of sweet potato were collected from 52 major potato growing regions of Orissa, West Bengal, Uttar Pradesh, Haryana, Bihar and Kerala. Spores were recovered from the soil following the wet sieving and decanting method (2). Morphological characteristics of the spores were studied and they were identified to the species level with the help of the Manual for the identification of VA mycorrhizal fungi by Shenck and Perez (3).

Table 1. Characteristics of identified AM species

AM species	Shape & Nature of spore	size (μm)	Compound wall thickness (μm)	No. of wall layers	Wall layer morphology	Dia. of hyphal attach ment (mm)	Nature of hypha
Acaulospora laevis	Globose, Yellow brown, smooth to dull roughened	119-520	2-4	3		1.5-3	Hypha below the spore attachment had branches
Acaulospora sp.	Globose, hyaline	125-150	3-4	2		6-8	1 Cylindrical
Gigaspora aurigloba	Globose, dul yellow	125-400	7-15	2	Inner wall thick	40-60	1 Bulbous
Gigaspora calospora	Globose, greenish yellow, smooth	220-325	4-7	2	Outer wall slightly thickened	25-35	1 Bulbous
Gigaspora gigantea	Globose, greenish yellow	300-550	3-7	2	Outer wall thin inner thick	40-45	1 Bulbous
Gigaspora gilmori	Globose, hyaline, white	150-210	7-8	2	Outer layer brittle and separable	25-40	1 Bulbous
Gigaspora gregaria	Globose to subglobose, red to dark brown with surface projections	150-230	10-15	2	Outer wall thick	35-47	1 Bulbous
Gigaspora margarita	Globose, smooth, spore contained oil droplets	150-280	10	2		57	1 Bulbous
Gigaspora rosea	Globose, hyaline	225-300	2.5-7.5	1		25-40	1 Bulbous
Gigaspora sp.	Globose, dull rough	200-400	3.5-8	1		30	1 Bulbous
Gigaspora aggregatum	Globose, orange brown, smooth	100-120	5	2	Outer layer thick	12	Hypha with straight attachment, pore open
Glomus caledonium	Subglobose, yellowish brown, smooth to dull rough	124-275	6	2	Almost equal	12	1 Cylindrical pore occluded by wall
Glomus constrictum	Globose, dark brown, shiny and smooth	130-300	7-12	2	Almost equal	18	1 Cylindrical, constricted, below the constriction the hypha branched dichotomously
Glomus fasciculatum	Globose to flattened, greyish brown, smooth, spore contents hyaline	110-120	7-8	2	Outer thin		1 Cylindrical
Glomus microcarpum	Globose, light yellow, smooth or appeared roughened from adherent debris	40-80	4-6	1	Thin	7-9.5	1 Cyindrical septum present at the base the spore attachment
Glomus mosseae	Globose to subglobose spore surface smooth	110-125	7	2	Outer wall thin	18	1 Cylindrical pore occuluded by plug
Glomus sp.	Globose, yellow to brown	150-170	8	2	Outer layer thin	10-13.5	1 Cylindrical

Table 1 contd...

Table 1 contd...

1	2	3	4	5	6	7	8
Sclerocystis rubiformis	Subglobose, dark brown spore wall laminate	200-300	3-7.5	1	Thick	10-12	1 Cylindrical thick walled
Sclerocystis sp.	Subglobose, brownish black, spore surface smooth to dull roughened	100-140	1.5	1	Outer layer thick	7-9	1 Cylindrical hyphal wall thicker than the spore wall
Scutellospora sp.	Globose to subglobose, yellowish brown with warts on the surface	300-380	3-5	2	Outer wall thick	40-45	1 Suspensor like hyphal attachment

Table 2. Distribution of AM species in the sweet potato soils of India

State	AM species
ORISSA	*Gigaspora calospora, Gigaspora gregaria, Gigaspora* sp., *Glomus fasciculatum, Glomus microcarphum, Glomus mosseae, Glomus* sp., *Sclerocystis* sp.
WEST BENGAL	*Glomus caledonium, Glomus mosseae, Glomus* sp.
BIHAR	*Glomus aggregatum, Glomus fasciculatum, Glomus microcarpum*
UTTAR PRADESH	*Acaulospora* sp., *Gigaspora gilmori, Gigaspora rosea, Glomus fasciculatum, Glomus microcarpum, glomus masseae, Glomus* sp.
HARYANA	*Gigaspora margarita, Glomus fasciculatum, Glomus microcarpum, Glomus* sp.
KERALA	*Acaulospora laevis, Acaulospora* sp., *Gigaspora aurigloba, Gigaspora gigantea, Gigaspora margarita, Gigaspora rosea, Gigaspora* sp., *Glomus aggregatum, Glomus caledenium, Glomus constrictum, Glomus fasciculatum, Glomus microcarpum, Glomus mosseae, Glomus* sp., *Sclerocystis rubiformis, Sclerocystis* sp., *Scutellospora* sp.

3. RESULTS AND DISCUSSION

Arbuscular mycorrhizal isolates belonging to the species of *Acaulospora, Gigaspora, Glomus, Sclerocystis* and *Scutellospora* were obtained from the sweet potato soils of India. The characteristics the identified species are presented in Table 1. Among the identified *Glomus* species, *G. microcarpum* was the most dominant species (19%) and *G. constrictum* was the least (1%). Among the species of *Gigaspora* the percentage of incidence was more for *G. gigantea* and the lowest percentage of distribution was recorded for *G. gregaria, G.rosea, G. gilmori, G. calospora* and G. aurigloba. Among *Sclerocystis* population *S. rubiformis* amounted to 1%. Out of the 2% population of *Acaulospora*, the two species *viz., A. laevis* and an unidentified one were equally distributed. *Scutellospora* formed only 1% of the total AM population in the soils assayed.

AMF species encountered in the sweet potato soils of various locations also differed (Table 2). Diversity in the composition of AMF species was more pronounced in the sweet potato soils of Kerala as compared with other regions. This could be due to the influence of agroclimatic differences (4). Similar reports on the differences in composition of AMF species in Indian soils have been reported in several other instances (5).

ACKNOWLEDGEMENT

The authors are thankful to Dr. G.T. Kurup, Former Director, CTCRI, Trivandrum for providing research facilities and encouragement.

REFERENCES

1. Sharma MP, Gaur A, Bhatia NP & Adholeya A, *Mycorrhiza*, 6 (1996) 441.

2. Gerdemannn JW & Nicolson TH, *Trans Br Mycol Soc*, 46 (1963) 235.

3. Schenck NC & Perez Y, *Manual for the Identification of VA Mycorrhizal Fungi*, University of Florida, Gainesville, USA (1987).

4. Harikumar VS, *Endomycorrbizal studies in sweet potato (Ipomoea batatas* (Lam.) L.) Ph.D. Thesis, University of Kerala, Trivandrum (1997).

5. Thapar HS, Kamala Uniyal & Verma RK, *Indian Forester*, 117 (1991) 1059.

CHAPTER – 7

EFFECT OF ALUMINUM ON MINERAL NUTRITION OF THE Ecto-*Mycorrhizal* FUNGUS *Cantharellus cibarius*

M. Sudhakara Reddy, Ajay Pandey, N. Tejoprakash and V. Ramamurthy*

Biotechnology Division, Thapar Centre for Industrial Research & Development, Patiala 147 001, Punjab.
*Dept. of Biotechnology, Himachal Pradesh University, Shimla - 171 005, H.P.

ABSTRACT

Aluminum is phytotoxic to many plant species at levels commonly found in moderate to high acidic soils. Low pH and nutrient availability enhance the competitive adsorption of Al over calcium and magnesium by their displacement from apoplast exchange sites. Mycorrhiza plays an important role in the acquisition of mineral nutrients in higher plants and also protect from detrimental effects of toxic elements. In the present study the effect of Al on the mineral nutrition of the edible ecto-mycorrhizal fungus *Cantharellus cibarius* was studied. Different concentrations of Al^{3+} (0, 10, 25, 50, 100 and 200 mg/l) were amended in MMN medium and the fungus was inoculated and incubated for 4 weeks at 25°C. After 4 weeks, the mycelia were filtered, dried and analyzed for Al, Ca and Mg content in the biomass using atomic absorption spectrophotometry. The growth of the fungus and the contents of Ca and Mg in the mycelium were reduced as a function of Al ion concentration. Al content in the mycelium had also increased in relation to its availability in the growth medium. Our results suggests that Al influences the uptake of Ca and Mg by the ecto-mycorrhizal fungus, *C. cibarius*

1. INTRODUCTION

Aluminum (Al) toxicity is generally considered to be the most important stress factor for plants grown in acid soils (1, 16). At a soil pH 5.5, Al forms complexes and can not be assimilated by the plant. Low soil pH and nutrient availability enhance the competitive adsorption of Al over calcium and magnesium (4) by their displacement from apoplast exchange sites (5). In addition, the uptake of P and K is inhibited by the presence of Al in the apoplast (2). Mycorrhiza plays a major role in forest ecosystem especially in mineral nutrition by higher plants(6). Although mycorrhizal symbiosis may not necessarily prevent the uptake of Al and toxicity to the host (5, 8), Al is accumulated within the mycorrhizal sheath and ectomycorrhizal trees often show better growth and improved resistance to unfavorable conditions than non-mycorrhizal trees(6). Al^{3+} can influence the growth of mycorrhizal fungi *in vitro* (17). The presence or absence of VA or ectomycorrhizal fungi can change the mineral content of plants treated with Al (7,8,10). The ectomycorrhizal fungus *C. cibarius* and its allied species produces the edible mushrooms which are appreciated throughout the world. In India *C. cibarius* has been reported in association with pines and spruce in Himalayan region. Inoculation of this fungus to planting stock can improve the growth and survival and also helps in production of the fruit bodies which are of great economic value. Most of the soil in Himalayan region are in acidic range, Al may have an influence on plant growth. Hence an attempt was made to study the influence of Al on the mineral content of the edible ectomycorrhizal fungus *C. cibarius in vitro*.

2. MATERIAL AND METHODS

The ecto-mycorrhizal fungus *C. cibarius* (ATCC 13228) was obtained from American Type Culture Collection, USA. Modified Melin-Norkans (MMN) medium(11) was prepared and 50 ml of the medium was added to each 250 ml conical flask and autoclaved for 15 min. at 121°C. Three mycelial disc (5 mm diameter) of the ecto-mycorrhizal fungus *C. cibarius* was inoculated to each flask, cut from the edge of the actively growing mycelium in petriplate. The stock solution of Al (2000 ppm) was prepared using $Al_2(SO_4)_3.16H_2O$ and filter sterilized. This solution was then added to each flask to achieve different concentrations of Al i.e. 0, 10, 25, 50, 100 and 200 mg/l after three days of the fungal inoculation. The flasks were incubated at 25° C for 4 weeks, filtered on a filter paper and the dry weight of the mycelium was noted. The dry mycelium along with the filter paper was digested in concentrated HNO_3 and the volume was made to 50 ml. By using atomic absorbtion spectrophotometry the contents of Al, Mg and Ca were determined in the digested mycelium

3. RESULTS AND DISCUSSION

The results showed that as the concentration of Al increased, the dry weight of the mycelium decreased. The highest biomass found when 10mg/l Al was added to the culture medium. The Al concentration increased in the mycelium as the concentration of Al increases in the culture medium. Mg content was drastically reduced as the concentration of Al increase in *C. cibarius* mycelium. Calcium content also decreased when the concentration of Al increased (Table 1).

Although effect of Al on plants are well known (13), few investigation have been done on mycorrhizal fungi. The growth of mycorrhizal fungi can be inhibited or stimulated by aluminum, depending upon the fungal species and the concentration of aluminum used (14). In the present study, growth of the fungus increased at lower concentrations of the aluminum. As the concentration increased, the biomass decreased.

The phytotoxicity of Al depends on the pH, organic matter content, concentrations of other cations, anions, total salts, plant species and individual genetic properties of the plants(18). Mycorrhizal fungi are known as great accumulators of heavy metals(3). In our work, the content of the Al in mycelia increased with increasing Al concentrations in the culture media. Ca

Table 1. Effect of Al on the mineral content of *Cantharellus cibarius*

Conc. of Al (mg/l)	Dry Weight of mycelium (g/l)	mg / gram tissue		
		Al	Mg	Ca
0	18.3	0.87	14.8	26.7
10	26.6	2.10	0.30	6.2
25	23.5	4.53	0.49	10.5
50	18.1	6.20	3.31	19.6
100	15.6	3.66	1.89	25.0
200	18.9	11.96	2.01	26.7

and Mg uptake are most frequently reduced in different woody species under the influence of aluminum(5, 12, 15). Zel and Bevc(18) reported the increase in concentrations of Al, Ca and Na contents and decrease in the concentrations of P, Mg and K in the mycelium when grown *in vitro* in presence of Aluminum. In the present study also both the Ca and Mg content were reduced in the mycelium as the concentrations of Al increase in the culture medium. From the results, it was concluded that Al influences the mineral content of ecto-mycorrhizal fungi. Mineral nutrition is one of the main functions of the fungal partner in mycorrhizal associations, so these changes can affect the availability and transport of nutrients also to the symbiotic higher plants. The axenic testing of ecto-mycorrhizal fungi for tolerance to Al may not be an accurate measure of metal tolerance. However, this may complement an inoculation program by assessing the relative tolerance of a suite of fungal isolates to Al, but more accurate evaluation of metal tolerance and increases in mineral uptake would be a trial of the symbiosis under native soil conditions.

ACKNOWLEDGMENTS

The authors wish to thank Prof. M. P. Kapoor, Director, TCIRD for encouragement and facilities.

REFERENCES

1. Baligar VC, Wright RJ, Fagaria NK & Foy CD, *J Plant Nutr* 11(1988) 549.
2. Bengtsson B, *Vegetario,* 101(1992)35.
3. Byrne AR, Ravnik V & Kosta L, *The Sci Total Environ,* 6(1976) 65.
4. Cronan CS, *Tree Physiol,* 84(191) 319.
5. Godbold DL, Fritz E & Hustermann A, *Proc Natl Acad Sci* USA, 85 (1988)3888.
6. Harley JL & Smith SE, *Mycorrhizal symbiosis,* Academic press, London(1983).
7. Hodson MJ & Wilkins DA, *New Phytol,* 118(1991)266.
8. Jentschke G, Schlegel H & Godbold DL, *Physiol Plant,* 82(1991)266.
9. Jones MD, browning MHR & Hutchinson TC, *Water Air and Soil Pollution,* 31(1986)441.
10. Koslowsky SD & Boerner REJ, *Environ Pollut,* 61(1989)107.
11. Marx DH, *Phytopathology,* 59(1969)153.

12. Raynal DJ, Joslin JD, Thoruton FC, Schaedle M.& Henderson GS, *J Environ Qual,* 19(1990)180.

13. Taylor GJ, In: *Metal ions in Biological System,* (Sigel H & Sigel A. eds). Publ. Marcel Dekker, New York, (1988)123.

14. Thompson GW & Medve RJ, *Appl Environ Microbiol,* 48(1984) 556.

15. Thornton FC, Schaedle M & Raynel DJ, *J Environ Qual,* 18(1989) 541.

16. Van Raij B, In: *Plant-soil interactions at low pH,*(Wright RJ, Baligar VC & Murrmann RP eds) Kluwer Academic Publisher(1991) 159.

17. Zel J, Blatnik A & Gogala N, *Water air and Soil Pollution,* 61(1992).

18. Zel J and Beve L, *Water Air and Soil Pollution,* 71(1993)271.

CHAPTER - 8

UTILIZATION OF REGION SPECIFIC *Cyanobacteria* AS BIOFERTILIZER FOR RICE — A CASE STUDY

Siba P. Adhikary

P. G. Department of Botany, Utkal University, Bhubaneswar -751 004, Orissa

ABSTRACT

Occurrence and distributional pattern of heterocystous cyanobacteria in the rice field soils of Orissa and tolerance of selected species to fertilizers viz., Urea, Ammonium sulphate, NPK, DAP, MOP and SSP; insecticides, Monocrotophos and Endosulfan and the herbicide, Hiltachlor have been presented. Agrochemical stress tolerant isolates were cultivated in galvanized iron trays and polybags in the outdoor for producing inocula for field application. Competition of inoculated cyanobacteria with indigenous species during the crop cycle and the regulatory role of fertilizers for their establishment in the soil based dried culture, fresh wet culture and urea supplemented plots was investigated. Nitrogenase activity of the soils inoculated with cyanobacteria was higher than the control and N-fertilizer supplemented plots. Most of the inoculated species competed successfully with the indigenous flora and established in the fields contributing higher amount of fixed N to the soils. Increase of grain yield by over 25 percent was obtained in plots algalized with fresh cyanobacterial cultures containing regional isolates which was much higher than the yield of the plots inoculated with soil based culture. Cyanobacterial biofertilizer technology can be improved by inoculating fresh cyanobacterial mixture containing regional isolates and also by identifying and alleviating the limiting factors for establishment of the inoculated strains during rice cultivation.

1. INTRODUCTION

The rice field ecosystem provides an environment favourable for the growth of cyanobacteria with respect to their requirement for light, water, temperature and nutrient availability. De (1) drew attention to the importance of these organisms in the sustenance of fertility of rice field soils. Since then, long-term fertility experiments and nitrogen fixation measurement have confirmed the importance of cyanobacteria in maintaining moderate but consistent rice production in fields receiving no N fertilizer, and many trials have been conducted to increase the rice yield by inoculation of these organisms (2-8). Venkataraman and his collaborators conducted comprehensive studies on cyanobacteria of rice fields and convincingly demonstrated their contribution to rice production and developed simple techniques for their use by rice farmers (9-10).

Reconsidering the agronomic utilization of cyanobacteria for rice cultivation, a case study was undertaken with the aim to study : (1) the occurrence of cyanobacteria in the rice fields of various agroclimatic zones of Orissa state and finding out cyanobacteria rich and deficient locations, (2) screening of agrochemical stress compatible isolates for using as inocula, (3) developing inoculum by cultivating in iron trays and in polybags and assessing their quality, (4) *in situ* nitrogenase activity of the rice field soils inoculated with cyanobacteria, (5) fate of inoculated cyanobacterial isolates during and after cultivation of rice crop and (6) contribution of cyanobacterial inoculation to paddy yield.

2. MATERIALS AND METHODS

(a) Experimental

Based on their better growth, higher nitrogenase activity and occurrence in different agroclimatic zones of the state, 9 species of which 2 species each belonging to the genera *Anabaena*, *Nostoc*, *Calothrix* and *Aulosira* and one species of *Cylindros permum* were selected for screening for their tolerance to various insecticides, e.g. Balawan (monocrotophos, 36 SL), Thiokill (endosulfan, 35 EC), herbicide, Hiltachlor (35 EC) and commercial fertilizers *e.g.* urea, ammonium sulphate, NPK (15:15:15), di-amino phosphate (DAP), muriate of potash (MOP) and single super phosphate (SSP). Five of these cyanobacteria that showed a wider tolerance to the agrochemicals were semi-mass cultured in 2 L conical flasks in the laboratory or in 1 L transparent polythene bags with cotton wool plug to be used as inocula in the field or for using to produce soil based material for subsequent field use. Soil based inocula were raised in 180 cm x 90 cm x 15 cm galvanized iron trays according to the method described by Venkataraman(9).

(b) Field Experiments

Field experiments were conducted in 5 m x 5 m plots with three replicates per treatment arranged on a randomized block design in the village Maniakati, Block Surada in the Ganjam district of Orissa state during Kharif 1995, 1996 and 1997. Experiments were conducted only during Kharif crop (July to December) due to the reason that 94.6% of total cultivated area of state are used for growing rice during Kharif. Further, the site of the experiment was located in a rain-fed rice cultivated area with low agrochemical input in the fields which is the characteristic feature of above 70 percent of rice fields of the state. The treatments were : algalization with (1) fresh inocula cultivated in BG 11 medium in polybags, (2) soil based inoculum cultivated in galvanized iron trays, (3) Urea (60 kg per hectare in two split doses, once before transplantation and than during panicle emergence), (4) 30 kg urea per hectare before transplantation and after 7 days fresh cyanobacteria from polybags, and (5) control without algalization or fertilizer application. High yielding high nitrogen responsive varieties of rice, one each of short duration and long duration varieties were used in the experiments. Phosphorus and Muriate of potash with or without 30 kg/ha of Urea as per the design of the experiment were added basally during field preparation before transplantation. Nursery grown, three weeks old rice were transplanted at 20 cm x 20 cm and the flood water was maintained at 3 to 4 cm. Cyanobacterial inoculation was done a week after transplantation by adding a mixture of freshly harvested cyanobacteria from polybags at the rate of 50 gram air dried weight or soil based inocula at 100 grams per plot. The fields were observed regularly for cyanobacterial growth throughout the paddy cultivation cycle and visible scums were examined under microscope.

In situ nitrogenase activity measurement was conducted as per Allimango and Yoshida (11) with slight modification. Glass test tube (18 mm x 150 mm) with opening at both ends were taken. One of the end was fitted with a rubber serum cap. The tube was placed 50 mm deep in the paddy soil between the rice hills. Ten per cent gas phase inside the tube was replaced with acetylene using a gas tight syringe. After incubation for 1 hour, gas samples were withdrawn into a pre-evacuated glass tube fixed with a serum cap. Aliquots of the gas sample of the vacutizer tube was injected into a AIMIL - NUCON 5765 Gas chromatograph with FID and fitted with 2 m stainless steel column with Porapak-R (oven temp. 75°C, injector temp. 100°C, detector temp. 110°C, carrier gas N flow, 30 ml/min). The ethylene formed from the reduction of acetylene was measured. ARA of the soil samples of the fields was also measured differently. The soil samples in between the rice hills were removed with a 1.5 cm diameter cork borer to a depth of 50 mm. Ten such soil cores were placed within an assay bottle fitted with serum cap. Acetylene was introduced to each tube and the reaction was terminated after one hour of incubation in the field with trichloroacetic acid (10%). Samples were brought to the laboratory for acetylene reduction assay. So for ARA measurement, minimum 20 samples were taken from each field and the average value was considered. ARA measurements was carried out always in the morning hours between 8 to 10 AM, as there are reports that maximum phototropic nitrogenase activity in the fields occurred during this period (12,13). A pinch of the soil samples used for ARA measurement was grown in enriched culture in the laboratory and the rest were air dried and weighed. Nitrogenase activity was calculated as P mol C_2H_4 produced per gram soil per hour. The cyanobacterial forms appeared in the culture of each sample was identified.

3. RESULTS AND DISCUSSION

(a) Screening of agrochemical stress tolerant *Cyanobacteria*

Capability of tolerance of nine different isolates of cyanobacteria has been presented in Table 1 and 2. The results showed that there exists a great deal of variation in the growth and nitrogen fixing capability among the species of different genera and also within the same genus, which also vary considerably in response to the insecticides, herbicide and nitrogenous and phosphatic fertilizers; capability to tolerate to various concentra- tions of certain commercial fertilizers vary up to about two order of magnitude within the species tested. Further, nitrogenase activity of all the species was inhibited in presence of much less concentration of the nitrogenous fertilizers in comparison to their recommended doses. Even recommended dose of the insecticide, endosulfan, inhibited ARA of *Calothrix* UU24112, *Anabaena* UU147 and *Cylindrospermum* UU245. Thus care should be taken for selecting the dose of these agrochemicals while applying to the fields inoculated with cyanobacteria.

Table 1. Inhibition of growth and nitrogenase activity of various *Cyanobacteria* isolated from rice fields in presence of commercial grade fertilizers

Commercial fertilizer	Tolerance to the fertilizers	Concentration range	Inhibition of nitrogenase activity	Concentration range	Recommended dose
Urea (46 N)	Aulosira 25118> Aulosira 29147> Cylindrospermum 245> Anabaena 147> Nostoc 24100> Anabaena >2493 Nostoc 29130> Calothrix 24112> Calothrix 29139>	100-5,000 ppm	Aulosira 25118 = Nostoc 29130 = Anabaena 147 = Cylindrospermum 245> Aulosira 29147 = Calothrix 2909> Anabaena 2493 = Nostoc 24110 >Calothrix 24112	5-25 ppm	200 ppm
Ammonium sulphate (21 N)	Nostoc 29130> Aulosira 25118> Aulosira 29147> Cylindrospermum 245> Nostoc 24100> Anabaena 147> Calothrix 29139> Anabaena 2493> Calothrix 24112>	100-10,000 ppm	Nostoc 29130 = Aulosira 25118 = Anabaena 147> Cylindrospermum 245> Aulosira 29147 = Calothrix 29139 = Nostoc 24100> Calothrix 24112 = Anabaena 2493	5-25 ppm	400 ppm
NPK (15:15:15)	Cylindrospermum 245> Aulosira 25118> Nostoc 29130> Anabaena 147> Calothrix 24112> Calothrix 29139> Anabaena 2493> Nostoc 24100> Aulosira 29147>	1000-5,000 ppm	Anabaena 147 >Aulosira 25118 = Nostoc 29130> Cylindrospermum 245> Calothrix 24112 = Calothrix 29139 >Aulosira 29147 >Nostoc 24100= Anabaena 2493	100-500 ppm	200 ppm
Di-amino Phosphate (DAP) (18 N, 46 P)	Nostoc 29130> Anabaena 147> Calothrix 29139> Nostoc 24100> Calothrix 24112> Aulosira 25118> Aulosira 29147> Anabaena 2493> Cylindrospermum 245>	1000-10,000 ppm	Anabaena >147 Nostoc 29130 >Aulosira 25118 >Calothrix 29139 = Cylindrospermum 245= Calothrix 24112 = Aulosira 29147 >Anabaena 2493 = Nostoc 24100	25-200 ppm	100 ppm
Muriate of Potash (MOP) (26 K)	Aulosira 25118 = Nostoc 29130 >Anabaena 147> Calothrix 29139 >Calothrix 24112 >Aulosira 29147 >Cylindrospermum 245 >Anabaena 2493 >Nostoc 24100	3000-20,000 ppm	Aulosira 25118 = Nostoc 29130 = Anabaena 147> Calothrix 29139 >Calothrix 24112 = Aulosira 29147 = Anabaena 2493 >Nostoc 24100= Cylindrospermum 245	2000-5000 ppm	100 ppm
Single super phosphate (SSP) (16 P)	Nostoc 29130 = Nostoc 24100 = Aulosira 25118> Anabaena 147 = Cylindrospermum 245> Calothrix 24112 >Calothrix 29139> Aulosira 29147 >Anabaena 2493	2500-15,000 ppm	Nostoc 29130 = Aulosira 25118 >Anabaena 147 = Nostoc 24100> Cylindrospermum 245> Aulosira 29147 >Calothrix 29139 > Anabaena 2493 = Calothrix 24112	1000-2000 ppm	300 ppm

Table 2. Inhibition of growth and nitrogenase activity of various *Cyanobacteria* isolated from rice fields in presence of commercial grade insecticides and herbicide

Insticide/ herbicide	Tolerance of pesticides	Concentration range	Inhibition of nitrogenase activity	Concentration range	Recommended dose
Thiokill (endosulfan, 35 EC)	Nostoc 29130> Aulosira 25118> Aulosira 29147> Calothrix 29139 = Anabaena 2493> Nostoc 24100= Anabaena 147 = Calothrix 24112 = Cylindrospermum 245	10-75 ppm	Nostoc 29130> Aulosira 25118> Anabaena 2493> Aulosira 29147> Calothrix 29139> Nostoc 24100> Calothrix 24112> Anabaena 147> Cylindrospermum 245>	4-75 ppm	4 ppm
Balwan (monocrotophos 36 SL)	Nostoc 29139 = Aulosira 25118 = Calothrix 29139 = Anabaena 2493 = Anabaena 147> Aulosira 29147 = Nostoc 24100 = Calothrix 24112 = Cylindrospermum 245	300-400 ppm	Aulosira 25118> Nostoc 29130> Anabaena 2493> Calothrix 29139> Anabaena 147> Nostoc 24100> Calothrix 24112> Aulosira 29147 = Cylindrospermum 245	300-400 ppm	4 ppm
Hiltachlor (50 EC)	Nostoc 29130 = Aulosira 25118> Nostoc 24100> Anabaena 147> Aulosira 29147 = Calothrix 29139 = Anabaena 2493 = Calothrix 24112> Cylindrospermum 245	100-400 ppm	Nostoc 29130> Aulosira 25118> Nostoc 24100> Anabaena 147> Calothrix 24112> Calothrix 29139> Anabaena 2493> Aulosira 29147> Cylindrospermum 245>	100-400 ppm	10 ppm

(b) Mass cultivation of selected *Cyanobacteria* and preparation of inocula

Results on the influence of commercial fertilizers applied to rice field soils on the multiplication of inoculated cyanobacteria reveal that when the organisms were cultivated with garden soil in the tray, the native algal species also grew alongwith the inoculated cyanobacteria. The native forms were predominantly *Aphoncothece* sp, *Anabaena* sp. or *Cylindrospermum* sp. and also species belong to green algal genera *Sirogonium*, *Spirogyra* and *Oedogonium*, besides diatoms and desmids. In the control containing soil without fertilizers, often *Anabaena* UU147, *Nostoc* UU29130 and *Aulosira* UU25118 competed successfully with the native flora and grew well forming higher biomass; air dried weight of the floating patches in the tray after 30 days of inoculation was 16 grams per sq. meter area. The biomass of algae increased when supplemented with various commercial fertilizers at the recommended dose of their field application. With urea and ammonium sulphate, *Aphonothece* sp. and green algal members dominated. Though few filaments of *Nostoc* UU29130 and *Aulosira* 25118 grew along with other algae, nitrogenase activity was completely inhibited (Table 3). In presence of SSP (300 μg/ml) also, growth of native cyanobacteria and green algae was encouraged, however, dominated by *Aulosira* 25118 and *Calothrix* 29139 and the ARA was about 30% less than the control. In presence of recommended dose of muriate of potash (MOP), the inoculated cyanobacteria grew and formed flakes as that of the control trays and the ARA of these organisms were almost equivalent to control (Table 3). These results showed that application of superphosphate encouraged the native algal population of the soils, and further, growth of inoculated cyanobacteria was

Table 3. Growth and acetylene reduction activity (ARA) of *Cyanobacteria* during outdoor cultivation in iron trays containing soils amended with commercial fertilizers

Soil 1 recommended dose of agrochemicals	Growth of cyanobacteria		Green algae ***	Air dried floating biomass (g/m^2) in 30 days	ARA n mol C_2H_4/g air dried mat/h
	Native*	Inoculated**			
Control	+ + 1 2	+ + + + + 3 4 5 6 7	+ 8	16.4 ± 3.8	3.9 ± 1.7
Urea (200 μg/ml)	+ 1	+ + 4 6	+ + + 8 9 10	24.5 ± 4.2	0
Ammonium sulphate (400 μg/ml)	+ 1	+ + 4 6	+ + + 8 9 10	21.9 ± 5.1	0
MOP (Potash) (100 μg/ml)	+ + 1 2	+ + + + + 3 4 5 6 7	+ 8	18.2 ± 3.6	4.3 ± 2.1
SSP (phosphate) (300 μg/ml)	+ + 1 2	+ + + 4 6 7	+ + 8 9	27.3 ± 5.8	2.7 ± 0.9

Values of biomass yield and ARA represent mean of five independent determinations ± S.D.

*The native cyanobacterial species present in soil which appeared during cultivation were: 1. *Aphanothece* sp. and 2. *Anabaena* sp.

**A mixture of cyanobacterial species e.g. 3. *Anabaena variabilis* UU 147, 4. *Nostoc carneum* UU 29130, 5. *Cylindrospermum* sp. UU 245, 6. *Aulosira* sp. UU 25118 and 7. *Calothrix* sp. UU 29139.

***Green algal forms appeared during cultivation under different treatments were: 8. *Sirogonium* sp., 9. *Spirogyra* sp. and 10. *Oedogonium* sp.

The kind of alga having highest biomass in the tray during cultivation is indicated by underlining against their respective number; + = one algal species.

inhibited in the presence of recommended dose of nitrogenous fertilizers. Once green algal forms and unicellular cyanobacteria dominate in the soils due to N-fertilizer application, they did not allow the inoculated cyanobacteria to get established.

(c) Fate and activity of *Cyanobacteria* introduced into rice field soils

Results of the experiments (Table 4) showed that the inoculated cyanobacterial forms bloom profusely a month after inoculation; the species which occurred in the field were *Nostoc, Aphonothece, Anabaena* and occasionally *Calothrix*. The cyanobacterial forms continued to grow in the field attached to rice plants as well as formed patches on the wet soils during 2nd month of the crop cycle. Quantity of cyanobacterial biomass in the inoculated plots was highest during October. Just after harvest, *Anabaena* UU147, *Aulosira* UU25118, *Cylindrospermum* UU245 and/ or *Calothrix* UU29139 form conspicuous patches on wet soils (Table 4). After harvest the fields were left fallow for a month and then the dried patches on the soil were examined in enriched culture. These mats were dominated by *Aulosira* UU25118 mixed with *Calothrix* UU29139 or native *Cylindrospermum* species. During the crop cycle, in the control plots though the native *Aphanothece* sp., *Anabaena* sp. and *Cylindrospermum* sp. occurred, in the algalized plots, the inoculated species dominated throughout. In urea treated plots, however, native cyanobacterial species and other algal forms occurred profusely throughout the crop cycle.

Table 4. Acetylene reduction rate in experimental plots with or without *Cyanobacterial* inoculation during Kharif cropping season, 1997

Date of measurement during crop cycle	Nitrogenase activity (p mol C_2H_4 /g soil/h in the soils of the experimental plots				
	Control	Inoculated with cyanobacteria*			
		Soil based dried culture	Fresh wet culture	Urea 60 kg/hectare	Urea 30 kg/ hectare + fresh wet culture
25 Aug. 1997 (after transplanting)	22 ± 8 (ND)	46 ± 13 (ND)	104 ± 41 (1,2)	0** (ND)	0** (ND)
22 Sept. 1997	87 ± 29 (6,7)	186 ± 42 (1,2,5,6)	217 ± 66 (1,2,5,6)	39+11 (6)	92 ± 24 (1,2,6)
28 Oct. 1997	233 ± 46 (6,7,8)	534 ± 71 (1,2,3, 5,7)	724 ± 55 (1,2,3,4,7)	98 ± 21** (6,7)	231 ± 57 (1,4,5)
4 Dec. 1997 (before harvest)	321 ± 61 (6,7,8)	494 ± 44 (3,4,5)	636 ± 59 (1,3,4)	286 ± 35 (6,7)	543 ± 43 (1,2,4,7)
11 Jan. 1998 (dried field after harvest)	254 ± 78 (6,8)	486 ± 56 (3,5,8)	574 ± 36 (1,3,4)	203 ± 42 (6)	498 ± 6 (1,4)

Cyanobacterial inoculation (soil based culture as well as fresh wet culture in polybags directly from the laboratory) was done one week after transplantation i.e. on 18 August, 1997.

** Urea at the rate of 30 kg/hectare was applied to the field before measurement of ARA.

- ARA activity was measured in the morning between 9 to 10 AM.
- The cyanobacterial inoculum contained mixture of 5 species, 1. *Anabaena variabilis* UU147, 2. *Nostoc carneum* UU 29130, 3. *Cylindrospermum* sp. UU 245. 4. *Aulosira* sp. UU 25118 and 5. *Calothrix* sp UU 29139. The other cyanobacterial species which appeared in the un-inoculated fields were: 6. *Aphanothece* sp. 7. *Anabaena* sp. and 8. *Cylindrospermum* sp.
- The cyanobacteria occurred in the ARA measured soil samples has been mentioned below the ARA value in parenthesis and and the organism showing highest biomass is indicated by underlining their respective number, ND - not detected.

The nitrogenous fertilizers when applied at the rate 60 Kg N/hectare or at half rate alongwith cyanobacterial inoculum, exhibited depressive effect on the nitrogenous activity (Table 4). ARA was not detected in the urea supplemented soils up to a week after inoculation; nitrogen fixation value was much less than to control in the urea supplemented plots up to one month after transplantation. Nitrogenase activity of the soils inoculated with cyanobacteria was maximum at the time of panicle emergence stage in the month of October, where as in control and fertilizer supplemented plots, maximum activity was obtained during ripening of the crop. Besides, in the later plots, a stimulation of growth of non-nitrogen-fixing algae was observed due to nitrogen supplementation which decreased the growth of nitrogen-fixing-forms by competition and antagonistic effect (Table 4). These results showed that phototropic nitrogenase activity of the soils of the experimental plots introduced with cyanobacteria was higher than the control and nitrogenous fertilizer supplemented plots, and further, the inoculated cyanobacterial species established in the field contributing higher amount of fixed nitrogen to the soils.

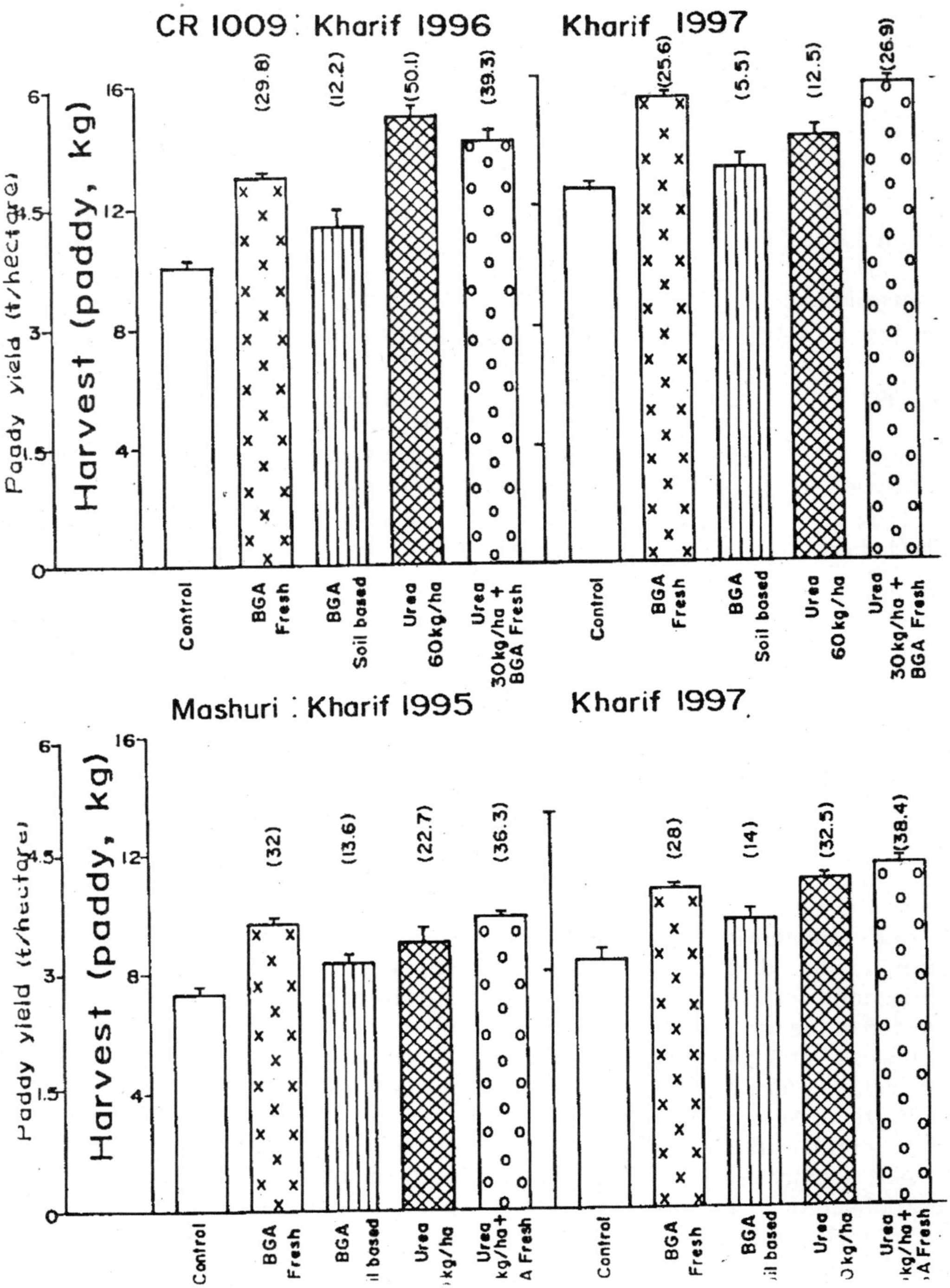

Fig. 1. Effect of algalization of on grain yield.

(d) Effect of algalization on grain yield

Results of the algalization experiments conducted in field during three Kharif seasons showed that by inoculating fresh cyanobacterial isolates from the rice fields of this region, the yield of Mashuri was increased by 32 and 28 per cent over control during 1995 and 1997 respectively (Fig. 1); by supplementing soil based inoculum, yield during both the years was increased by about 14 per cent over control. Yield increased by 22.7 and 32.5 percent during 1995 and 1997 in the 60 kg/hectare urea treated plots. The yield was further increased to 36.3 and 38.4 percent when half of the urea dose (30 kg/hectare) and fresh inoculum was supplemented. The yield data of CR 1009 rice variety during Kharif 1996 and 1997 was quite different from the results obtained with Mashuri rice cultivation during Kharif 1995 and 1997 (Fig. 1). In the soil based cyanobacteria inoculated fields, the yield was higher by 12.2 and 5.5 per cent over control during 1996 and 1997 respectively. Paddy yield further increased to 29.8 and 25.6 per cent over control during the same years when fresh cyanobacteria inoculum was supplemented. With 60 kg /hectare urea treatment, yield increased over 50 per cent during 1996, which was the maximum yield and such a value was not obtained even when half of the urea dose and fresh cyanobacterial inoculum was supplemented together; yield in the latter treatment was about 10 per cent higher than the value obtained from fresh cyanobacteria inoculated plots. But during 1997, yield in the 60 kg/hectare urea supplemented plots was considerably reduced to 12.5 percent but increased over control; with 30kg urea per hectare + fresh inoculum, the yield was comparable to the yield obtained in the fresh cyanobacteria inoculated plots.

All these results showed that in fresh cyanobacteria inoculated fields the yield of Mashuri as well as CR 1009 rice increased by 25.6 and 32 per cent over the control during the three years of study. In addition, when fresh inoculum and half the dose of urea (30 kg/hectare) was supplemented together, there was an additional benefit in the yield though it varied with the rice variety and year of cultivation. Further, though the soil based inoculum increased the yield of both the rice varieties, it was always considerably less than their respective fresh cyanobacteria inoculated plots. The results that fresh cyanobacterial inoculum either singly or in combination with reduced dose of nitrogenous fertilizer increase the yield of short duration (Mashuri) as well long duration (CR 1009) rice varieties by over 25 per cent as compared to control suggest the potentiality of cyanobacterial inoculants containing regional isolates for use as biofertilizer for rice cultivation.

Most of the studies on algalization was focussed only on grain yield and little emphasis has been paid on the ecological and microbial processes that might influence this. Algalization trials have been carried out using inocula containing foreign strains selected in the laboratory and almost none except that of Reddy and Roger (14) have paid attention to the establishment of inoculated strains. There are reports that foreign strains inoculated in a soil rarely establish (15, 16), which possibly was the reason of not obtaining significant benefits through algalization in certain trials (17, 18). However, results of the present study showed that identifying and alleviating limiting factors which regulates the establishment of inoculated strains, and by inoculating fresh cyanobacteria isolated from that particular region, cyanobacterial biofertilizer technology can be substantially improved.

ACKNOWLEDGEMENTS

The work was possible due to the financial assistance from the Department of Biotechnology, Government of India (BT/R&D/ 02/06/93 and BT/PR 1338/ SPD/11/90/98). I thank Mr. J.K. Sahu, Mr. H. Nayak, Mr. A. Mishra Dr. Mrs. Jayanti Sahu and Mr. Maheswar Sahu for their help in the experiments.

REFERENCES

1. De, P.K. 1939. Proc R Soc Lond 127B 121-132.
2. Sankaram, A., M.J. Mudholkar & M.N. Sahay 1967. *Ind. J. Microbiol.* 7 57-62.
3. Goyal S.K. & G.S. Venkataraman 1970. Phykos. 9 137-138.
4. Goyal, S.K. & G.S. Venkataraman 1971. Phykos 10 32-33.
5. Aiyer, R.S., S. Salahudeen & G.S. Venkataraman 1972. Ind J Agric Sci 42 380-383.
6. Huang, C.Y. 1978. Bot. Bull. Acad. Sin. 19 41-52.
7. Goyal, S.K. 1985. Phykos. 24 149-151.
8. Mandhare, V.K., P.L. Patil & D.B. Bhanvase. In: Cyanobacterial nitrogen fixation (B.D. Kaushik ed.), Associated Publishing Co., New Delhi p 313-320, (1990).
9. Venkataraman, G.S. 1978. Algal Biofertilizer for Rice, Informative Manual, Shiv Narain Printers, New Delhi .
10. Venkataraman, G.S. 1981. Blue-green algae for rice production a manual for its promotion, FAO field document no. 2 (RAS/75/004), FAO, Rome.
11. Allimagno, B.V. & T. Yoshida 1977. Plant Soil 47 239-244.
12. Roger, P.A. & I. Watanabe 1982. IRRI research paper series 78, The International Rice Research Institute, Manila, Philippines.
13. Anand, N. & V.S.R. Murugesan 1996. Curr Sci 70 119-120.
14. Reddy, P.M. & P.A. Roger 1988. Biol Fertil Soils 6 14-21.
15. Roger, P.A. & S.A. Kulasooriya 1980. Blue green algae and rice, The International Rice Research Institute, Manila, Philippines.
16. Roger, P.A. In. Biological nitrogen fixation associated with rice production (Dutta S.K. & C. Sloger Eds.), Oxford & IBH Publishing Co., New Delhi. pp. 119-137, (1991).
17. Reddy, P.M., P.A. Roger, W. Venture & I. Watanabe 1986. Phil Agri 69 629-632.
18. Kulasooriya, S.A. & W.K. Hirimburegma. In: Biological nitrogen fixation associated with rice production (Dutta S.K. & C. Sloger eds.), Oxford & IBH publishing Co, New Delhi p 191-210 (1991).

CHAPTER – 9

ISOLATION AND SCREENING OF *Azospirillum* STRAINS FOR DEVELOPING BIOFERTILIZER FOR WHEAT (*Triticum aestivum* L.)

R.P. Singh, G.R. Lawanaya and A. Vaishmpayan

Department of Genetics and Plant Breeding, Institute of Agricultural Sciences
Banaras Hindu University, Varanasi.

ABSTRACT

Biofertilizers are less expensive, ecofriendly, preparation containing live or latent cells of efficient nitrogen fixing, phosphate solublizing organisms used for improving the grain yield and soil texture for long term sustainable agriculture. *Azospirillum* bacteria, out of many nitrogen fixing organisms, have gained much importance for their associative nitrogen fixing capacity and secretion of growth promoting substances in the vicinity of roots or in the roots of cereals and vegetables, since these bacteria inhabits root rhizosphere soil and cortical cells or xylem vessels of roots etc. Ten *Azospirillum* strains were isolated from the roots of wheat plants collected from various places in Varanasi district. The strains were purified and tested for their effect on plant vigour, grain yield and total biomass production of wheat variety in pots. The two effective strains were further tested on four wheat varieties (HUW 234, K 9006, HUW 12 and HP 1731) in pots during 1997-98 along with standard strain of *Azospirillum* from IARI. Inoculation of *Azospirillum* (both local culture and IARI culture) increased the plant vigour, grain yield and total biomass in all wheat varieties in comparison to control, however, the local culture II exhibited comparatively more increase in yield attributing traits in all varieties, except, HP 1731.

1. INTRODUCTION

Biofertilizers are highly efficient, less expensive and socially acceptable microbial preparations of live and latent cells of nitrogen fixing and phosphate solublizing organisms used for improving soil texture, biological and chemical properties of soil for long term sustainable agriculture for increased grain yield production (11).

Azospirillum bacterium, out of many nitrogen fixing organisms has gained much importance as biofertilizer due to its associative nitrogen fixing ability in roots of cereals, vegetables and grasses (12). Further reports are available which suggest that *Azospirillum* secretes growth promoting substances in the roots which improve the plant vigour (6). Since *Azospirillum* inhabits close proximity of roots and the cortical cells/xylem vessels of roots, the nitrogen fixed and growth promoting substances secreted by it, might be utilized more efficiently by growing crop (3). Dobereiner and Boddey (3) reported that different cereal species or cultivars exhibit different ability in promoting the association with *Azospirillum* strains.

The present study was carried out to develop an efficient Azospirillum biofertilizer for wheat crop and also to screen wheat genotypes and efficient Azospirillum strains for better association to achieve increased grain yield production.

2. MATERIALS AND METHODS

(a) Isolation of *Azospirillum*

Root samples of wheat containing rhizosphere soil were collected from various places of Varanasi district. The freshly collected roots were washed thoroughly in running tap water to remove soil particles. The roots were cut into small pieces [0.5-1.0 cm length]. Root pieces were surface sterilized in 0.1% $HgCl_2$ for 5 minutes followed by quick dipping in 70% ethanol. Root pieces were washed for five times with sterile distilled water. The sterilized root pieces were macerated and transferred to culture tubes containing slant of nitrogen free Dobereiner's bromothymol blue [NFB] semisolid medium [2]. The culture tubes were kept in incubator at 30 ± 2°C temperature. The culture tubes were observed regularly for appearance of bacterial growth and change of colour of the medium around root pieces. The cultures were purified on Potato Infusion Agar Medium. Pure pinkish wrinkled colonies were picked up with the help of inoculating needle and transferred to culture tubes containing slant of NFB medium for maintenance and in liquid culture medium in culture flasks for their growth and preparation of *Azospirillum* inoculum.

The strains of *Azospirillum* were isolated following above procedure. These locally isolated strains along with *Azospirillum* strain from I.A.R.I. were grown on semisolid malate agar medium for four days and their nitrogen fixation ability was estimated quantitatively by microkjeldahl method.

(b) Pot Experiment

Two high nitrogen fixing local strains and one strain from I.A.R.I. were tested with four varieties of wheat (K 9006 HP 1731, HUW 12 and HUW 234) in pots to find out best association of wheat genotypes with bacterial strains.

(c) Details of experiment

Treatment : Four

T_1 = Control

T_2 = *Azospirillum* strain from IARI coated around seeds.

T_3 = Local culture I coated around seeds (strain No 3).

T_4 = Local culture II coated around seeds (strain No. 8).

Variety : Four

V_1 = K 9006 V_3 = HUW 12

V_2 = HP 1731 V_4 = HUW

Replication : Three, Pot area = 0.18 Sq. m approx.

Weight of soil per pot : 10 kg

Chemical fertilizer level : 60 : 40 : 40 kg N: P : K/ha.

Table 1a. Composition of Nitrogen free Malic acid medium (NFB).

Chemical compounds	Amount per litre
Malic acid	5.0 g
KOH	4.0 g
K_2HPO_4	0.5 g
$FeSO_4.7H_2O$	0.05 g
$MnSO_4.H_2O$	0.01 g
$MgSO_4.7H_2O$	0.10 g
NaCl	0.02 g
$CaCl_2$	0.01 g
$Na2MoO_4$	0.002 g
Distilled water	1000 ml
Bromothymol blue (0.5% alcholic soln)	2.0 ml
Agar for liquid medium	1.75 g
Agar for semisolid medium	15-18 g/l
pH	6.8

Table 1b. Potato infusion Agar medium for purity check.

Potato/Chemical compound	Amount in one litre
Potatoes (peeled)	200.0 g
Malic Acid	2.5 g
KOH	2.0 g
Sucrose	2.5 g
Biotin	0.1 g
Bromothymol blue 0.5%	2 drops
Distilled water	1000 ml

Amount of fertilizer applied per pot = 25g urea (15+ 10) in all pots including control, 12g M O P, 35g, SSP.

(d) Seed treatment

0.5 ml *Azospirillum* suspension culture was coated around 20 g wheat seeds with the help of 1.0 ml sticker solution (25 g Gur/sugar) and 1.0 g Gum Acacia in 50 ml water). The mixture was shaken properly for good adherence of bacteria around seeds. The control seeds were shaken in 1.5 ml sticker solution only.

3. RESULTS AND DISCUSSION

Isolation of Azospirillum strains : The colour of the medium around root pieces changed from yellowish green colour to bluish to brilliant blue colour after four days of transfer due to growth of the Azospirillum bacteria. White pellicles were formed on the surface just below the lower meniscus of the medium layer after four days of transfer of root pieces. Positive cultured tubes showing *Azospirillum* growth were collected. Ten *Azospirillum* strains were isolated from the wheat roots collected from various places. These strains were cultured on slant containing NBF medium for maintenance and the liquid medium for inoculum preparation.

Nitrogen fixation of Azospirillum strains: Significant variation was found in amount of nitrogen fixed by ten strains. The amount of nitrogen fixed varied from 14.3 mg/g of Na-malate to 30.2 mg/g Na-Malate [Table 2]. The two best nitrogen fixing strains No 3 and 8, out of ten strains isolated, and one IARI strain were selected for pot experiment.

Table 2. *Azospirillum* isolates and their ability to fix nitrogen

Strain No.	Nitrogen fixation mg/g Na-malate
1. Local strain	16.4
2. Local strain	22.0
3. Local strain	26.4*
4. Local strain	25.0
5. Local strain	23.1
6. Local strain	24.5
7. Local strain	14.3
8. Local strain	30.2**
9. Local strain	20.7
10. Local strain	18.0
11. IARI strain	26.0
CD (at P = 0.05)	

*3 Local strain used as LC I;
**8 Local strain used as LC II

The strains exhibited great variation in their nitrogen fixation ability *in vitro*. Similar variations in nitrogen fixation with strains of *Azospirillum* were reported (13).

(a) Pot experiment

It was observed that inoculation of *Azospirillum* increased the vigour of wheat plants after 25-30 days of sowing. Bacteria containing seeds showed early germination by one day in comparison to untreated control. *Azospirillum* inoculation improved the plant colour from yellowish green (control) to green. The improvement in crop vigour and plant colour may be attributed to the availability of uptakable nitrogen and growth promoting substances in close vicinity of plant roots (6). *Azospirillum* secreted growth promoting substances viz., GA, IAA and kinetin etc, which promote germinability and plant growth (8,9).

(b) Plant height

The plant height showed significant increase with the treatments and also with varieties. However, the interaction between treatment and variety was non significant. Further, mean plant height was found more in T_4 followed by T_2 and T_3 (Table 3). The variety V_1 (K 9006) exhibited maximum height followed by HUW 234 in treatment 4. On the other hand V_2 (HP (1731), a short duration variety, exhibited less increase in plant height in all treatments. HUW 234 showed maximum percent increase in plant height (4.68) followed by K 9006 (4.18) and HUW 12 (3.9) (Table 4). T_4 showed best effect on plant height increase in all varieties with their respective controls. The increase in plant height with *Azospirillum* application occurred due to increased availability of uptakable nitrogen and secretion of growth promoting substances. Similar results have been reported in pearlmillet (6,7).

(c) Biomass

Azospirillum application increased the total biomass of the plants/pots in all wheat varieties in comparison to control. Increase in biomass was statistically significant for both treatments and varieties.

The varieties have shown different response to different treatment (Table 3). Variety K 9006 exhibited maximum mean biomass in response to local culture II (T_4) followed by variety V_4 (HUW 234) in T_4 and variety K 9006 in T2. In contrast HP 1731 resulted low biomass/pot.

Table 3. Effects of *Azospirillum* on different quantitative characters of wheat (based on mean value of three replicates).

Treatment	Characters		
	Plant height (cm)	Biomass/ pot t (g)	Grain Yield/ pot (g)
T_1V_1	101.2	119.56	36.91
V_2	89.0	106.54	34.98
V_3	94.5	114.99	37.42
V_4	99.6	115.73	37.26
T_2V_1	103.6	124.54	40.81
V_2	90.6	108.84	36.28
V_3	96.3	116.79	80.93
V_4	102.4	118.76	41.25
T_3V_1	102.9	120.22	38.46
V_2	90.2	110.94	36.98
V_3	96.0	117.43	39.04
V_4	102.2	116.38	38.74
T_4V_1	105.4	125.16	39.63
V_2	90.4	109.65	36.55
V_3	98.2	118.22	39.23
V_4	104.3	122.20	44.04
CD (P= 0.05)	1.63	0.8	1.61
SEDM	0.78	1.32	0.65

Table 4. Effect of *Azospirillum* inoculation on plant height (cm) in wheat varieties (values are sum of three replicates).

Treatment	Varieties				Mean
	V1	V2	V3	V4	
T1	303.5 (0)*	267.6 (0)	283.5 (0)	298.5 (0)	228.35 (0)
T2	310.8 (2.40)	272.9 (1.98)	288.9 (1.89)	307.9 (2.81)	294.95 (2.29)
T3	308.8 (1.74)	270.6 (1.1)	288.0 (1.58)	306.6 (2.61)	293.5 (1.78)
T4	316.9 (4.18)	271.2 (1.38)	294.6 (3.9)	312.8 (4.68)	298.7 (3.5)
Mean	309.8	270.6	288.75	306.35	

CD (P=0.05) Tr x V=3.25; SEDM Tr x V=1.59; CV = 2.0%

*Value in parenthesis denotes per cent increase over respective control.

T_4 was found most efficient to increase total biomass in all varieties in comparison to other treatments, however treatment with IARI culture (T_2) was found equally good to increase total biomass of variety K 9006. HUW 234 had shown maximum per cent increase in biomass (5.61) in response to T_4 followed by variety K 9006 (4.70) in response to T_4 and (4.2) in response to T_2 (Table 5).

Table 5. Effect of *Azospirillum* inoculation on biomass (g) in wheat varieties (values are sum of three replicates).

Treatment	Varieties				Mean
	V1	V2	V3	V4	
T1	358.68 (0)*	320.82 (0)	344.97 (0)	347.19 (0)	342.9 (0)
T2	373.75 (4.20)	326.52 (1.77)	350.37 (1.57)	356.28 (2.62)	351.7 (2.54)
T3	360.66 (0.55)	332.82 (3.72)	352.29 (2.21)	349.14 (0.57)	348.7 (1.68)
T4	375.48 (4.7)	328.95 (2.5)	354.09 (2.6)	366.67 (5.61)	356.79 (3.90)
Mean	367.1	327.3	350.43	354.82	

CD (P=0.05) Tr x V=3.21; SEDM Tr x V=1.57; CV=1.65%

*Value in parenthesis denotes per cent increase over respective control.

Increase in total biomass was found due to increased vigour of wheat plants in response to *Azospirillum* inoculation which increased the nitrogen availability. Secondly some growth promoting substances released by bacteria in root tissues increased the crop vigour, thus increased vigour of plants enhanced the total biomass (7, 9).

(d) Grain yield

Application of *Azospirillum* increased the grain yield in all wheat varieties with their respective controls. Grain yield per pot was found significant for varieties and treatments. Variety V4 exhibited maximum grain yield in T4 followed by T2 (Table 3). Treatment with IARI culture (T2) increased the grain yield of K 9006 more than T4. The variety HUW 234 showed highest per cent increase in grain yield over control in T4 whereas variety K 9006 showed maximum per cent increase in grain yield (10.56) over control in response to T2 (Table 6). Minimum per cent increase in grain yield was found in variety HUW 12 and HP 1731.

Table 6. Effect of *Azospirillum* inoculation on grain yield per pot (g) in wheat varieties (values are sum of three replicates).

Treatment	Varieties				Mean
	V1	V2	V3	V4	
T1	110.73 (0)*	104.94 (0)	112.26 (0)	111.78 (0)	109.9 (0)
T2	122.43 (10.56)	108.84 (3.7)	116.79 (4.0)	123.75 (10.45)	117.2 (6.57)
T3	115.38 (4.2)	110.94 (5.7)	117.12 (4.3)	116.22 (3.95)	114.9 (4.5)
T4	118.8 (7.34)	109.65 (4.5)	117.69 (4.8)	132.15 (18.13)	119.6 (8.73)
Mean	116.86	168.6	116.0	121.0	

CD (P=0.05) Tr x V=2.65; SEDM Tr x V = 1.3; CV = 4.13%

*Value in parenthesis denotes per cent increase over respective control.

The grain yield increase in response to *Azospirillum* application was due to cumulative factors induced by bacterial inoculation viz., increase in number of effective tillers (10), availability of more uptakable nitrogen fixed by *Azospirillum* (1, 12), increased production of nitrogen assimilatory enzymes in plants (5), increased number and length of roots (7) and secretion of growth promoting substances in roots (7, 9). All these factors produced additive effect to increase the grain yield in treated pots. The increase in grain yield in response to *Azospirillum* inoculation has been reported in many crops viz., wheat (8), maize (4) sorghum (11) and *Brassica juncea* (1).

Wheat genotypes exhibited different association with *Azospirillum* strains which is in conformity with the reports of Dobereiner and Boddey (3). Variety HUW 234 and variety K 9006 showed efficient association with *Azospirillum* local culture II and IARI strain respectively, these associations increased the grain yield significantly.

REFERENCES

1. Chauhan D.R., Paroda S., Kataria O.P. & Singh K.P. 1995. Indian J Agron 41 : 86.
2. Dobereiner J., Marriel I.E. & Nery M. 1976. Can J Microbiol, 22 : 1664.
3. Dobereiner J. & Boddey R.M. 1981. In Current Perspectives in Nitrogen fixation (Gibson AE & Newton WE eds) Austration Acad Sci., 305.
4. Follik E. & Okon Y. 1996. World J Microbiol & Biotech, 12 : 511.
5. Gunasekharan S. & Shivakumar S. 1996. Indian J Microbiol, 36 : 223.
6. Kundu B.S., Sangwan P., Sharma P.K. & Nandwal 1997. AS Indian J Plant Physiol, 2 : 101.
7. Kundu B.S. & Tauro P. In Current status of Biological Nitrogen Fixation (Singh R ed) HAU Hissar 183.
8. Mertens T. & Hess D. 1984. Plant & Soil, 82 : 87.
9. Purushothaman D. & Menon P. 1984. Indian J Microbiol., 24 : 225.
10. Sawarkar S.D. & Goydoni D.M. 1996. Indina J Agron., 41: 409.
11. Subba Rao N.S. 1982. Biofertilizer in Agriculture Oxford and IBH Publishing Co New Delhi. 126.
12. Subba Rao N.S. 1984. Current Developments in Biological Nitrogen Fixation Oxford and IBH Publishing Co New Delhi. 227.
13. Subba Rao N.S., Singh C.S. & Tsuru, S. 1984. IBL Mocrobiol, 39 : 607.

CHAPTER - 10

NEGATIVE (STIMULATORY) ALLELOPATHY OF *Lantana camara* L. ON LINSEED *var.* KIRAN

P. Oudhia, S.S. Kolhe and and R.S. Tripathi,

Department of Agronomy, Indira Gandhi Agricultural University, Raipur 492 001 - India.

ABSTRACT

Lantana camara L. is one of the common problematic weed of Chattisgarh plains of India having allelopathic property. In order to find out the allelopathic potential of *Lantana camara* L. on germination and seedling vigour of linseed var. Kiran, an experiment under controlled condition was carried out at Department of Agronomy, IGAU, Raipur (India) during 1995-96. The study revealed that 3DAS, root extract of 120 hours and at 11DAS, root extract of 168 hours produced maximum germination. Maximum root and shoot elongation were noted under leaf and stem+leaf extract of 264 hours, respectively. The study indicated the possibility of utilizing stem+leaf extract of 264 hours for presowing soaking treatment of linseed seed in order to achieve higher shoot elongation without affecting its germination. This stimulatory allelopathy was the result of interaction between micro-organisms and weed material in the presence of water. The study suggested the products of this beneficial interaction can be utilized for developing eco-friendly crop growth regulators.

1. INTRODUCTION

The term allelopathy refers to all bio-chemical interactions positive or negative between organisms including micro-organisms (1). The negative (stimulatory) allelopathic potential of any weed can be utilized for promoting seedling vigour, dry weight and finally high crop production (2-6). The negative allelopathy producing extracts of weed parts can also be utilized for tissue culture and biotechnological operations and developing growth promoting substances (7). Similarly, the positive (inhibitory) allelopathy of weed towards other weeds can be utilized for developing ecofriendly and cheaper "Gren Herbicides". *Lantana camara*, locally known as Gotiphool, is one of the common problematic weeds of Chhattisgarh region (8). Due to non-availability of any information regarding

Table 1. Allelopathic effect of *Lantana camara* L. on germination (%) of linseed

	Decaying Period			
	120 hours	168 hours	216 hours	264 hours
Source		**3 DAS**		
Root	66.20 (54.45)	43.30 (41.15)	46.00 (42.71)	18.60 (25.56)
Stem	40.00 (39.23)	28.70 (32.38)	23.30 (28.86)	62.00 (46.15)
Leaf	25.20 (30.13)	2.70 (9.40)	27.80 (31.82)	61.40 (51.59)
Stem + Leaf	42.00 (40.40)	40.0 (39.23)	40.60 (39.58)	29.30 (32.71)
Control (Water)	52.00 (46.15)			
LSD (0.05)	6.11			
		5 DAS		
Root	72.00 (58.05)	76.30 (60.67)	66.30 (54.51)	60.70 (51.18)
Stem	70.40 (57.04)	75.90 (60.60)	79.00 (62.72)	78.00 (62.03)
Leaf	46.00 (42.71)	67.50 (55.24)	60.70 (51.18)	61.40 (51.59)
Stem + Leaf	42.00 (40.40)	45.30 (42.30)	63.20 (52.65)	72.70 (58.50)
Control (Water)	67.90 (55.45)			
LSD (0.05)	4.97			
		7 DAS		
Root	72.00 (58.05)	82.10 (64.87)	68.90 (56.11)	65.40 (53.97)
Stem	73.00 (58.69)	76.70 (61.14)	79.00 (62.72)	78.00 (62.03)
Leaf	50.00 (45.00)	74.80 (59.84)	65.80 (54.21)	61.40 (51.59)
Stem + Leaf	43.30 (41.15)	56.20 (48.56)	64.70 (53.55)	72.70 (58.50)
Control (Water)	72.50 (58.37)			
LSD (0.05)	5.36			
		9 DAS		
Root	72.00 (58.05)	85.40 (67.54)	72.80 (58.56)	68.00 (55.55)
Stem	80.70 (63.94)	85.10 (67.29)	80.10 (63.57)	84.80 (67.05)
Leaf	52.70 (46.55)	88.00 (69.73)	67.10 (55.01)	58.70 (50.01)
Stem + Leaf	46.60 (43.05)	68.00 (55.55)	68.70 (55.98)	78.30 (62.24)
Control (Water)	82.10 (64.87)			
LSD (0.05)	8.38			
		11 DAS		
Root	72.00 (58.05)	88.90 (70.54)	74.10 (59.41)	70.10 (56.85)
Stem	80.70 (63.94)	88.70 (70.36)	80.10 (63.57)	84.80 (67.05)
Leaf	82.00 (51.94)	88.00 (69.73)	70.30 (56.98)	60.70 (51.18)
Stem + Leaf	49.20 (44.54)	69.30 (56.38)	70.10 (56.83)	80.20 (63.58)
Control (Water)	82.10 (64.87)			
LSD (0.05)	7.18			

Note : Figures in parentheses indicate angular value.

Table 2. Allelopathic effect of *Lantana camara* L. on seedling vigour of linseed

	Decaying Period			
	120 hours	168 hours	216 hours	264 hours
Source		**SHOOT LENGTH (cm./plant)**		
Root	3.13	6.35	6.67	8.26
Stem	4.91	6.83	5.34	8.76
Leaf	6.70	6.93	7.55	9.20
Stem + Leaf	5.73	6.69	7.26	10.20
Control (Water)	7.43 (1.24)			
LSD (0.05)	1.24			
		ROOT LENGTH (cm./plant)		
Root	3.69	5.82	4.36	4.13
Stem	6.11	4.05	4.17	5.38
Leaf	8.89	3.99	5.10	5.07
Stem + Leaf	5.00	6.73	3.35	5.57
Control (Water)	4.08			
LSD (0.05)	1.25			

allelopathic potential of this obnoxious weed on linseed an experiment under laboratory condition was carried out at Depatment of Agronomy, IGAU, Raipur, India during 1995-96.

2. MATERIALS AND METHODS

Roct. stem, leaf and stem+ leaf of *Lantana camara* L. were allowed to decay for the period of 120, 168, 216 and 264 hours as per treatment in water in the ratio of 1:10 w/v. At different time periods the extracts were prepared and applied in specific amount (15 ml.) on 50 representative seeds of linseed through seed soaking treatment. the experiment was laid out in Completely Randomized Design with a control (water). The experiment was replicated thrice and repeated twice. At 3, 5, 9, 7 and 11 DAS (Days After Sowing) the germination was recorded and as 11 DAS, root and shoot elongation were mesured. Linseed variety Kiran was used as test crop.

3. RESULTS AND DISCUSSION

The extracts of different parts of *Lantana camara* produced significant effect on germination and seedling vigour of linseed. At 3DAS, root extract of 120 hours produced maximum germination whereas lowest germination was noted in leaf extract of 168 hours. Leaf extract of 264 hours resulted in comparable germination with root extract of 120 hours. Stem extract of 216 hours and root extract of 168 hours produced maximum germination at 5 and 7 DAS, respectively. Stem + leaf extract of 120 hours lowered the germination at 5, 7, 9 and 11 DAS. At 11 DAS, stem extract of 120 216 and 264 hours, leaf extract of 168 hours, stem+leaf extract of 264 hours and root extract of 168 hours produced comparable germination with control (water) (Table 1).

Maximum root elongation was noted under leaf extract of 264 hours whereas stem+ leaf extract fo 264 hours resulted in maximum shoot elongation. Stem+ leaf extract of 216 hours and root extract of 120 hours lowered the root and shoot elongation to the minimum (Table 2). The negative (stimulatory) allelopathy of *Lantana camara* as produced by different treatment combinations in this experiment was also observed indifferent crops in early studies (9). As the weed parts decayed in water for different time periods, the interaction between weed parts and micro-organisms might have produced some allelochemicals which was resulted in stimulatory germination and seedling vigour. The study revealed that the promising extracts resulted in stimulatory allelopathy *viz.*, stem+leaf extract of 264 hours, may be utilized for achieving higher root and shoot elongation without affecting germination, through presowing soaking treatment of crop seed in these extracts. This extracts may also be utilized as a growth media in order to promote the initial growth of crop seeds in commercial laboratories. As the weed, *Lantana camara* is available in plenty throughout India. farmers can prepare these extracts by dipping weed parts in water for specific time period and treating their seeds with it. The study also revealed that decaying of different parts of different time periods, a common phenomenon in nature, can alter the allelopathic potential of that particular weed. These observations were in line with those reported in earlier studies on different crops. (10 & 11). Furthermore this work under field condition would provide better understanding of alletopathic phenomenon of this weed on linseed.

REFERENCES

1. Narwal SS. Allelopathy in crop production. *Scientific publication* Jodhpur (1994) pp. 145.
2. Oudhia P, Kolhe SS & Tripathi RS. *Weed News* 2(172) (1995) 15.
3. Oudhia P, Kolhe SS & Tripathi RS *Agril Biol Res*, 12(1&2) (1996) 12.
4. Oudhia P, Kolhe SS & Tripathi RS *Indian J Plant Physiol*, 2(4) (1997a) 327.
5. Oudhia P, Kolhe SS & Tripathi RS. *Indian J Weed Sci*, 29(1&2) (1997b) 4.
6. Oudhia P, Kohle SS & Tripathi RS. *Weed Oryza* 35(2) (1998a) (In Press).
7. Oudhia P, Kohle SS & Tripathi RS. *Leg Res* (1998b) (In Press).
8. Oudhia Pankaj and Dixit Anil, *Weed News* (1994) 19.
9. Oudhia P, Kohle SS & Tripathi RS. *Crop Res* 16(1) (1998b) (In Press).
10. Oudhia P, Kohle SS & Tripathi RS. *World Weeds* 4 (1998c) 109.
11. Oudhia P, Kohle SS & Tripathi RS. *Agri Sci Dig* (1998c) (In Press).

CHAPTER - 11

GENE TRANSFER THROUGH SOMATIC HYBRIDIZATION IN *Brassicas*

*R.P. Singh, B.K. Prasad, S.P. Singh and B.D. Singh**

Department of Genetics and Plant Breeding Institute of Agricultural Sciences
*School of Biotechnology
Faculty of Science, Banaras Hindu University, Varanasi.

ABSTRACT

Genetic variability is of utmost importance for crop improvement and it could be created through gene transfer. Conventional gene transfer could be achieved through sexual hybridization but it is confined to closely related sexually compatible species only, which limits the gene flow from other species possessing desirable genes. Protoplast fusion enables the recovery of hybrid between sexually incompatible species. Many such hybrids are partially fertile and also possess desirable traits. Hybrids produced through protoplast fusion in Brassicas are being used in breeding programme. Similarly cybrids can also be produced through normal protoplasts fusion or from fusion between protoplast and cytoplast. These cybrids possess mitochondria of one species and protoplast of second species. Through somatic hybridization and proper selection, C3 plants could be transferred into C4 plants with higher photosynthetic efficiency leading to higher grain yield. The various applications of somatic hybridization technology in Brassicas are discussed using suitable example.

1. INTRODUCTION

Genetic variability is prerequisite for any crop improvement programme and it could be created through gene transfer. Although, conventionally, it could be created through sexual hybridization but it is limited to closely related sexually compatible species which limit the gene flow from other species possessing desirable genes. Wild relatives of cultivated species have contributed remarkably to the improvement of cultivated species and in various cases, the crop retained their commercial status [5]. However sexual incompatibility hampers the utilization of various sources of desirable genes in crop improvement programme, but the same may be overcome through somatic hybridization (Table 1). Although the genetic transformation represent the ultimate strategy for gene transfer but it is quite expensive and technically most exacting because it requires identification of genes, isolation, cloning of concerned gene, transfer of gene through suitable vector and selection of transformed cells/plants. However, some characters of interest governed by two or more genes, transfer of such characters may pose many difficulties. Similarly transfer of cytoplasmic organelles is not possible through genetic transformation. Thus transfer of cytoplasmic organelles and traits govern by more than one gene can be readily achieved through somatic hybridization.

2. PROTOPLAST ISOLATION AND FUSION

Plant protoplasts can be routinely isolated by enzymatic digestion of cell wall with cellulase and pectinase. Leaf mesophyll and embryogenic suspension culture protoplasts are quite useful for somatic hybridization [1, 14].

Protoplasts are generally fused by a treatment with 20-40% solution of polyethyleneglycol (PEG) of 1500-1600 MW, followed by dilution with high calcium (Ca) and high pH elution medium. PEG induces adhesion of protoplasts belonging to any plant species, the protoplast fusion, however, occur during elution [1]. Some PEG preparations may be toxic to plant cell; this problem can be resolved by deionization of PEG solution or by using lower concentration alongwith a few per cent of dimethylsulphoxide (DMSO [14]).

Alternatively, a very short time (few micro-seconds) field of high voltage (500-1000 V/cm) may be used to induce fusion between protoplasts that are first brought in close contact by a non uniform alternating field between two electrodes. Electrofusion has been mostly used with members of Solanaceae often with very high rate (<50%) of fusion [14].

Table 1. A summary of the various approaches available for gene transfer in plants.

Problem/objective of gene transfer	Approaches	
	Conventional	Biotechnological
A. *Intervarietal Crosses*		
Nonflowering, sterile parents	None	Somatic hybridization (SH)
B. *Interspecific Crosses*		
Hybrid sterility	Chromosome doubling	Somatic hybridization
Sexual incompatibility	Bridging species	Somatic hybridization
1. Endosperm abortion	None	Embryo rescue, SH
2. Prefertilization barriers	Stigma/style manipulation etc.	*In vitro* pollination, SH
3. Very wide crosses	None	SH, Genetic transformation
4. Combining mitochondria of one parent with chloroplastsof the other	None	Somatic hybridization
5. Production of recombinant mtDNA and cpDNA	None	Somatic hybridization
C. *Entirely unrelated organisms*		
Animals, bacteria viruses etc.	None	Genetic transformation

In a modification of electrofusion, protoplast pairs are individually transferred, using a micromanipulation set up, into microfusion chambers, where they are induced to fuse by a single or multiple, negative DC-pulse (50μs; 800-1800 V/cm) after mutual dielectrophoresis (1 MHz; 65-80 V/cm). Protoplasts of a *Nicotinana tabacum* line were microfused with those of another *N. tabacum* cms line; the fusion products, contained individually in 60 μl droplets were transferred into microculture chambers, where they produced macroscopic colonies. A majority of the plants regenerated from these calli were somatic hybrids, while many were cybrids. Most of the plants had chloroplasts of one of the other fusion parent, while few of them were variegated. The mitochondrial genome of the plants was different from those of the fusion parents, indicating an extensive mt DNA rearrangement [19]. This appears to be the common status of mitochondria and chloroplasts in somatic hybrids and cybrids.

3. SELECTION OF SOMATIC HYBRIDS

In any fusion experiments, the heterokaryon/hybrid cell formed by fusion of a single protoplast from each of the two fusion partners is, ordinarily, the only desirable cell type, and it forms only a small proportion of the population. Several strategies have been devised for selection and isolation of such heterokaryons/hybrid cells. Initially, these strategies were based on complementation in the somatic hybrid of natural or induced deficiencies of the fusion parents [1]. At present, however, different selectable marker transgenes are transferred into the two fusion parents, which allows an efficient selection for the hybrid cells (see later). Isolation of heterokaryons/hybrid cells or colonies may be based on natural morphological differences between the fusion parents. Detectable differences can, however, be more readily produced by staining the nuclei of protoplasts with distinct dyes, which may even be adopted to automatic cell sorting leading to the recovery of large numbers of heterkaryons rather rapidly [1, 14]. However, many workers prefer to plate the entire protoplast population after the fusion of treatment and to screen the resulting colonies/plants to the somatic hybrids/cybrids [1, 9].

4. SYMMETRIC SOMATIC HYBRIDS

When normal protoplasts of two species are fused, the resulting somatic hybrid plants may retain the somatic chromosome complement of both the fusion parents. Such somatic hybrids, called symmetric hybrids, immediately give rise to a new species, A novel species *Brassica naponigra* was produced by fusion of the protoplasts of *B. napus* and *B. nigra*, two sexually incompatible species. *B. nigra* offers valuable sources for resistance to *Phoma lingam*, the causal agent of black-leg disease, to which *B. napus* accessions show, at best; only moderate resistance. *B. naponigra* was fertile and resistant to black-leg. Ingergenomic recombination would be required to transfer the black-leg resistance gene from the 'B' genome of *B. nigra* into the 'C' or 'A' genome of *B. napus*. There is some evidence that such recombination does take place. This somatic hybrid (*B. naponigra*) is being used in breeding programmes [18].

B. napus has been resynthesized by fusing the protoplasts of *B. oleracea* var. *botrytis* (a dwarf cauliflower with 62.9% erucic acid) and *B. campestris* var. *rapifera*, or var. *oleifera* (53-54% erucic acid). Most of the somatic hybrids had reduced plant height, large seed size, and nonshatering pods, while some had good tolerance to downy mildew; the erucic acid content of the hybrids was 57.4% [6]. Similarly, *B. napus* resynthesized from *B. oleracea* var. *botrytis* and *B. campestris* strains having low linolenic acid contents (4.6% in *B.*

oleracea and 4.1% in *B. campestris* respectively), showed 3.5 to 7.9% linolenic acid contents; one progeny showed 3.5% linolenic acid in Ro and R1 generations. The synthetic *B. napus* lines can be used in breeding programmes as additional sources of useful genes [7].

Xanthomonas campestris pv campestris is the most serious pathogen of crucifers throughout the world; it causes black rot of *B. oleracea*, which does not have lines with adequate levels of resistance. Certain lines of *B. napus* are highly resistant to black rot but crosses between *B. napus* and *B. oleracea* are extremely difficult, success depending on the genotypes involved and on embryo rescue. Four of the seven somatic hybrids obtained from fusion of *B. napus* and *B. oleracea* protoplasts were fertile and highly resistant to the disease. The hybrid was successfully back crossed to *B. oleracea* 'line 15' using embryo rescue. The resulting BC1 progenies back crossed easily to 'line 15' without embryo rescue and several black rot resistant BC2 plant were isolated [4].

Somatic hybridization offers the unique opportunity for combining mitochondria of one species with the chloroplasts from another; this may be highly desirable in some cases. A cms version of *B. juncea* cv RLM 198 was developed through a back cross programme using a *B. napus* cms line ('ogura' cytoplasm) and the nonrecurrent female parent. The ogura cytoplasm was discovered in a wild population of radish (*Raphanus sativus*) and was subsequently transferred into *B. napus*. The cms version of RLM 198 showed a high degree of leaf chlorosis at low temperatures and several floral deformities. Protoplasts of cms (ogura) RLM 198 were fused with those of normal male fertile RLM 198. A total of 123 plants were regenerated of which 4 plants remained dark green at 8°C and were male sterile. These parents were back crossed thrice to RLM 198, and normal green cms plants (2n=36) showing 90% female fertility were isolated. The parents had chloroplasts of *B. juncea* and novel recombinant mitochondria which retained the ability to produce male sterility. In addition to the correction of chlorosis, these plants showed increased vigour and vastly improved female fertility over the earlier cms version of RLM 198 [9].

5. ASYMMETRIC HYBRIDS

Intensive somatic hybridization is being carried out in *Brassica* to generate new source of cms or to transfers genes for disease resistance and other traits. Gene transfer is more likely to be achieved by producing asymmetric hybrids in which almost all the chromosomes of one fusion partner are eliminated. Such hybrids are produced due to a preferential chromosome elimination of one of the fusion parents from normal somatic hybrid/cells either spontaneously or in response to an irradiation treatment. Asymmetric hybrids resulting from spontaneous chromosome elimination often provide examples of useful gene transfers during *in vitro* culture. Six plants regenerated from fusion products of *B. juncea* and *Trachystoma ballii* protoplasts showed 2n=36 (18" at MI) and intermediate morphology. These plants were back crossed 3 times to *B. juncea*, and in BC3 some plants were identified that were able to restore fertility in *B. juncea* cms lines having *T. ballii* cytoplasm. Restoration seemed to be due to a single dominant gene [11]. The cms (*T. ballii* cytoplasm) line was obtained by five back crosses to *B. juncea* of the 4 symmetric hybrids obtained by fusion of *B. juncea* and *T. ballii* protoplasts [10].

In an effort to transfer black-leg resistance to *B. napus*, irradiated mesophyll protoplasts of *B. juncea*, *B. nigra* of *B. carinata* (all are resistant to black-leg) were fused with non irradiated protoplasts of *B. napus* (stained with 5 (6)-carboxyfluo-

Table 2. Some selected examples of symmetric and asymmetric somatic hybrids and cybrids.

Fusion partner: Recipient	Fusion partner: Donor	Trait Transferred	Remarks	References
		Symmetric hybrids		
Brassica juncea +	*Trachystoma ballii*	Cytoplasmic male sterility (by repea ted Back crossing)	*B. juncea* cms line	Kirti et al. (1995b)
Brassica napus +	*B. nigra*	Black-leg (Phoma lingam) resistance	Included in breeding programme	Sjodin and Glimelius (1989a)
B. napus +	*Moricandia arvensis*	C_3-C_4 intermediate features (Kranz leaf anatomy and CO_2 compensation point)	—	O'Neill et al.(1996)
B. napus+	*Raphanus sativus*	Resistance to beet cyst nematode	Reduced fertility, unable to backcross with *B. napus*	Lelivelt and Krens (1992)
B. napus+	*Thlaspi perfoliatum*	High nervonic acid content	Partial fertility on back cross	Fahleson et al. (1994)
B. oleracea var. *botrytis* + PI 320999	*B. campestris* var. *Oleifera* Pl 347607	Low linolenic acid *B. napus* (3.5%)	Stably inherited in R_1 and R_2	Health and Earle (1997)
B. oleracea var. *botrytis* +	*B. campestris*	High erucic acid content (57.4%)	Short height, non shattering pod, large seeds, downy mildew tolerance	Health and Earle (1995)
		Symmetric hybrids		
B. oleracea +	*B. napus*	Black rot resistance	Black rot resistance BC2 plants, single dominant gene	Hansen and Earle (1995)
B. oleracea +	*Raphanus sativus*	Resistance to club root	Partial fertility on selfing and back crossing	Hagimori et al. (1992)
		Asymmetric hybrids (spontaneous origin)		
B. juncea +	*T. ballii*	Domionant restorer gene	Restorer lines R 72 & R 74	Kirti et al. (1997)
		Asymmetric hybrids (irradiation of donor)		
B. napus + + +	*B. juncea** *B. nigra** *B. carinata**	Black leg (Phoma lingam) resistance	Seed set on back cross with *B. napus*	Sjodin and Glimelius (1989b)
		Cybrids (spontaneous origin)		
B. nigra (*B. juncea* cms) +	*B. Campestris* var. *oleifera***	Atrazine resistant cy-toplasmic male sterile	CMS broccoli plants	Christey et al. (1991)
B. juncea cv. + RLM 198 (cms ogura)	*B. juncea* cv. RLM 198**	RLM 198 chloroplast and cms of ogura (mt DNA recombination)	Increased vigor and female fertility	Kirti et al. (1995a)
B. oleracea (Ogura cms) +	*B. napus* (atrazine resistance)**	Atrazine resistance	—	Jourdan et al. (1989)
		Cybrids (irradiation of donor)		
B. oleracea var. + *broccoli* cv. Green commet	*B. napus** sterility	'Polima' cytoplasmic male sterility	One plant set seed on back crossing broccoli	Yarrow et al. (1990)
		Cybrids (cytoplast fusion)		
B. napus cv. Westar +	*Raphanus sativus* cv. Kosena (CMS)@	Cytoplasmic male sterility	One cms plant bac kcross with *B.napus*	Sakai and Imamura (1990)

*The protoplasts were irradiated. **Protoplasts of both the parents were fused.
@ Cytoplasts of this parent were used for fusion.

rescein diacetate. Heterokaryons from among fusion products were selected by flow sorting, and cultured on 8 pm medium having 3 5M sirodesmin PL, a toxin produced by *Phoma lingam.* Both symmetric and asymmetric hybrids were obtained. All the symmetric hybrids and 19 of the 24 asymmetric hybrids produced via toxin selection were resistant to *P. lingam*, while and nonselected asymmetric hybrids were susceptible. The pollen fertility ranged from 0.5-8%, and the plants set seed when back crossed with *B. napus* [17].

6. CYTOPLASM TRANSFER

Cytoplasmic-genetic male sterility provides, at present, the most successful tool for large scale hybrid seed production. Many male sterile cytoplasm have been obtained from related species. Therefore, transfer of cytoplasm in an effort to develop cms lines of crop plants has been the focus of several protoplast fusion experiments (Table 2). A general approach for the transfer of cytoplasm is to irradiate the protoplasts of the cytoplasm donor species and to treat the protoplast of the recipient species with iodacetic acid.

Irradiated protoplasts of a *B. napus* cms lines fused with iodoacetate-treated protoplasts of *B. oleracea* (broccoli). Four cybrids were recovered which had nuclear genome of broccoli and *B. napus* mitochondria and chloroplasts. However, only one of these plants set some seed which back crossed with the broccoli parent; surprisingly, seed set remained low even in the second back cross [21].

Cybrids have been often produced by fusion of protoplasts of the recipient species with the cytoplasts isolated from protoplast preparation of the donor parent. Cytoplasts can be prepared by percoll/mannitol density gradient ultracentrifugation. The density gradient may be prepared by overalyering three iso-osmotic solutions of different densities, viz., from bottom to top: I m each of 50% percoll diluted with 0.9 M mannitol, 20% percoll diluted with 0.6 M mannitol, and 0.6 M mannitol). The cytoplasts isolated after ultracentrifugation are mixed with the protoplasts of the recipient species and subjected to fusion either by PEG treatment or electrofusion. Cybrids have been produced in several cases mainly to transfer cytoplasmic organelles (Table 2).

Cytoplasts of a *Raphanus sativus* cms line fused with iodacetoamide-treated *B. napus* protoplasts. Of the seventeen plants regenerated, ten plants had diploid chromosome number of *B. napus*; one of these plants was male sterile and had novel mt DNA pattern [16].

Cytoplast fusion offers advantages over irradiation procedure for cybrid production as it eliminates the need for back crossing of the cybrids [16], as well as the risk of unwanted nuclear gene transfer from the donor parent. Spontaneous origin of cybrids from normal protoplast fusion experiments may also be accompanied with nuclear gene transfers from the donor parent [11].

7. CONCLUSION

Somatic hybridization allows transfer of cytoplasmic organelles in a single generation and offers unique opportunities for combining in a single hybrid the mitochondria of one species with chloroplasts from another species. This capacity may permit improvement of characteristics of certain cms lines [9], which may to their commercial exploitation. In addition, even nonflowering lines can be utilized in breeding programmes [15].

At present bulk of studies on various strategies of somatic hybridization are concerned with species of *Brassica, Nicotiana, Solanum,* and *Lycopersicon.* In case of potato, the technique is already being used in commercial breeding programmes of the Netherlands. In India, several *Brassica* lines derived from somatic hybridization are being ordered into breeding programmes.

REFERENCES

1. Bhojwani SS & Razdan MK 1983. Plant Tissue Culture : Theory and Practice. Elsevier, Publication, Amsterdam, 237.
2. Fahleson J, Eriksson I, Landgren M, Stymne S, glimelius K 1994. Theor Appl Genet 87 795.
3. Hagimori M, Nagaoka M, Kato N & Yoshikawa H 1992. Theor Appl Genet 84 819.
4. Hansen LN & Earle ED 1995. Theor Appl Genet 91 1293.
5. Harlan JR 1976. Genetic resources in wild relatives of crops Crop Sci 16 329.
6. Health DW & Earle ED 1995. Theor Appl Genet 91 1129.
7. Health DW & Earle ED 1997. Euphytica 93 339.
8. Jourdan PS, Earle ED & Mustschler MA 1989. Theor Appl Genet 78 71.
9. Kirti PB, Banga SS, Prakash S & Chopra VL 1995a. Theor Appl Genet 91 517.
10. Kirti PB, Mohapatra T, Baldev A, Prakash S & Chopra VL 1995b. Plant Breeding 114 434.
11. Kirti PB, Baldev A, Gaikwad K, Bhat SR, Kumar VD, Prakash S & Chopra VL 1997. Plant Breeding 116 259.
12. Lelivelt CLC & Krens FA 1992. Theor Appl Genet 83 887.
13. O'Neill CM, Murata T, Morgan CL & Mathias RJ 1996. Theor Appl Genet 93 1234.
14. Pelletier G 1993. In : Plant Breeding Principles and Prospects (Hayward MD, Bosemark NO & Romagosa I Eds) Chapman & Hall, London, 93.
15. Rasmussen JO, Nepper JP & Rasmussen OS 1996. Theor Appl Genet 92 403.
16. Sakai T & Imamura J 1990. Theor Appl Genet 80 421.
17. Sjodin C & Glimelius K 1989b. Theor appl Genet 78 513.
18. Sjodin C & Glimelius K 1989a. Theor appl Genet 77 651.
19. Spangenberg G, Osusky M, Oliveira MM, Freydl E, Nagel J, Pais MS, & Potrykus I. 1990. Theor Appl Genet 80 577.
20. Spangenberg G, Freydl E, Osusky M, Nagel J & Potrykus I 1991. Theor Appl Genet 81 477.
21. Yarrow SA, Burnett LA, Wildeman RP & Kemble RJ 1990. Plant Cell Reports 9 185.

CHAPTER - 12

SELECTION OF *Azotobacter* STRAINS FOR GROWTH AND SURVIVAL UNDER STRESS CONDITIONS

*Nirmala Agrawal and H.P. Singh**

Soil Microbiology Laboratory, Deptt. of Soil Science College of Agriculture, G.B.P.U.A. & T., Pantnagar, 263 145 (U.P.), INDIA

ABSTRACT

Twenty two strains of *Azotobacter* isolated from oat and wheat rhizosphere were tested *in vitro* to find out the ability to tolerate common occurring stress conditions such as salinity/sodicity, acidity/alkalinity and temperature. The nitrogenase activity of different strains measured in terms of total nitrogen fixed greatly differed and varied from 0.83 to 6.57 mg N/25 ml of broth culture due to strain PWN-6 and PWN-3. Out of 22 strains 3, 11 and 8 were recorded as low, medium and high nitrogen fixing strains. All strains differed widely in their tolerance to salinity/ sodicity, acidity/ alkalinity and temperature.

Strain PWN-14 was most sensitive while strain PWN-10, PWN-11, PWN- 17, PWN-19, PWN-20, PWN-21 were highly tolerant to sodium chloride. Low, medium, and high tolerant strains were 10, 7 and 5 for sodium sulphate and 4, 1, 17 for magnesium sulphate, respectively. About half of the strains recorded as highly resistant against the concentration of ≤ 2.0% $CaCO_3$.

All the strains grew between pH 6.0 to 8.0, none at pH 3.5 and only 4 strains at pH 10.0. *Azotobacter* strains which showed their growth at acidic pH of 5.0 to 6.0 and 8.5 to 9.5 were nearly 77% and 55%, respectively. All strains grew at temperature range between 25 to 40°C, none grew at 50°C and strain PWN-4, PWN-5, PWN-6, PON-13, PWN-18, PWN-21 and PWN-22 shown their growth at 45°C.

1. INTRODUCTION

A three dimensional system of soil-plant-microbe plays an important role in supplying adequate amount of nutrients to plant to improve our agriculture. Among

* Author for correspondance

beneficial microbes, biological nitrogen fixers both symbiotic such as *Rhizobium* as well as asymbiotic viz., *Azotobacter* account for about 65 percent of nitrogen currently utilized in agriculture and will be increasingly important in future crop productivity especially for sustainable system, small scale operations and marginal land utilization (1).

Azotobacter has been reported to increase plant growth, nutrient uptake and yield of cereals and forage grasses because of ability of asymbiotic nitrogen fixation, synthesis and secretion of various amino acids, growth promoting substances (2) vitamins. antibiotics and anti-pathogenic substances (3). Though, it has been reported to have extreme demands such as carbon substrate, major and micro- nutrients, pH, temperature, aeration, salinity/ sodicity, etc for its growth, multiplication and survival in a particular condition, all the species do not have identical requirement particularly under stress conditions. Therefore, an attempt has been made in the present investigation to select strains of *Azotobacter* for growth and survival under stress conditions.

2. MATERIALS AND METHODS

(a) *Azotobacter* Strains

Twenty two strains of Azotobacter viz. PWN-1, PWN-2, PWN-3, PWN- 4, PWN-5, PWN-6, PWN-7, PWN-8, PWN-9, PWN-10, PWN-11, PON-12, PON-13, PWN-14, PWN-15, PWN-16, PWN-17, PWN-18, PWN-19, PWN-20, PWN-21 and PWN-22 isolated from rhizosphere of Oat and Wheat plants grown on different types of soils were used.

(b) Preparation of Broth Culture

Jensen,s (1954) broth medium containing sucrose 20g, agar 15g, $CaCO_3$ 2.0g, $MgSO_4$ 0.5g, NaCl 0.5g, $K_2HPO_4$1.0g, $FeSO_4$ 0.1g, sodium molybdate trace, and water 1000 ml was prepared and pH was adjusted to 7.0± 0.2. Five ml of broth medium was transferred to each test tube and plugged with cotton wool, The test tubes were sterilized at 1.0Kg/cm^2 pressure for 30 minutes. After cooling, each tube containing broth medium was inoculated aseptically by a loopful of individual strain of *Azotobacter*. The tubes were shaked on a mechanical shaker for a week at 28± 2°C and microbial population of broth culture recorded as PWN-1 (108.25), PWN-2 (11.32), PWN-3 (81.33), PWN-4 (48.59), PWN-5 (28.94), PWN-6 (35.21), PWN-7 (47.81), PWN-8 (59.85), PWN-9 (44.99), PWN-10 (84.48), PWN-11 (53.390, PON-12 (50.95), PON-13 (71.56), PWN-14 (53.87) PWN-15 (38.27), PWN -16 (94.78), PWN-17 (78.82), PWN-18 (28.02), PWN-19 (41.76), PWN-20 (74.56), PWN-21 (39.86), PWN-22 (41.95 10^6CFUs/ml).

(c) Efficacy of Nitrogenase

To determine the efficacy of nitrogenase enzyme of each strain, 25ml Jensen,s broth medium was transferred into each 150ml capacity conical flask and plugged carefully with cotton wool. After autoclaving, the flasks were cooled and inoculated in triplicate by adding 0.1ml. broth culture of each strain. The flasks were shaked on a rotatory shaker for 7 days at 28± 2°C. From each flask 5 ml. of broth culture was used to determine the total nitrogen by the microkjedahl method as described by Jackson (5).

(d) Tolerance to Salt

Salt tolerance of different *Azotobacter* strains was determined against sodium chloride (NaCl), sodium sulphate (Na_2SO_4), magnesium sulphate ($MgSO_4$) and calcium carbonate ($CaCO_3$). The concentrations of different salts used were 0.5, 1.0, 1.5, 2.0, 3.0, 4.0, and 5.0 per cent for NaCl; 1.0, 2.0, 3.0, 4.0, 5.0, 6.0, and 7.0 per cent for Na_2SO_4; 0.1, 0.2, 0.5, 1.0, 2.0, 3.0 per cent for $MgSO_4$; and 0.1, 0.2, 0.5, 1.0, 2.0 and

3.0 per cent for $CaCO_3$.These concentration were made on the volume basis in Jensen medium. Jensen,s medium of different concentrations of different salts as described above were prepared and autoclaved as usual. Jensen medium plates were prepared by pouring about 12-15 ml medium in each petriplate and allowed to solidify. Filter paper discs of about 6 mm were cut and sterilized, after wetting in an autoclave at 1.0 Kg/cm^2 pressure for 30 minutes. The discs were dried in an oven at 70°C for 24 hours followed by dipping in broth culture of each strain and spotted onto the medium in petriplate with the help of sterilized forcep. Thus, eight strains at one concentration were tested in each plate. The plates were inoculated at 28± 2°C for 3 days. After inoculation, the plates were observed for presence or absence of growth of Azotobacter.

(e) Tolerance to Acidity/Alkalinity

Jensen,s agar medium of different pH namely 3.5, 4.0, 4.5, 5.0, 5.5, 6.0, 6.5, 7.0, 7.5, 8.0, 8.5, 9.0, 9.5, and 10.0 were prepared by adjusting the pH of the medium by dilute solutions of HCl or NaOH as per requirement. Each medium was sterilized at 1.0 kg/cm^2 for 30 minutes in an autoclave. Medium plates were prepared. A loopful of each strain was taken out with the help of inoculating needle and streaked onto agar medium in a plate under aseptic conditions. Thus, all the strains were tested at different concentrations against all the pH levels. Plates were for 3 days at 28±2°C and at the end of incubation period the presence or absence of growth was recorded.

(f) Tolerance to Temperature

Jensen, s agar medium plates were prepared. A loopful of each strain streaked onto medium in triplicates was inoculated at 15°C, 20°C, 25°C, 30°C, 35°C, 40°C, 45°C and 50°C for 7 days and observed for the presence or absence of its growth.

3. RESULTS AND DISCUSSION

The nitrogen fixing capacity of nitrogenase enzyme of 22 strains of *Azotobacter* isolated from different rhizospheric soil of oat and wheat showed a wide variation among themselves and ranged from 0.003% to 0.026% nitrogen and 0.83 to 6.57 mg N/25 ml broth culture (Table 1). The highest nitrogen fixing strain PWN-3 fixed about 6 times more nitrogen than the lowest nitrogen fixing strain PWN-6. The popularity of free living nitrogen fixing bacteria *Azotobacter* is because of its importance in nitrogen economy in agriculture due to its most important character to fix atmospheric nitrogen. The wide variations in nitrogen fixing capacity of different strains of *Azotobacter* have also been reported in literature (6-8).

Table 1. Nitrogen - Fixing capacity of different *Azotobacter* strains

Azotobacter strains	%N	N-Fixed (mgN/25 ml)
PWN-1	0.011	2.79
PWN-2	0.023	5.78
PWN-3	0.026	6.57
PWN-4	0.016	3.97
PWN-5	0.009	2.27
PWN-6	0.003	0.83
PWN-7	0.009	2.49
PWN-8	0.007	1.80
PWN-9	0.008	2.44
PWN-10	0.014	3.66
PWN-11	0.014	3.56
PWN-12	0.007	1.74
PWN-13	0.008	1.97
PWN-14	0.008	1.88
PWN-15	0.014	3.22
PWN-16	0.005	1.33
PWN-17	0.008	1.88
PWN-18	0.012	3.03
PWN-19	0.022	5.54
PWN-20	0.011	2.80
PWN-21	0.004	0.11
PWN-22	0.012	3.06
SEm±	0.0012	0.555
CD (0.01)	0.492	2.119
(0.05)	0.0040	1.585

The different strains of *Azotobacter* greatly differed in their ability to grow and to tolerate various concentrations of sodium chloride in growth medium (Table 2). Low (≤ 1.0%), medium (≥ 1.0 to ≤ 4.0%) and high (≥ 4.0%) NaCl salt tolerant strains of *Azotobacter* were 8, 3and 11, respectively, These finding are in accordance with findings of Keutner (9) who reported growth and nitrogen fixation by *Azotobacter* even at 8% sodium chloride in culture medium. The high concentrations of NaCl has been reported to inhibit the catalytic activity of the molybdenum-containing nitrogenase complex of *Azotobacter* as a dead end inhibitor without altering the number of MgATP hydrolyzed/ electron transferred to substrate (10).

Different strains of *Azotobacter* differed to a great extent in their ability to grow in the medium containing various concentrations of sodium sulphate salt. The low (≤ 2.0%), medium (≥ 3.0 to ≤ 5.0%) and high (≥ 6.0%) Na_2SO_4 salt tolerate strains were10, 7 and 5, respectively. The majority of the *Azotobacter* strains were resistant at high concentration of $MgSO_4$. Strain PWN-1 and PWN-2 were found to be susceptible as these could only be grown at 0.2% concentration of the salt. The tolerance of high concentration of sulphate, salts by majority of *Azotobacter* strains are in

Table 2. Growth pattern of different strains of *Azotobacter* as influenced by various concentration of Sodium Chloride and Sodium Sulphate

Azotobacter strains	NaCl (%)							Na_2SO_4 (%)						
	0.5	1.0	1.5	2.0	3.0	4.0	5.0	1.0	2.0	3.0	4.0	5.0	6.0	7.0
PWN-1	+	+	–	–	–	–	–	–	–	–	–	–	–	–
PWN-2	+	+	–	–	–	–	–	+	+	–	–	–	–	–
PWN-3	+	+	+	+	+	+	–	+	+	–	–	–	–	–
PWN-4	+	+	–	–	–	–	–	+	–	–	–	–	–	–
PWN-5	+	+	–	–	–	–	–	+	+	–	–	–	–	–
PWN-6	+	+	+	+	+	+	–	+	+	–	–	–	–	–
PWN-7	+	+	+	+	+	+	–	+	+	+	+	+	-	–
PWN-8	+	+	+	+	+	–	–	+	+	+	+	+	+	+
PWN-9	+	–	–	–	–	–	–	+	+	+	+	+	+	+
PWN-10	+	+	+	+	+	+	+	+	+	+	–	–	–	–
PWN-11	+	+	+	+	+	+	+	+	+	+	+	–	–	–
PWN-12	+	+	+	+	+	+	–	+	+	+	+	–	–	–
PWN-13	+	–	–	–	–	–	–	+	–	–	–	–	–	–
PWN-14	–	–	–	–	–	–	–	+	+	–	–	–	–	–
PWN-15	+	–	–	–	–	–	–	+	+	+	+	–	–	–
PWN-16	+	+	+	+	+	+	–	+	+	+	+	+	+	+
PWN-17	+	+	+	+	+	+	+	+	+	+	+	+	+	–
PWN-18	+	+	–	–	–	–	–	–	–	–	–	–	–	–
PWN-19	+	+	+	+	+	+	+	+	+	+	–	–	–	–
PWN-20	+	+	+	+	+	+	+	+	+	+	+	+	+	+
PWN-21	+	+	+	+	+	+	+	+	+	–	–	–	–	–
PWN-22	+	+	+	–	–	–	-	+	+	–	–	–	–	–

corroboration with the research findings of Jensen (11) who reported high requirement of magnesium for *Azotobacter.*

As expected, all the strains of *Azotobacter* grew well on the medium containing 0.2% $CaCO_3$ generally used in *Azotobacter* medium. The $CaCO_3$ concentration resulted in variable growth pattern (Table 3). About half of the strains were highly resistant against ≥ 2.0% concentration of calcium carbonate. Nine strains were sensitive to ≤ 0.5% concentration of this salt. The differential requirement for calcium salt for different species of *Azotobacter* have also been reported. The results obtained in the present investigation are further substantiated with the results of Zaitseva *et al.* (12) who noted high requirement of calcium foe most species of *Azotobacter.*

Based on growth pattern on Jensen,s agar medium, all the strains grew between pH 6.5 to 8.0 (Table 4). At pH 10.0 only four strain namely, PWN-14, PWN-17, PWN-21 and PWN-22 could shown their growth. The occurrence of growth of different strains of *Azotobacter* under alkaline conditions are in the conformity with the results reported by Mohmoud *et al.* (13) who observed growth of 26 strains of *Azotobacter chroococcum* between pH 6.5 to 9.5. None of the strain was able to grow at pH 3.5. However,

Table 3. Growth pattern of different strains of *Azotobacter* as influenced by various concentration of $MgSO_4$ and $CaCO_3$

Azotobacter strains	$MgSO_4$ (%)						$CaCO_3$ (%)					
	1.0	2.0	0.5	1.0	2.0	3.0	0.1	0.2	0.5	1.0	2.0	3.0
PWN-1	+	–	–	–	–	–	–	+	+	–	–	–
PWN-2	+	–	–	–	–	–	+	+	–	–	–	–
PWN-3	+	+	+	+	+	–	+	+	+	+	+	+
PWN-4	+	+	+	+	–	–	+	+	+	+	+	+
PWN-5	+	+	+	+	+	–	+	+	+	+	+	+
PWN-6	+	+	+	+	+	–	+	+	–	–	–	–
PWN-7	+	+	+	+	+	+	+	+	+	+	+	+
PWN-8	+	+	+	+	+	+	+	+	+	+	+	–
PWN-9	+	+	+	+	+	–	+	+	–	–	–	–
PWN-10	+	+	+	+	–	+	+	+	+	+	+	+
PWN-11	+	+	+	+	+	+	+	+	+	+	+	+
PWN-12	+	+	+	+	+	–	+	+	+	+	+	+
PWN-13	+	+	+	–	–	–	+	+	–	–	–	–
PWN-14	+	+	–	–	–	–	+	+	–	–	–	–
PWN-15	+	+	+	+	+	–	+	+	+	+	+	+
PWN-16	+	+	+	+	+	–	+	+	+	+	+	+
PWN-17	+	+	+	+	+	+	+	+	+	+	+	+
PWN-18	+	+	–	–	–	–	+	+	–	–	–	–
PWN-19	+	+	+	+	+	+	+	+	–	–	–	–
PWN-20	+	+	+	+	+	+	+	+	+	+	+	+
PWN-21	+	+	+	+	+	+	+	+	+	+	+	+
PWN-22	+	+	+	+	+	-	+	+	–	–	–	–

Table 4. Growth pattern of different strains of *Azotobacter* at different pH levels.

Strains	pH													
	3.5	4.0	4.5	5.0	5.5	6.0	6.5	7.0	7.5	8.0	8.5	9.0	9.5	10.0
PWN-1	–	–	–	+	+	+	+++	+++	+	+	+	+	–	–
PWN-2	–	–	–	+	+	+	+++	+++	+	+	–	–	–	–
PWN-3	–	–	–	+	+	+	+++	+++	+	+	–	–	–	–
PWN-4	–	–	–	+	+	+	+++	+++	+	+	–	–	–	–
PWN-5	–	–	+	+	+	+	+++	+++	+	+	–	–	–	–
PWN-6	–	–	–	+	+	+	+++	+++	+++	+++	+++	+	+	–
PWN-7	–	–	–	–	–	+	+	+++	+++	+	+	–	–	–
PWN-8	–	–	–	+	+	+	+++	+++	+++	+	+	+	–	–
PWN-9	–	+	+	+	+	+	+++	+++	+++	+	+	–	–	–
PWN-10	–	–	–	+	+	+	+++	+++	+++	+	+	+	+	–
PWN-11	–	–	–	+	+	+	+++	+++	+++	+	+	+	–	–
PWN-12	–	–	–	+++	+++	+++	+++	+++	+	+	+	–	–	–
PWN-13	–	–	–	+++	+++	+++	+++	+++	+	+	+	–	–	–
PWN-14	–	–	–	–	+	+	+	+	+++	+++	+++	+	+	+
PWN-15	–	–	–	–	–	–	+	+++	+++	+++	+	–	–	–
PWN-16	–	–	–	–	–	–	+	+	+++	+	+	+	–	–
PWN-17	–	–	–	–	–	–	+	+	+++	+++	+	+	+	+
PWN-18	–	+	+	+++	+++	+++	+++	+++	+	+	+	–	–	–
PWN-19	–	–	–	+	+	+	+	+++	+++	+	+	+	+	–
PWN-20	–	–	–	+	+	+	+	+++	+++	+	+	+	–	–
PWN-21	–	–	–	+	+	+	+	+++	+++	+	+	+	+	+
PWN-22	–	–	–	+	+	+	+	+++	+++	+	+	+	+	+

strain PWN-5, PWN- 9 and PWN-18 at pH 4.0 and strain PWN-5, PWN-9, PWN-11, PON-12, and PWN-18 at pH 4.5 showed their growth on Jensen,s agar medium. Though, majority of the strains of *Azotobacter* have been reported to be grown in alkaline range of pH, strains which are able to grow in acidic conditions as low as pH 3.0 has also been reported elsewhere (14).

The different strains of *Azotobacter* had wide variability in the ability of temperature tolerance (Table 5). All the 22 strains of Azotobacter grew at temperature range between 25 to 40°C, 5 strains recorded to grew at 15°C temperature, 7 strains namely, PWN-4, PWN-5, PWN-6, PON-13, PWN-18, PWN-21 and PWN-22 tolerated temperature of 45°C. The temperature tolerance capacity of different strains of *Azotobacter* has also been reported by Danilova *et al.* (15) at low temperature, by Raj Kumar and Lakshman (16) at mesophilic temperature and by Dilworth *et al.* (17) at thermophilic temperature range.

Table 5. Survival of different strains of *Azotobacter* at different temperatures

Strains	Temperature (°C)							
	15	20	25	30	35	40	45	50
PWN-1	–	+	+	+	+	+	–	–
PWN-2	–	+	+	+	+	+	–	–
PWN-3	+	+	+	+	+	+	–	–
PWN-4	–	–	+	+	+	+	+	–
PWN-5	–	+	+	+	+	+	+	–
PWN-6	–	–	+	+	+	+	+	–
PWN-7	–	–	+	+	+	+	+	–
PWN-8	+	+	+	+	+	+	–	–
PWN-9	–	–	+	+	+	+	–	–
PWN-10	–	+	+	+	+	+	–	–
PWN-11	+	–	+	+	+	+	–	–
PWN-12	–	+	+	+	+	+	+	–
PWN-13	+	–	+	+	+	+	–	–
PWN-14	–	+	+	+	+	+	–	–
PWN-15	–	–	+	+	+	+	–	–
PWN-16	–	–	+	+	+	+	–	–
PWN-17	–	+	+	+	+	+	+	–
PWN-18	+	–	+	+	+	+	–	–
PWN-19	–	+	+	+	+	+	–	–
PWN-20	–	+	+	+	+	+	–	–
PWN-21	–	+	+	+	+	+	+	–
PWN-22	–	–	+	+	+	+	+	–

REFERENCES

1. Ladha J.K. & Peoples M.B. (ed.) Management of BNF for the development of more productive and sustainable Agricultural systems (1994). Kluwer Academic publishers, London.
2. Shende S.T. *Azotobacter* inoculation-Anumeration input for agricultural crops. Natl. symp. on Biological nitrogen fixation (1977) 60 p. IARI. New Delhi.
3. Huyer Marianne & Page William J. *Appl Environ Microbiol* 54 (11) (1988) : 2625 - 2631.
4. Jensen H.L. *Res Soc Appl Bacteriol* 74 (1) (1954) : 89-94.
5. Jackson M.L. Soil chemical analysis. Prentice Hall of India. Pvt. Ltd., New Delhi, (1958).
6. Sanoria C.L. & Sundara Rao W.V.B. *Indian J Agric Sci* 45 (1975): 224.
7. Mahmoud S.A.Z., Sawy M-El, Ishac, Y.Z., Safty M.M-El, El- Sawy-M. & El-Safty M.M. Ecological Bulletins - Stockolm, 26 (1978) : 99-107.
8. Tippannavar C.M. & Reddy T.K.R. *J Fmg System* 3 (3-4) (1987) : 67-71.

9. Keutner N. *Exp Sta Rev* 17 (1907) : 118.

10. Deits T.L. & Howard J.B. *J Biol Chem* 265 (7) (1990) : 3859 - 3867.

11. Jensen H.L. *Acta Agric Scand* 4(2) (1954b) : 224 - 236.

12. Zaitseva G.N., Khmel I.A. & Belozerskii A.N.*Dokl Akad Nauk SSSR* 141 (1961) : 1961 740.

13. Mahmoud S.A.Z., El-Sawy M., & El-Safty M.M. *Egyptian J Soil Sci Spl Issue.* (1975) 453.

14. Kirakosyan A.V., Ananyan L.G. & Melkonyar Zh. S. Mikrobiologii, (3), Erevan, Izd-vo Akad. Nauk Arm SSR, (1966) 13-23.

15. Danilova I.V., Botvinko I.V. & Egorov N.S. *Microbiol* New York 61 (6) (1992): 660 - 670.

16. Raj Kumar K. & Lakshman M. *Indian J Microbiol* 35(1) (1995) : 25-30.

17. Dilworth, M.J., Eldridge M.E. & Eady, R.R. *Biochem J London* 289(2) (1993): 395 - 400.

CHAPTER - 13

A HIGHLY EFFICIENT MICROPROPAGATION PROTOCOL OF TEAK (*Tectona grandis* L.)

SK Tiwari, EA Siril and KP Tiwari

Forest Genetics, Biotechnology & Tree Improvement Division State Forest Research Institute, Jabalpur (M.P.) 482 008 INDIA

ABSTRACT

A highly efficient micropropagation protocol for the Teak (*Tectona grandis*) has been developed. Nodal explants cultured on MS medium supplemented with 22.2 μM benzyl aminopurine and subjected to a serial transfer after inoculation in 12 hrs., 24 hrs, 48 hrs and 72 hrs to MS medium containing 22.2 μM-BAP facilitated maximum explant establishment (76.8%). Explants retained in initial culture medium over 12 hrs significantly reduced explant establishment. Seasonal variation affected explant response. Explants collected in May showed maximum (76.8%) response. *In vitro* developed shoots were remultiplied over 18 times in every 4 weeks interval without any loss of vigor. MS medium supplemented with 22.2 μM benzyl aminopurine and 0.57 μM indole-3- aceticacid resulted in the maximum (5.76) number of shoots. *In vitro* raised microshoots subjected to an *ex vitro* rooting by giving a dip in indole-3-butyricacid (9.8 μM) for 2 minutes and followed by planting them in polyethylene pots containing soil-vermiculite (1:1 v/v) mixture resulted in 77.9% survival. They were weaned in a glass house and finally moved to an agro-net shadehouse. An estimation of efficiency of the protocol revealed that a single explant can produce over 2000 hardened plants through 18 times multiplication cycle.

1. INTRODUCTION

Tectona grandis L. (Family Verbanaceae) is one of the most valuable structural timber tree of the tropics. It is indigenous to Burma, Thailand and Central and south India. Teak has been introduced in all agroclimatic regions of Central India and has naturalized in almost all regions. The tree is characterized by tall clear bole with rounded crown. Its leaves are broadly elliptical or obovate. Worldwide attraction of this species is due to its quality timber in regard to mechanical and physical properties. Wood is resistant to termite attack due to the presence of sesquiterpenes (1,2). In Central and South India the species has been recognized as the prime candidate for tree cultivation by public and private sector enterprises putting an annual target to plant in 50,000 ha. per year. Plant material for the above target however, cannot be met by conventional method alone. Teak reproduced conventionally through seeds. Germination is difficult due to the hard seed coat often hampers production of large number of seedling in a defined time. Possibility of propagation of teak via cuttings is reported (3). But method has several limitations ultimately provide a few number of propagules from a selected plus individual. As the species is a valuable source of timber tissue culture of this the species has been started in the seventies (4). Further to improve establishment multiplication of this species has gained attention (5). However, adult tree micropropagation of teak remained problematic with sparse culture establishment, reduced rate of multiplication of teak remained problematic with sparse culture establishment, reduced rate of multiplication and poor rooting percentage. Poor explant response and rapid explant bronwning are major hurdles to successful establishment of teak *in vitro*. This paper describes a method for the Micropropagation of phenotypically identified plus individuals of teak using nodal segments as explants.

2. MATERIALS AND METHODS

(a) Culture establishment

The explants were collected from known phenotypic plus tree source (Seoni, M.P.). Fresh shoot of current season's growth were collected and swabbed well with alcohol (50% v/v) soaked muslin cloth. They were cut into 3-4 cm long pieces having a single node and subjected to teepol (1%) washing for 30 minutes. They were surface sterilized successively in ethyl alcohol (80% v/v) for 30 seconds and 5 minutes in 0.1% mercuric chloride solution containing few drops of twee-20. This was followed by five rinses in sterile distilled water. After trimming both ends explants were placed aseptically in 100-ml bottles containing 25-ml medium. As preliminary trials resulted in immediate media browning and subsequent death of explants an alternative, serial transfer technique was adopted. The technique includes culture of materials in MS medium supplemented with 3% sucrose and 0.8% agar. Explants initially cultured on MS medium containing 2.2 μM BAP and were subjected to different serial transfer methods (Table 1). First transfer of explant was after 120-hrs. (T1), 24-hrs (T2, 48-hrs (T3) or 72- hrs (T4). Explants cultured on MS medium containing 22.2 μM BAP and not subjected to serial transfer served as control (T5).

In *T. grandis* under natural conditions, fresh sprout appears in early summer (Mar.-April). They grow rapidly throughout the summer (May-June) and monsoon (Jul-Aug) and thereafter growth ceases by early winter and leaf shedding starts in early winter (December). To determine the most suitable month for culture establishment, explants were collected in every month from March to February and cultured using a five-step serial transfer technique (T1).

(b) Multiplication

Auxiliary bud sprouts collected from T1 treatment were used for further multiplication. The exudation of phenolic was noticed even in the first subculture and, therefore, new shoot segments with one node each were kept in sterile double distilled water immediately after cuttings for 15 minutes and were blotted well with

Table 1. Effect of various transfer treatments on explant establishment of *T. grandis*

Treatment*	Transfer sequence (hrs)	Percentage establishment
T1	12 → 24 → 36 → 48 → 72	76.80 a
T2	24 → 36 → 48 → 72	6.60 b
T3	48 → 72	20.00 c
T4	72	4.50d
T5	Direct	0.00d

Means within column followed by same letters are not significantly ($p<0.05$) different as determined by F-LSD test
*Medium used : MS + 22.2 µM BAP

sterile filter paper. They were placed in MS medium added with different concentrations (4.44, 11.1, 22.2 or 33.3 µM) BAP with or without 1AA (0.0, 0.57 or 1.43 µM). However, second subculture onwards phenolic exudation was less thus explants were directly placed in fresh multiplication medium.

(c) Rooting

Shoots (>2.5 cm) harvested from IVth subculture product onwards were used for rooting. Microshoots were thoroughly washed in running tap water. After blotting the cut ends shoots were dipped in IBA (4.9, 9.8 or 24.5 µM) solution prepared in 50% ethyl alcohol for different duration (30 sec, 1 min. 2 min.). They were planted in plastic pots containing soil: vermiculite mixture enriched with one fourth MS salt solution and placed in hardening chamber (32± 2°C RH 90%). Initially plants were covered with microholed polyethylene bags. When plants completely wined they were moved to shadehouse. Three-month-old plants were planted in the field and performance is being observed.

(d) Experimental design and data collection

All experiments were conducted by using a completely randomized block design (CRBD). Each treatment represented three replications of twelve explants. Significance was determined by analysis of variance (ANOVA) and the least significant ($p<0.01$) difference among mean values was estimated using Fisher's LSD test. Data presented in percentages were subjected to arcsin transformation before analysis, and then converted back to percentage for presentation in Table 2.

3. RESULTS AND DISCUSSION

(a) Culture establishment

Primarily explants showed rapid exudation of phenolics, *in vitro* and explant immediately turned reddish-brown within 24 hrs. Explants subjected to a five-transfer (T1) treatment (Table 1) found most effective to establish culture. Explants retained in initial culture medium over 24 hrs. significantly ($p<0.01$) reduced percentage of establishment (Table 1). The explants not subjected to a serial transfer failed to respond in vitro. Survived cultures starts producing new shoots from the node and grew into 4.0-cm size within 6 weeks. Each new shoot had 4-5 nodes (Plate 1-A). Although several workers (12) have suggested frequent transfer of explants to fresh medium for the establishment of cultures the schedule for the transfer is not yet been elucidated well. In a previous report (5) on tissue culture of *T. grandis* three continuous passage to MS medium containing 4.44 µM each BAP and 4.65 µM kinetin along with adenine sulphate has been suggested. Explants kept over 12 hrs. in initiation medium caused formation of autotoxic substance and subsequent death of explants.

Plate 1. Micropropagation of adult Teak (*Tectona grandis*)

A. *In vitro* shoot development from nodal segment cultured on MS + 22.2 μM BAP

B. Shoot multiplication and subsequent elongation (left) in MS + 22.2 μM BAP + 0.25 μM IAA, (right) elongation suppressed shoots in MS + 33.3 μM BAP

C. *Ex vitro* rooting of microshoots four weeks after potting

D. Hardened plants (3-month-old) growing in shadehouse

Table 2. Effect of plant growth regulators on shoot proliferation of *Tectona grandis*

Cytokinin BAP (μM) +	Auxin IAA (μM)	Shoot No. ± S.E.	Shoot length (cm) ± S.E.
4.44	0.00	1.46 ± 0.086 h	3.9 ± 0.057 bc
4.44	0.57	1.36 ± 0.066 h	4.3 ± 0.063 a
4.44	1.43	1.06 ± 0.032 h	4.4 ± 0.087 a
11.1	0.00	2.70 ± 0.057 e	3.8 ± 0.063nb
11.1	0.57	2.80 ± 0.057 e	4.4 ± 0.063 a
11.1	1.43	2.26 ± 0.032 f	4.0 ± 0.032 b
22.2	0.00	5.63 ± 0.087 a	3.2 ± 0.087 d
22.2	0.57	5.76 ± 0.032 a	3.7 ±0 .066 c
22.2	1.43	5.00 ± 0.057 b	3.3 ± 0.120 d
33.3	0.00	3.53 ± 0.033 d	2.0 ± 0.033 ef
33.3	0.57	3.80 ± 0.057 c	2.2 ± 0.063 e
33.3	1.43	2.03 ± 0.087 g	1.8 ± 0.087 f

ANOVA Summary Table

Source	Df(n-1)	Shoot number		Shoot length	
		MS	F	MS	F
Replication	2	0.0185	1.62ns	0.002	0.16 ns
Cytokinin	3	27.302	239.02*	8.920	500.54
Auxin	2	2.519	220.78*	0.625	35.05*
Auxin x Cyto	6	0.346	30.33*	0.075	4.20*
Error	22	0.011		0.178	
Total	35	30.197		9.640	

Means within column followed by same letters are not significantly ($p<0.05$) different as determined by F-LSD test

* Significant at 0.01 level

Table 3. Effect of different IBA treatments on rooting of micropropagated shoots

Treatment	IBA concentration (μM)	Rooting percentage
30 sec	4.9	0.0 d
	9.8	4.5 d
	24.5	16.3 c
1 Min	4.9	16.3 c
	9.8	20.3 c
	24.5	19.3
2 Min	4.9	39.8 b
	9.8	77.9 a
	24.5	46.6 b

Means with column followed by same letters are not significantly ($p<0.05$) different as determined by F-LSD test.

Explants collected in May gave maximum response as measured by percentage establishment *in vitro* (Fig. 1). In Central India, May is the middle of the summer season and is the active growth period of Teak in nature. Better response of

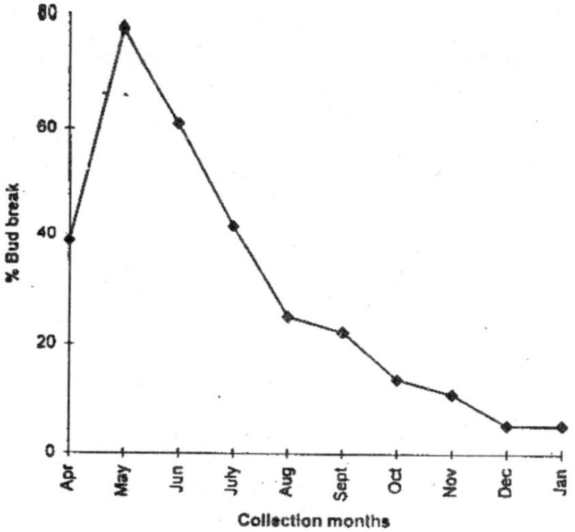

Fig. 1. Effect of explant collection months on culture establishment

actively growing shoot buds over dormant buds is in agreement with several other woody perennials (13). Devi *et al.* (5) have reported December as the best month for the establishment *T. grandis* of cultures from field grown trees. Establishment of cultures during active growth phase of tree requires an efficient phenolic elimination strategy. Adopted serial transfer technique is considerably reduced death of cultures due to phenolic toxicity.

(b) Multiplication

Explants cultured on different cytokinin supplemented media resulted in multiple shoot proliferation. ANOVA revealed a significant ($p<0.01$) effect of various cytokinin concentrations tried. Role of auxin in regard to shoot multiplication and elongation was significant ($p<0.01$). Media supplemented with 22.2 μM BAP and 0.57 μM IAA resulted in the production of significantly ($p<0.01$) high number (5.76) of shoots. Shoots elongated into 3.7-cm size within a four-week culture period. Auxin- cytokinin interaction significantly ($p<0.01$) controlled shoot formation and its subsequent elongation. Increasing concen- tration of auxin in multiplication medium, however, adversely affected new shoot formation. Addition of low levels of auxin alongwith cytokinin has been known to increase shoot number and also shoot elongation in other trees like *Gmelina arborea* (6) *Sapium sebiferum* (7). Presence of BAP at high concentration was inhibitory to growth of auxiliary buds, support finding of Hu and Wang (8).

(c) Rooting

In vitro developed microshoots were planted in potting mixture showed emergence of new leaves within ten days. Shoots survived after one month of planting assumed as rooted micro shoots and rooting percentage was estimated with respect to treatment. ANOVA revealed significant ($p<0.01$) effect of auxin concentration and treatment duration. There was an interaction ($p<0.01$) of both concentration and duration which ultimately determines rhizogenesis of micro cuttings. Shoots treated with 9.8 μM IBA for 2 min, resulted in 80.0% rooting. *Ex vitro* rooting attracts micro propagators because of simultaneous rooting and hardening of plants (6, 9). It also cuts the cost of production significantly. Further such plants gave high percentage of field survival (10). As previous reports (11) on micropropagation a Teak a two-step *in vitro* rooting procedure was necessary.

Major improvement over previous protocols (5, 11) are increased percentage of culture establishment, high rate of multiplication and *ex vitro* rooting which can beneficially merge the steps in involved in micropropagation of Teak. In brief the present protocol has the following advantages; high cloning success in regard to *in vitro* establishment of culture, an easily adaptable multiplication procedure and *ex vitro* rooting with high percentage of soil establishment of plants. The protocol will effectively meet the challenges of high demand of Teak cloned planting material.

ACKNOWLEDGEMENTS

The authors are grateful to Dr. M.C. Sharma for providing suggestions. Laboratory support from Mr. A.P. Agrawal and Mrs. M. Parihar are thankful acknowledged.

REFERENCES

1. Anon. The Wealth of India Vol X, (1996) Pub. CSIR, New Delhi, India.
2. Maslekar AR. Forestor's companion. India Forestry Handbook Jugal Kishore & Co., India (1981).
3. Nautiyal S, Singh U & Gurumurthi K. Indian Forester 118 (1992) 112.
4. Gupta PK, Nadgir AJ, Mascarenhas AF & Jaganathan V. Plant Science Lett, 17 (1980) 259.
5. Devi YS, Mukherjee BB & Gupta S. Indian Journal of Exp Bio, 32 (1994) 668.
6. Tiwari KP, Sharma MC & Tiwari SK. SFRI Technical bulletin No.29, State Forest Research Institute, Jabalpur, India (1997).
7. Siril EA & Dhar U. Plant Cell Reports, 16 (1997) 637.
8. Hu Cy & Wang PJ, Evan DA, Sharp WR, Ammirato PV & Yamada Y etd. Macmillan Publ Co Pvt Ltd New York (1983) pp.177.
9. Nemeth G & Bajaj YPS ed. In : Biotechnology in Agriculture and Forestry Vol. I, Trees 1. Springer Verlag, Berlin Germany (1986) 49.
10. Moncousin CH & Bajaj YPS ed. In : Biotechnology in Agriculture and Forestry Vol. 17. High-Tech and Micropropagation, Springer - Verlag, Berlin, Heidelberg, Germany (1991).
11. Mascarenhas AF, Kendurkar SV, Gupta PK, Khusepe SS & Agrawal DG, Bonga JM & Durzan DJ eds. In : Cell and Tissue Culture in Forestry Vol.3. Martinus ub. Dordrecht, Netherlands (1987) 300.
12. Gill, RJS & Gill SS. Indian Forester, 120 (1994) 504.
13. Chalupa V, Bonga JM & Durzan eds. In Cell and Tissue Culture in Forestry. Vol.3, Martinus NiJhoff, Dordrecht (1987) pp. 224.

CHAPTER - 14

PESTICIDE TOLERANCE BY COMMON NON-HETEROCYSTOUS FILAMENTOUS *Cyanobacteria* OF RICE FIELDS OF UTTAR PRADESH

O.N.Tiwari and G.L.Tiwari

Department of Botany, University of Allahabad, Allahabad - 211002 U.P.

ABSTRACT

Cyanobacteria show high variations in their tolerance limits to the effect of different pesticides. Cyanobacterial species occur abundantly in rice fields of tropical and subtropical countries where agricultural technologies are involved in extensive use of pesticides for selective elimination of pests and weeds of rice crops. In the present study, non-heterocystous filamentous Cyanobacteria which differ from heterocystous forms for lack of akinetes, and heterocysts were selected. Ten species viz., *Phormidium fragile* (AU 246), *Phormidium foveolarum* (AU 150), *Microcoleus paludosus* (AU 189), *Microcoleus chthonoplastes* (AU 381), *Oscillatoria acuminata* (AU 66), *Oscillatoria brevis* (AU 377), *Lyngbya palmarum*(AU 2), *Lyngbya spiralis* (AU 111), *Plectonema boryanum* (AU 143), and *Plectonema nostocorum* (AU 97) were treated with pesticides *viz.*, 2,4-D, dimecron and malathion which are commonly used in rice fields.

1. INTRODUCTION

The widespread use of pesticides in rice fields can be harmful to both, the crops and the associated Cyanobacteria. Different species of cyancbacteria are known to vary in their sensitivity to a different pesticides (1). Sharma and Gaur (2) reported, increased growth in some strains of *Aulosira fertilissima* and *Nostoc* sp. on treatment with pesticide combinations, but nevertheless, decreased nitrogen fixation in all the strains. Agro- chemicals such as herbicides, fungicides and insecticides are either stimulatory or inhibitory or neutral to cyanobacterial growth and nitrogen fixation. A greater understanding on the impact of a wide range of these chemicals is necessary to appreciate their environmental bearing on the ecology of these organisms, particularly under improved soil and crop management systems. Extensive use of pesticides and herbicides in agriculture has created its problem of their effect on soil micro-organisms. Cyanobacteria are amongst the major beneficial nitrogen fixing micro-organisms in the soil used in rice cultivation (3,4). It is therefore desirable to know the effect of some commonly used herbicide on their survival and growth of these organisms. In the present study, variation of the pesticide tolerance by non-heterocystous cyanobacteria at generic and species level has been investigated.

2. MATERIALS AND METHODS

Ten wild non-heterocystous filamentous Cyanobacterial species belonging to 5 genera *viz., Phormidium, Microcoleus, Oscillatoria, Lyngbya* and *Plectonema* commonly occurring in rice fields of Uttar Pradesh, were isolated from different localities and used as experimental organisms. Axenic cultures of each species were maintained in BG-11 nitrogenous medium (5). Fifteen ml culture tubes of borosil were loaded with 10 ml of nitrogenous inorganic medium containing different doses of pesticides *viz.,* 2,4-D, malathion and dimecron in triplicate. One ml of diluted suspension of each organisms was inoculated in cotton stoppered hard glass test tubes and incubated at temp 30 ± 2°c and pH 7.5 under daily cycle of 14 hrs. light (4K) and 10 hrs. of darkness. Effect of pesticides on organisms was evaluated at the exponential phase of growth. Growth in terms of total chlorophyll was measured by measuring O.D. at 665 nm through systronics (UV-VIS spectrophotometer, and calculation was made as per the method described by (6).

3. RESULT AND DISCUSSION

In the present study, five genera and ten species of the order *Oscillatoriales viz., Phormidium* (2), *Microcoleus* (2), *Oscillatoria* (2), *Lyngbya* (2), and *Plectonema* (2) have been identified upto the species level.

4. RESPONSE OF DIMECRON (50% EC)

Maximum biomass was observed at 5 ppm concentration of dimecron in *Phormidium foveolarum, P. fragile* and *Oscillatoria brevis* and growth increase over the control was recorded as 27.6%, 29.26% and 26.96% respectively. While others viz., *Microcoleus paludosus, M. chthonoplastes, Oscillatoria acuminata, Lyngbya palmarum, L. spiralis, Plectonema boryanum* and *P. nostocorum* produced maximum biomass at 1 ppm concentration of dimecron and growth increased over the control as, 28.09%, 25.08%, 7.31%, 7.06%, 22.0%, 15.0% and 3.41% respectively. This observation indicated that the most suitable dose of the dimecron is 1 ppm because most of the investigated taxa produced maximum biomass in this concentration. The present study revealed that the most common species viz., *Phormidium foveolarum, P. fragile* and *Oscillatoria brevis* were the strains that produced good biomass and also exhibited more tolerance to pesticide dimecron than the other investigated organisms. Earlier there has been no work on the effect of dimecron on the growth of non-heterocystous filamentous cyanobacteria. Anand and Veerappan (7) reported that higher concentration of dimecron are toxic to *Anabaena, Nostoc* and *Cylindrospermum.* They also observed that *Anabaena* and *Nostoc* are more resistant than *Cylindrospermum.*

5. RESPONSE OF MALATHION (50% EC)

Effect of malathion (50% EC) indicated that *Phormidium foveloarum, P. fragile,*

Microcoleus paludosus, M. chthonoplastes and *Oscillatoria acuminata* were produced maximum biomass at 25 ppm concentration and growth increase over the control was as 18.10%, 37.84%, 55.89%, 85.81% and 24.25% respectively. Out of the 5 taxa, examined *Oscillatoria acuminata* produced maximum biomass (2.306 μg/ml), and the maximum growth with the increase over the control as 85.81% in *Microcoleus chthonoplasts* at 25 ppm concentrtion of dimecron. Three taxa viz., *Oscilltoria brevis, Lyngbya palmarum* and *L. spiralis* produced highest biomass at 50 ppm concentration of dimecron and growth increase over the control was as, 40.95%, 55.92% and 64.22% respectively. *Plectonema boryanum* and *P. nostocorum* produced higher biomass at 10 ppm concentration and growth increase over the control was as 41.66%, 21.19% respectively. Present observation indicated that *Oscillatoria brevis, Lyngbya palmarum* and *L. spiralis* were tolerant to higher concentration while *Plectonema* sp. was tolerant to the lower concentration of malathion.

Earlier Torres and O'Flaherty (8) observed reduction in the Chlorophyll content at 1 ppm concentration of malathion in unicellular and filamentous green algae. In *Nostoc hatei* Gangawane and Chaporker (9) reported stimulatory effect of malathion on growth (dry weight and total nitrogen) in the presence of 1 ppm to 10 ppm concentration. Inhibitory and lethal doses were found to be 50 ppm and 500 ppm respectively. However, in *Chlorococcum hypnosporum* (green alga), chlorophyll production was not inhibited in lower concentration (0.55 to 0.01 ppm) and it was found stimulatory in *Chlorella, Tribonema* and *Oscillatoria* sp. Dikshit and Tiwari (10) recorded inhibition in Chlorophyll a pigment of *Anabaena variabilis, Aulosira implexa, A. fertilissima, Scytonema chiastum* and *S. stuposum* in presence of 100 ppm of malathion, while the growth of all the above forms was absent in presence of 200 ppm concentration except

Table 1. Effect of pesticides on 10th day growth of non-heterocystous filamentous *Cyanobacteria*

Experimental organisms	Dimecron			Malathion			2,4-D		
	Conc. in ppm	Chl. a (μg/ml)	Growth increase over control	Conc. in ppm	Chl. a (μg/ml)	Growth increase over control	Conc. in ppm	Chl. a (μg/ml)	Growth increase over control
Phormidium foveolarum	5	0.957	27.6%	25	0.685	18.10%	50	0.996	22.96%
Phormidium fragile	5	0.596	29.26%	25	0.510	37.84%	50	0.584	35.81%
Microcoleus paludosus	1	1.076	28.09%	25	1.297	55.89%	25	1.314	55.50%
Microcoleus chthonoplastes	1	1.182	25.08%	25	1.598	85.81%	25	1.417	50.74%
Oscillatoria acuminata	1	1.644	7.31%	25	2.306	24.25%	25	2.140	22.20%
Oscillatoria brevis	5	1.794	26.96%	50	1.863	40.95%	25	1.715	41.38%
Lyngbya palmarum	1	1.487	7.06%	50	2.214	55.92%	25	3.496	146.20%
Lyngbya spiralis	1	1.480	22.00%	50	1.813	64.22%	25	2.814	154.89%
Plectonema boryanum	1	0.830	15.00%	10	1.214	41.66%	25	1.112	52.54%
Plectonema nostocorum	1	0.440	3.41%	10	0.406	21.19%	25	0.874	153.33%

Scytonema chiastum. Although results are comparable with other workers, but useful comparison can not be derived because most of the previous observations were based on heterocystous forms.

6. RESPONOSE OF 2,4 DICHLORO-PHENOXY ACETIC ACID (18% W.P.)

2,4-D is a weedicide as well as growth promoting hormone so its use in rice fields is more useful than other pesticides. Effect of 2,4-D (18% WP) on growth of non-heterocystous filamentous cyanobacterial population varied to the genus level. The effective dose in *Phormidium* sp. was 50 ppm and maximum biomass was recorded at this concentration. *Microcoleus, Oscilla toria, Lyngbya* and *Plectonema* were the organisms which produced maximum biomass at 25 ppm concentration. Amongst 10 investigated taxa, *Lyngbya palmarum* recorded highest biomass (3.496 µg/ml) and growth increase over the control was as, 146.20% in 25 ppm concentration. Present observation indicated that 25 ppm concentration was more suitable in all the wild strains and only *Phormidium* sp. showed good interaction at 50 ppm concentration. Nitrogen fixing cyanobacteria, which dominated the water logged paddy fields, were also reported to tolerate much higher concentration of 2,4-D, although lower concentrations were stimulatory to their growth (11-13). The aim of present work was to investigate the biological implications of common pesticides which are normally used in rice field ecosystems dominated by cyanobacterial population.

ACKNOWLEDGEMENTS

Authers thank to the Head, Department of Botany, University of Allahabad for providing laboratory facilities. They also thank the Department of Biotechnology New Delhi for financial support.

REFERENCES

1. Sharma, V.K. and Gaur, Y.S. *Pesticides* : **12** (1982) : 5-6
2. Sharma, V.K., and Gaur, Y.S. In: Sheshadri, C.V., Thomas, S. and Jeeji Bai, N. (eds.). *Proceedings of national workshop on algal systems.* Indian society of Biotechnology, New Delhi, illustr. **273** (1980) pp. 47-50.
3. Singh, R.N. Role of Blue-green algae in nitrogen economy of Indian agriculture. ICAR New Delhi (1961) pp. 170.
4. Stewart, W.D.P. *Plant and Soil* : **32** (1970) : 555-588.
5. Rippka, R., Deruelles, J., Waterburry, J.B., Herdman, M. and Stainer, R.Y. *J. Gen. Microbiology* : **111** (1979): 1-61.
6. Mackinney, G. *J.Biol.Chem.* **140** (1941): 315-322.
7. Anand, N. and Veerappan, B. *Phykos* : **19** (1980): 210-212.
8. Torres, A.M.R. and O'Flaherty, L.M. *Phycologia.* **15**(1976) : 25-36.
9. Gangawane, L.V. and Chaporkar, C.B. *Indian Jour. Microbiology.* **23** (1983) : 198-199.
10. Dikshit, G. and Tiwari, G.L. Effect of pesticides on certain nitrogen fixing Cynaobacteria. In. Kaushik, B.D., (ed.) *Proc. Nat. Symp. Cyanobacterial nitrogen fixation,* (1990) pp, 495-500, IARI, New Delhi.
11. Venkataraman, G.S. and Rajyalakshmi, B. *Curr. Sci.* **40(6)** (1972): 143-144.
12. Singh, P.K. *Arch.microbiol.* : **97**(1974): 69-72.
13. Da Silva, E.J., Henriksson, L.E. and Henriksson, E. *Arch. Environ. Contnm. Toxicol* : **3** (1975): 193-204.

CHAPTER - 15

Eucalyptus pauciflora — A POTENTIAL SOURCE OF SUSTAINABLE, ECOFRIENDLY STORAGE PESTICIDE

Amritesh Chandra Shukla, S.K. Shahi and Anupam Dikshit

Biological Product Laboratory, Department of Botany, University of Allahabad, Allahabad - 211002, (U.P.)

ABSTRACT

With a view to develop harmless and effective natural fumigant to control storage pests of food commodities, a detailed *in vitro* pesticidal screening of some aromatic plants against the dominant storage fungai *viz., Aspergillus flavus, Penicillium italicum* and *Alternaria alternata* as well as insect pests *viz., Rhizopertha dominica, Trogoderma granarium, Sitophilus oryzae* the weevils and *Corcyra cephalonica, Ephestia cautella*, the moth, the oil of *Eucalyptus pauciflora* Sleb (Family - Myrtaceae) was found to be strongest toxicant. The oil at 5% concentration killed the tested fungal pests within 4-6 hrs. but against insect pests it ranged from 40-60 minutes only, when used as a contact toxicant. However, as a fumigant, killing time was found to be 7 hrs. against *R. dominica* and *S. oryzae*, 5 hrs. against *T. granarium* and 6 hrs against *C. cephalonica* and *E. cautella* respectively. Besides a broad pesticidal spectrum including some spp. of *Aspergillus, Penicullium, Fusarium, Alternaria* etc. the oil tolerated heavy doses of inoculum. The toxicity of the oil was found to be thermostable and persisted up to 36 months.

After testing their non-phytotoxicity on wheat and groundnut (the host plant), the maximum phytotolerable concentration as well as the minimum effective dose of the fumigant, formulation were determined as 9 gm at 50% concentration cub meter. The fumigated commodities withstood organoleptic test as well as their pharmacological parameters did not show any adverse effect. Besides, formulated fumigant was found to be superior over tested synthetics, it was also economically viable. As such, after successful multilocational trial, the oil of *E. pauciflora* can be commercialised as an ecofriendly, sustainable storage pesticide in the form of '*Biopauci*'.

1. INTRODUCTION

The major issue of the space age is world hunger. The main cause of food problem is the enormous increase in the world population which is growing as never before. It took until 1830 to build up the first billion people on earth. The second billion was added in just 100 years, the third came in by 30 years and by the end of the century it will exceed six billions. But the available volume of food has not expanded in pace with population. Besides this, in some of the countries inspite, of the use of all available means for plant protection about one-third of the yearly harvested food commodities (worth about 7.5 million dollars as well as a quantity good enough to food for millions of people) is destroyed by storage pests. Therefore, it becomes essential to check them. However, to control these storage pests, a number of synthetic pesticides are available in the market but their ill effects are well known. Recently, some plant origin fumigant biopesticides have been projected as an alternative to the synthetic pesticides in pest management (1-6) due to their desirable characteristics.

The present study was undertaken to formulate a renewable, biodegradable, non-phytotoxic biopesticide to control the storage loss of wheat (*Triticum vulgare*) and groundnut (*Arachis hypogea*), the two major representatives of cereal and oil seed crops.

The volatile secondary mutabilities (essential oil) was isolated from the selected plant *Eucalyptus pauciflora* Sieb and tested against the dominant storage fungal pests *viz.*, *Aspergillus flavus* Link, *Penicillium italicum* Wehmeyer and *Alternaria alternata* (Fr) Keissler as well as dominant storage insect pests *viz.*, *Rhizopertha dominica* Fab, *Sitophilus oryzae* Linn, *Trogoderma granarium* Everts, the weevils and *Corcyra cephalonica* Staint, *Ephestia cautella* Cadra and Walker, the moths. Besides the investigations on various physico-chemical, pesticidal and phytotoxicity, the oil was also subjected to various physical factors, comparison with synthetic pesticides, maximum phytotolerable concentration as well as minimum effective dose, organoleptic tests and pharmacological investigations.

2. MATERIALS AND METHODS

On pesticidal screening the volatile toxity of some aromatic plant spp. against dominant storage fungal pests by Inverted Petri Plate Method of Bocher (7), as well as against insect pests following the method of Krishnaswami *et. al.* (8) and Chakravary (9) with a slight modification, *Eucalyptus pauciflora* showed the strongest toxicant, inhibiting the test pests completely. Consequently, the plant was selected for detailed investigations. The essential oil was extracted from the fresh leaves of *E. pauciflora* (1 Kg.) which were chopped and subjected to hydrodistillation through Clevenger apparatus (10). A clear light yellow coloured oily layer was obtained on the top of the aqueous distillate, which was separated and dried over anhydrous sodium sulphate. The oil was characterized and standardized by gas liquid chromatography as well as by other physico-chemical properties which were determined by the methods of Langenau (11). The oil thus obtained was evaluated at different concentration and subjected for the determination of minimum killing time (MKT) of the oil against the storage fungal pests as well as storage insect pests. In case of fungal pests, the Mycelial Disc Killing Technique of Shahi *et. al.*, (12) with a slight modification suggests by Shukla *et. al.*, (13) has been followed. Two treatment sets were maintained, one with pure oil (i.e. 100% cone) other with 5% concentration of the oil. The treatment set at 5% conc. of the oil was prepared by mixing the required quantity of the oil samples in acetone (2% of the total quantity of the medium) and then added in pre-sterilized potato dextrose agar medium (pH 5.6). In control, sterilized water (in place of the oil) and acetone were used in the medium in appropriate amounts. Mycelial discs of 5 mm diameter, cut out from the periphery of 7 day old culture of the test fungal pests, *A. alternata, A. flavus* and *P. italicum* were aseptically placed in the culture tubes containing the treatment and control sets. These mycelial discs were taken out of the culture tubes at different time intervals and washed immediately in the washing solution (i.e. acetone and sterilized water in 1:2 ratio) to remove the treatment solution. These washed mycelial discs were aseptically transferred upside down to the potato dextrose agar medium in the petri plates. The same procedure was followed with the control sets. The inoculated petri plates were incubated at 27±1°C. After 7 day of the treatment the observation was recorded on the basis of 5 replicates, repeated twice (Table-1). The percentage of fungal pest-growth inhibition (FpGI) was calculated as per formula :

$$\text{FpGI (\%)} = \frac{dc - dt}{dc} \times 100$$

where,

"dc" indicates colony diameter of control disc, and "dt" indicates colony diameter of treated disc.

The nature of toxicity i.e. static/cidal of the oil was determined by the method of Garber and Houston (14). The effect of inoculum density of the fungal pests on the pesticidal toxicity of the oil at 5% conc was determined following the procedure outlined by Dikshit and Dixit (15) with a slight modification of Shukla *et. al.*, (13). The treatment (at 5% conc.) as well as control sets were prepared in potato dextrose broth (pH 5.6) as per usual methods. Mycelial discs of 5 mm diameter, cut out from the periphery of 7 days old culture of the test fungal pests, were aseptically inoculated in culture tubes containing PDB with 5% conc. of the oil. Controls, were run simultaneously. The number of fungal discs were increased progressively up to 25 disc in multiple of 5 mm diam. Absence of mycelial growth on the seventh day exhibited the oil's potential to withstand heavy discs of inocula of the fungal pests. The observations were recorded on the basis of an average value of 5 replicates, repeated twice (Table 1). The per cent fungal inóculum density (FID %) was calculated using the formula:

$$\text{FID (\%)} = \frac{FGc - FGt}{FGc} \times 100$$

where,

FGc indicates fungal growth in control set, and

FGt indicates fungal growth in treatment set

(i.e. either in volume or in area)

$$FG = VAG - VBG$$

where,

VAG indicates volume after growth and VBG indicates volume before growth

But, in case of insect pests, the treatment set at 5% conc. of the oil was prepared in distilled water as per usual methods and were gently incorporated in different petri plates (7.5 cm diam) containing the sterilized filter papers. Similarly, controls were also maintained. Five inoculates of each healthy insect pests *viz., R. dominica, T. granarium, S. oryzae, C. cephalonica* and *E. cautelia* were introduced with the treatment as well as control sets of the petri plates, separately and incubated at different time intervals.

Table 1. Pesticidal properties of the oil of *E. pauciflora* against fungal pests

Properties studied	Fungal-pest growth inhibition (FpGI)%					
	Alternaria alternata		*Aspergillus flavus*		*Penicillium italicum*	
	100%C	5%C	100%C	5%C	100%C	5%C
Minimum Killing Time (MKT)						
02 sec	100	—	80	—	100	—
05 sec	100	20	100	—	100	15
04 hrs	100	100	100	85	100	100
06 hrs	100	100	100	100	100	100
Inoculum Density (max. no. of fungal disc, each of 5 mm diam.)						
25	100	100	100	100	100	100
Thermostability (oil kept for 1 hr upto)						
80°C	100	100	100	100	100	100
Effect of storage (Storage upto)						
36 months	100	100	100	100	100	100

— indicates not considered

Table 2. Pesticidal properties of the oil of *E. pauciflora* against insect pests

Properties studied	Insect-pest lethality (IpL)%									
	Corcyra cephalonica		*Ephestia cautella*		*Sitophilus oryzae*		*Trogoderma gramarium*		*Rhizopertha dominica*	
	100%C	5%C	100%C	5%C	100%C	5%C	100%C	5%C	100%C	5%C
AS A CONTACT TOXICANT										
Minimum Killing Time (MKT)										
04 min	80	—	60	—	60	—	100	—	80	—
08 min	100	—	100	—	100	—	100	—	100	—
40 min	100	100	100	100	100	60	100	80	100	60
50 min	100	100	100	100	100	80	100	100	100	80
60 min	100	100	100	100	100	100	100	100	100	100
AS A FUMIGANT TOXICANT										
Minimum Killing Time (MKT)										
05 hrs	—	60	—	80	—	—	—	100	—	—
06 hrs	—	100	—	100	—	80	—	100		80
07 hrs	—	100	—	100	—	100	—	100	—	100
Inoculum Density (max. no. of insect populations)										
100	100	100	100	100	100	100	100	100	100	100
Thermostability (oil kept for 1 hr, upto)										
80°C	100	100	100	100	100	100	100	100	100	100
Effect of Storage (storage upto)										
36 months	100	100	100	100	100	100	100	100	100	100

— Indicates not considered

After the required intervals, the insect pest spp. were gently transferred to the petri plates containing sterilized dry filter paper. The same procedure was followed with the control set. The data recorded was mean of the five replicates, repeated twice (Table 2).

The percentage of insect pest lethality (IpL) was calculated as per formula:

$$IpL\ (\%) = \frac{Lc - Lt}{Lc} \times 100$$

where, Lc indicates lethality in control set, and

Lt indicates lethality in treatment set

The effect of inoculation density of the insect pests on the pesticidal toxicity of the oil at 5% conc. was determined by increasing the insect inocula at minimum lethal time. The number of insect pests were increased progressively up to 100 and each in multiple of 10 in numbers. The same procedure was followed with the control set. Absence of any living insect at the minimum lethal time exhibited the oils potential to withstand heavy population of the insect inocula. The observations were recorded on the basis of an average value of 5 replicates, repeated twice (Table-2).

The per cent insect inoculum density (IID %) was calculated using the formula:

$$IID\ (\%) = \frac{PPc - PPt}{PPc} \times 100$$

PPc indicates pest population in control set, and

PPt indicates pest population in treatment set.

The pesticidal spectrum of the oil at 5% conc. against the fungal pests *viz.*, *Aspergillus parasiticus* Speare, *Cladosporium cladosporioides* (Fresenius) de vitres, *Curvularia lunata* (Wakker) Boedijn, *Collectotrichum capsici* (Syd.) Bulter & Bisby, *C. falcatum* Went, *Fusarium oxysporum* Schlecht, *F. udum* de vires, *Helminthosporium maydis* Nisikado & Miyakel, *H. oryzae* Breda de Haan, *Penicullum implicatum* biourge and *P. minio-luteum* Dierckx as well as *Tribolium castaenum* Herbst. and other insect pests were also determined by the usual Poisoned Food Technique. The expiry of toxicity of the oil was also determined by storing the oil at room temperature and testing their pesticidal activity at their respective minimum lethality at a regular interval of 6 months for a total period of 36 months (Table 1).

Effect of temperature on the pesticidal toxicity of the oil was determined by incubating the oil at 40-80° in an incubator for one hour and then tested at their minimum lethality by the usual method (Table 1). The phytotoxic property of the oil as a contract toxicant at 5% conc. with seed germination and seedling growth of wheat and groundnut were studied following the method of Tripathi *et. al.*, (4) with a slight modification. The treatment set at 5% conc. as well as the control sets were prepared as per usual method. Seeds of groundnut and wheat (ten to twenty number) were soaked separately in the treatment solution as well as in control solution for 24 hrs. and then placed in pre sterilized petri plates containing two layer of whatman filter paper no.1, soaked in sterilized distilled water. All the plates were kept in an incubator maintained at 30±1°C. The per cent seed germination was recorded after 3,4 and 5 days, using the formula.

Seed germination (%) =

$$\frac{\text{Total number of germinating seed}}{\text{Total number of seeds}} \times 100$$

Besides this, the radical and plumule lengths were measured regularly up to one week and it was observed that the oil have no ill effect on seed germination and seedling growth of the host plant of wheat and groundnut (6).

But, for commercial viability of the biospesiticide of plant origin, the formulation should be cheaper, with broad pesticidal spectrum, reutilizable, eco-friendly as well as easier to handle. Therefore, for convenience and feasibility one should foresee that until unless active secondary metabolites (i.e. essential oil) are transformed into suitable formulations, they will not be accepted by the users. As such, keeping these facts in mind, further investigations have been carried out with special reference to the *'fumigation technology'*.

The fumigant activity of the essential oil of *E. pauciflora* was determined following the method of Krishnaswami *et. al.*, (8) with a slight modification. The oil at 5% conc. was prepared in distilled water as per usual method and was taken separately in a petri plate (7.5 cm diam) which was placed in different petri plates (15 cm diam), containing the five inoculants of each insect pest *viz., R. dominica. S. oryzae, T. granarium, C. cephalonica* and *E. cautelia* separately, and plates were sealed with grease. Similarly, controls were also maintained. After a regular interval of an hour the insect pest spp. were gently transferred to the fresh exposed petri plate. The data recorded was mean of the 5 replicates, repeated twice (Table 2). The percentage of insect pest lethality (IpL) was calculated by the usual formula. The effect of inoculum density of the insect pests on the pesticidal toxicity at minimum lethality of the oil was also determined. The temperature treatment, expiry of toxicity and phytotoxicity was also observed just to access their confirmational test as in case of contact toxicant. As such, the maximum phytotolerable concentration was determined following the method of Krishnaswami *et. al.*, (1974) with slight modification of Shukla *et. al.*, (13). The requisite quantity of the oil samples loaded on some 'tiller compounds' were used to prepare the formulations in variable concentrations of 10%, 20%, 30%, 40% and 50%.

To fumigate the seeds, petri plate containing 10-20 gm seeds of wheat groundnut were placed first in 5 liter desiccator separately and thereafter, formulations of variable concentrations were introduced separately and the desiccators were sealed with vacuum grease. In control set, seeds of wheat and groundnut were given similar treatment with fumigant-free formulations using only their compounds. After 10 days of the treatment, seeds were stored for six to twelve months to see the effect on seed germination, seedling growth and general health of the plants as per usual method (Table 3).

The minimum effective dose (MED) i.e. the ratio between *concentration* (i.e. 50%, 40%, 30%, 20% or 10%), *quantity* (i.e. 3 gm, 6gm, or 9gm) and *volume* (i.e. 5 liter volume of desiccator) of the formulated fumigant biopesticides, the technique of Krishnaswami *et. al.*, (8) was followed with a slight modification (13). Seeds of wheat and groundnut, 300 gm each, were fumigated following the usual method (as used in the maximum phytotolerate conc.) with formulations of 3, 6 and 9 gm each of 50%, 40% and 30% of conc. in 5 litre desiccator, separately. To find out the best container, seeds after 10 days of the fumigation were stored for twelve months in polythene, cotton and gunny bags at the rate of 100 gm each. On the basis of Pest analysis, Phytotoxicity, Organoleptic test and Pharmacological investigation the minimum effective dose (MED) was determined. The pest analyst i.e. insect pest as well as fungal pests of the fumigated and twelve month stored seeds of wheat and groundnut was done following the method of Chakraverty (9) as well as

Samson *et. al.*, (17) and the observations were recorded in Table 4 and 5. The phytotoxicity test was done as per usual method. The organoleptic test was calculated using the technique of Tiwari and Dixit (18) with slight modification. The previously treated and 12 months stored seeds of wheat and groundnut were investigated for organoleptic test with special reference to appearance,

Table 3. Maximum phytotolerable concentration of *E. pauciflora* oil

(i) Seed Germination (%)

Fumigant concentrations (%)	Germination period (days) 3		4		5	
	W	GN	W	GN	W	GN
0.0	80	70	82	72	86	76
10	78	70	82	76	84	76
20	78	68	84	70	84	78
30	80	70	80	72	86	78
40	80	72	80	72	86	76
50	82	70	84	72	86	78

W = indicates wheat
GN = indicates groundnut

(ii) Seedling Growth (mm)

(A) Wheat

Fumigant concentrations (%)	Seedling Growth of Wheat (mm) 5 days		6 days		7 days	
	R	P	R	P	R	P
0.0	16	15	23	20	31	25
10	15	15	24	22	33	24
20	18	14	22	22	32	26
30	16	14	22	20	31	27
40	16	16	24	20	31	25
50	16	15	24	20	32	27

(ii) Seedling Growth (mm)

(B) Groundnut

Fumigant concentrations (%)	Seedling Growth of Groundnut (mm) 5 days		6 days		7 days	
	R	P	R	P	R	P
0.0	08	11	12	18	18	24
10	08	09	14	16	19	26
20	09	13	14	16	19	24
30	10	12	12	19	20	25
40	09	12	11	18	20	25
50	09	12	13	18	20	26

R = indicates radicle & P = indicates plumule

Table 4. Pest analysis of Fumigated and 12 months stored seeds of Wheat

Isolated		TREATMENTS																	
		BIOPESTICIDE									SYNTHETIC PESTICIDES						Control		
		Eucalyptus pauciflora									$AlPH_3$			EDB					
		50%			40%			30%			56%			83%					
Pests	Bag	3gm	6gm	9gm	3gm	6gm	9gm	3gm	6gm	9gm	3gm	6gm	9gm	3gm	6gm	9gm	3gm	6gm	9gm
FUNGAL PESTS																			
Alternaria alternata	P																		
	C																+	+	+
	G													+			+	+	+
Aspergillus flavus	P										+			+	+	+	+	+	+
	C				+						+	+	+	+	+	+	+	+	+
	G	+	+		+	+	+	+	+	+	+	+	+	+	+	+	+	+	+
A. fumigatus	P																+	+	+
	C																+	+	+
	G	+			+			+	+	+	+			+	+		+	+	+
A. niger	P													+	+	+	+	+	+
	C													+	+	+	+	+	+
	G	+			+			+	+		+	+	+	+	+	+	+	+	+
A. parasiticus	P																+	+	+
	C																+	+	+
	G										+			+	+	+	+	+	+
A. versicolor	P																		
	C																+	+	+
	G																+	+	+
Fusarium cerealis	P																+	+	+
	C													+			+	+	+
	G	+			+	+		+	+	+	+			+	+		+	+	+
F. oxysporum	P																		
	C																+	+	+
	G										+			+	+		+	+	+
Penicillium gresiofulvum	P																		
	C																+	+	+
	G	+			+	+		+	+		+			+			+	+	+
P. italicum	P																+	+	+
	C	+			+			+	+		+			+	+		+	+	+
	G	+			+	+		+	+	+	+	+		+	+	+	+	+	+
Rhizopus spp.	P	+			+	+		+	+	+	+	+	+	+	+	+	+	+	+
	C	+	+		+	+	+	+	+	+	+	+	+	+	+	+	+	+	+
	G	+	+	+	+	+	+	+	+	+	+	+	+	+	+	+	+	+	+
INSECT PESTS																			
Rhizopertha dominica	P																		
	C																+	+	+
	G				+			+						+			+	+	+
Trogoderma granarium	P																		
	C																+	+	+
	G							+									+	+	+
Sitophilus oryzae	P																		
	C																		
	G																+	+	+
Tribolium castaneum	P																		
	C																		
	G																+	+	+
Ephestia cautella	P																		
	C																		
	G							+						+			+	+	+
Sitotroga cerealella	P																		
	C																		
	G																+	+	+

+ indicates presence of pests. P = polythene bag. C = cotton bag. G = gunny bag.

Table 5. Pest analysis of Fumigated and 12 months stored seeds of Groundnut

Isolated		TREATMENTS																	
		BIOPESTICIDE									SYNTHETIC PESTICIDES						Control		
		Eucalyptus pauciflora									$AlPH_3$			EDB					
		50%			40%			30%			56%			83%					
Pests	Bag	3gm	6gm	9gm	3gm	6gm	9gm	3gm	6gm	9gm	3gm	6gm	9gm	3gm	6gm	9gm	3gm	6gm	9gm
FUNGAL PESTS																			
Aspergillus flavus	P							+	+								+	+	+
	C				+	+		+	+	+				+	+	+	+	+	+
	G	+	+		+	+	+	+	+	+	+	+	+	+	+	+	+	+	+
A. parasiticus	P							+									+	+	+
	C							+	+					+			+	+	+
	G	+			+	+	+	+	+	+	+	+		+	+	+	+	+	+
A. tereus	P																		
	C																+	+	+
	G													+	+		+	+	+
A. versicolor	P																		
	C													+			+	+	+
	G				+			+	+		+	+	+	+	+	+	+	+	+
Fusarium oxysporum	P																+	+	+
	C													+			+	+	+
	G				+	+		+	+	+	+			+	+		+	+	+
Penicillium italicum	P																+	+	+
	C							+									+	+	+
	G				+			+	+					+	+	+	+	+	+
Rhizopus spp. *(Oryzae)*	P	+			+	+		+	+	+	+	+	+	+	+	+	+	+	+
	C	+			+	+		+	+	+	+	+	+	+	+	+	+	+	+
	G	+	+	+	+	+	+	+	+	+	+	+	+	+	+	+	+	+	+
INSECT PESTS																			
Corcyra cephalonica	P																		
	C																+	+	+
	G				+			+	+					+			+	+	+
Sitotroga cerealella	P																		
	C																		
	G							+									+	+	+

+ indicates presence of pests. P = polythene bag. C = cotton bag. G = gunny bag.

tenderness, flavour, taste and general acceptability.

The observations were made to their respective consumable forms. The samples of consumable forms were served to a batch of 30 persons comprising equal number of males and females of different age groups *viz.*, teen-agers (10), middle age group (10) and old age groups (10). The experiment was repeated twice. The persons were asked to score the aforementioned parameters of organoleptic properties, out of the maximum of ten with significant remarks, if any. The mean scores were recorded.

Just to assess the mammalian toxicity of the formulations of *E. pauciflora* if any, the seeds of wheat and groundnut treated with their minimum effective doses (Table 4 & 5) and stored for different periods, were given in their respective consumable forms to the human volunteers. The serum sample of such volunteers were taken after 48 hrs. of consumption of aforementioned treated commodities, each of 500 gm and subjected to the following preliminary pharmacological investigations *viz.*, body weight, alkaline phosphatases in serum, cholesterol in serum, glucose in serum, urea in serum, SGOT in serum and SGPT in serum. The analysis of these pharmacological parameters was done using the kits from Bayer Diagnostic India

Table 6. Synthetic Pesticides and Biopesticide : A comparison

Name of fumigants	Synthetic pesticides		Biopesticide
Characteristic features	Aluminium phosphide - 56%	Ethylene dibromide 83%	Biopauci-50% (*E. pauciflora)*
1. Chemical formula	AIPH3	$C_2 H_4 Br_2$	An aldehyde
2. Active constituent	Phosphine	Ethylene	Eucalyptol
3. Normal state	Solid	Liquid	Solid
4. Odour	Garlicky	Pungent almond like	
5. Penetration	Excellent	Low	Good
6. Flammability (% by vol. in air)	1-79	Non-inflammable	Non-inflammable
7. Solubility in water (g/100 ml)	Very slighly soluble	0.43 at 30°C	Insoluble in water
8. Minimum effective dose	2.25 gm/m3	22 gm/m3	9 gm/m3
9. Exposure duration	5-7 days	7-10 days	10-15 days
10. Physical changes	Solid to powdery form	Liquid to waxy form	No change
Remarks	* Grain fumgigant * Phytotoxic at high dosese * Utilizable once * Strictly prohibited	General fumigant Phytotoxic at high doses Utilizable once Strictly prohibited for oil seeds	General fumigent Non-phytotoxic upto 50% conc. Reutilizable Antifeedent

Ltd. except for alkaline phosphatases where the kits of Recon Diagnostic Pvt. Ltd. has been used. Further the samples have been analyzed on RA-50 "AMES" analyzer with the help of an expert pathologists. The range of toxicity at minimum effective dose was also determined up to the volume of per cubic meter. Comparative studies of the formulated fumigant biopesticide with some synthetic fumigants were carried out by comparing their pesticidal activity as well as their characteristic features (Table 4 to 6). The preliminary comparison of cost benefit analysis, while using the 'formulated fumigated biopesticides' with that of synthetic pesticides have been recorded in Table 7.

3. RESULTS

The findings suggests that *Eucalyptus pauciflora* oil can be exploited as a potential source of sustainable ecofriendly storage pesticide because of its following pesticidal activity.

(a) Oil as contact toxicant

The pure oil (i.e. 100% conc.) as a contact toxicant killed *Alternaria alternata, Aspergillus flavus* and *Penicillium italicum* in 2 sec, 5 sec and 2 sec respectively while, at 5% conc. the oil required 4 hrs. 6 hrs and 4 hrs to kill the same fungal pests respectively (Table 1). The toxicity of the oil at 5% conc. inhibited heavy doses of inocula of the test fungal pests (Table 1). However, in case of insect pests the pure oil killed the *Rhizophertha dominica, Sitophilus oryzae, Corcyra Cephalonica* and *Ephestica cautella* in 8 min. and *Trogoderma granarium* in 4 min, while at 5% conc. it required 60 min, 60 min, 40 min, 40 min and 50 min respectively against the same test insect and withstood heavy inoculum density (Table 2). The oil showed wide range of activity by inhibiting the 11 fungal pests

Table 7. Cost benefit Analysis : An overview

Formulated Biopesticide and trade name of the Synthetic fumigant pesticides	Pesticide concentration (%)	Cost/gm (Rs.)	Adverse effects	Expiry duration	Environmental impact
Biopauci (*E. pauciflora*)	50% w/v		No adverse effects	Not expire	Eco-friendly,
Economic product		0.63		even upto 36 months	renewable biodegradable
Luxury product		0.75			
Aluminum Phosphide					
($AlPH_3$)	56% w/w	0.50	Several side effects	24 months	Non-renewable, non-biodegradable and residual toxic.
Ethylene Di-Bromide					
(E.D.B)	83% w/v	1.17	Several side effects	24 months	Non-renewable, non-biodegradable and residual toxic.

as well as some insect pests completely. The toxicity of the oil persisted up to 36 months of storage and remained thermostable up to 80°C, the maximum units taken into consideration (Table 1 & 2). The oil proved non-phytoxic to wheat and groundnut (the host plant) showing no adverse effect on seed germination and seedling growth.

(b) Oil as fumigant toxicant

The oil is a fumigant toxicant, at 5% concentration killed the insect pests *R. dominica*, and *S. oryzae* in 7 hrs. *T. granarium* in 5 hrs and required 6 hrs against the *C. cephalonica* and *E. cautella* respectively at which the oil's toxicity withstood heavy population of the insect pests (Table 2). The temperature treatment, expiry of toxicity and phytotoxicity remained constant as in case of contact toxicant. The maximum phytotolerable concentration of the oil was found up to 50% (the maximum units taken into consideration showing no adverse affect on seed germination and seedling growth of wheat and groundnut (Table-3). For determining the minimum effective dose, the observations on the following parameters were under taken.

(i) **Pest analysis** of the fumigant free formulations (i.e. control set) showed that a total of 11 fungal pests and appeared on wheat when stored in gunny and cotton bags, while *A. alternata, A. versicolor, F. oxysporum* and *P. gresiofulvum* did not occur in isolation from seeds stored in polythene bags. However, in case of insect pests only *R. dominica, S. oryzae, T. granarium, T. castaneum, S. cerealella* and *Ephestia cautella* appeared in gunny bags, while no spp. of insect pests could be recovered from the seeds stored in polythene bags. However, in the treatment sets except the *Rhizopus* sp. all the fungal pests as well as insect pests were controlled from the seeds stored in all the bags at 9 gm dose of 50% concentration as compared to synthetic fumigants, where more or less only 50% pests could be controlled.

In control set of groundnut, 7 fungal pests have been isolated from the seeds stored in cotton and gunny bags. *A. tereus* and *A. versicolor* did not appear on seeds stored in polythene bags. However, in case of insect pests only *S. cerealella* and *C. cephalonica* were recovered from the seeds stored in polythene bags. However, in the treatment sets neither fungal pests nor insect pests could be isolated from the seeds stored in all the bags at 9 gm of 50% conc. as compared to synthetic fumigants, where more or less only 50% pests could be controlled.

(ii) **Phytotoxicity Test** on wheat and groundnut showed no adverse effect on seed germination, seedling growth and general health of the plants. However, in some cases the growth of preliminary radicle and plumule length of the fumigated seeds were healthier than the control sets.

(iii) **Organoleptic test** indicated that the seeds treated with the formulations of *E. pauciflora* were more acceptable and preferred by the consumers of both the sexes of all age groups.

(iv) **Pharmacological investigations on** mammalian toxicity of the formulated fumigant biopesticides reflected the similar results to that of control ones, indicating that the treated commodities can be safe and fearlessly consumed.

The minimum effective dose was observed as 9 gm at 50% conc. and that was measured in terms of the volume of per cubic meter. Polythene bags were found to be the best container as the seeds stored in them showed least pest association in comparison to cotton and gunny bags.

On comparing the pesticidal activity of the formulated biopesticides with that of synthetic pesticides *viz.*, AlpH and EDB, it was observed that the biopesticide has been found more or less superior than the synthetic ones, in controlling the tested fungal and insect pests (Table 4 & 5).

The characteristic features including the cost benefit analysis of the formulated fumigant biopesticide of *E. pauciflora* has also an edge over synthetic pesticides (Table 6 & 7).

4. DISCUSSION

Although the essential oils from various *Eucalyptus spp.* have been reported to contain antimicrobial activity (19-23) to our knowledge the detailed studies *viz.*, minimum killing time, maximum phytotolerable concentration, minimum effective dose, organoleptic tests, pharmacological tests etc. as of the essential oil of *E. pauciflora* is being reported here for the first time.

For determining the minimum killing time (MKT) of essential oil, treated mycelial discs were washed using distilled water by various workers (24-25), however, in the present investigation instead of water, washing solution consisting of acetone and water in 1 : 2 ratio was definitely a refinement over the previous technique, and was found good.

Similarly, regarding the insect pests, the methods which have been taken into consideration have also an edge over the previous methods. According to Wellman (24) a pesticide must be stable to the extremes of temperature and long period of storage. However, the effects of storage and temperature on pesticidal activity of the oils have received little attention. Chaturvedi *et. al.*, (25) found *Adenocalymma allicea* oil thermostable up to 100°C and toxic for 21 days. Singh *et. al.*, (26) reported *Peperomia pellucida* oil toxic up to 80°C and up to 150 days of storage. However, in the present study, the pesticidal toxicity of the oil of *E. pauciflora* was thermostable up to 80°C and toxic up to 36 months of storage.

In contrast to the oil of *Raphanus* (27); Juniperus (28), *Psoralea corylifolia* (29) and *Feronia elephantum* (30) which showed a narrow pestidal spectrum, the *E. pauciflora* oil exhibited a broad pesticidal spectrum like that of *Curcuma aromalica* (31), *Caesulla auxiliaries* (32); *Nepeta hindustana* (9) and *Eucalyptus citriodora* (13) and *Trachyspermum ammi* (6).

Earlier literature showed that generally essential oils are non phytotoxic for practical tise (1, 5, 6, 32, 33) which was found to be five for *E. pauciflora* oil also. Most of the treatments, under *in vivo* conditions, highly effective in preventing the fungal decay of the commodities were non-toxic to the pests *in vitro* under similar conditions. Conversely some treatments effective against the pests *in vitro* were completely ineffective in protecting the commodities from the fungal invasions (34). As such, in the present study the oil of *E. pauciflora* in the form of fumigant formulations has been tested for its pesticidal activity under both *in vitro* and *in vivo* condition and found quite effective in protecting the stored food commodities up to the volume of per cubic meter completely, containing the standard organoleptic as well as pharmacological features.

Besides the superiority of the 'formulated fumigant biopesticide' over some synthetics in controlling the tested fungal and insect pests, the characteristic features of the formulations of *E. pauciflora* have also an edge over the synthetic pesticides.

Thus, on the basis of above findings it can be safely said that 'formulated fumigant biopesticide' may prove to be better substitutes to the hazardous synthetic pesticides, which are responsible not only for injurious to the health of the human beings but also pollute the ecosphere. As such, the effective products recommended to agrochemical firms for their exploitation as herbal pesticide may be 'BIO PAUCI'.

ACKNOWLEDGEMENT

Thanks are due to Head, Department of Botany, University of Allahabad for providing the facilities; to Indian Agricultural Research Institute, New Delhi for providing cultures of storage fungi and their identification, to Prof. A.V.N. Paul, Div. of Entomology, IARI, New Delhi for providing authentic cultures of the insect pests and their identification; to Shri S.N. Pandey, District Manager, FCI Allahabad for providing their invaluable help; to Dr. V.C. Saksena, Prof. & Head, Pharmacology Department of M.L.N. Medical College, Allahabad to Dr. Veenu Saksena, Pathologist, Allahabad, for their invaluable help in pharmacological investigations of the samples and to Cuncil of Science and Technology, Lucknow, U.P. for financial assistance.

REFERENCES

1. Mishra AK & Dubey NK. *Applied and Environmental Microbiology*, US 60(4) (1994) 1101.
2. Singh M, Srivastava S, Srivastava RP and Chauhan SS. *Plant Foods for Human Nutrition*, 47(2) 1995) 109.
3. Bishop CD & Thornton IB. *Journal of Essential Oil Research, USA* 9(1) (1997) 77.
4. Shaaya E, Kostjukvshi M, Eilberg J & Sukprakarn C. *Journal of Stored-product Research*, U.K.V. (1997) 7.
5. Tiwari R. *Proc. national Symposium on Plant Diversity Pattern and bioresources Technology*, Gorakhpur, India 26 (1998).

6. Shukla AC, Shahi S & Dikshit A. Biotrachy an ecofriendly storage pesticide : progress and prospects. *Proc. 7th International Working Conference on Stored-product Protection*, China (1998) (Accepted).

7. Bocher, OF. *Z. Hyg. Infektion Skrankh*, 121 (1938) 166.

8. Krishnaswami MA, Patel JD, Nair KKS & Mathur H. *Indian Drugs*, 11 (1974) 6.

9. Chakraverty A. 3rd ed. *Oxford & IBH Publishing Co. Pvt. Ltd., N. Delhi (1995).*

10. Clevenger JV. *J. Am. Pharm. Assoc. 17 (1928) 346.*

11. Langenau IFF. The examination and analysis of essential oils, synthetics and isolates (Guenther, E, ed. 1948). The essential oil. *Robert E. Krieger Publishing Co.*, Huntington, New Yori (1948).

12. Shahi S, Shukla AC, Dikshit S and Dikshit A. *Proceedings of the 4th International Symposium on Diagnosis and Identification of Plant Pathogens*, Khuwer Academic Publishers, Netherlands, (1997) 257.

13. Shukla AC, Shahi S & Dikshit A. Efficacy nd potentiality of *Eucalyptus citriodora as a biopesticide. Proceedigns of the T.C.D. International Wrokshop on Application of biotechnology in Biofertilizers and Bipesticides*, New Delhi (1997) 81.

14. Garber RH & Houston BR *Phytopathology*, 49 (1959) 449.

15. Dikshit A & Dixit SN. *Indian Perfumer* 26(2-4) (1982) 216.

16. Tripathi NN, Dikshit A, Tripathi RD & Dixit SN. *International J Crude Drug Res* 21(2) (1983) 67-71.

17. Samson RA, Hoekstra ES, Frisvad JC & Filtenborg O, 4th ed. Introduction to food borne fungi. *Ponsen and Looyen*, Wageningen, The Netherlands (1995).

18. Tiwari R & Dixit V. *Indian Perfumer*, 39(2) (1995) 67.

19. Egawa A, Tsulsui O, Tatasuyama K & Halta T. *Experientia*, 33 (1977) 889.

20. Slavens J & Razinskaite D. *TSR Moks. Acad. Darbal. Ser. C.* 2 (1962) 63.

21. Suri RK, Nigam SS & Thind TS. *Drugs. Pharm. India*, 14(3) (1979) 35.

22. Deshmukh SK, Gupta DK & Tiwari OP. *Indian drugs*. 19 (1982) 370.

23. Mishra DN, Mishra AK & Dixit V. *Vegtos*, 3 (1990) 182.

24. Dikshit A. *Ph. D. Thesis*, University of gorakhpur, Gorakhpur, India.

25. Chaturvedi R, Dikshit A & Dixit SN. *Trop. Agric.* (Trinidad), 64 (1987) 318.

26. Wellman RH Plant Pathology Problem and Progress, Hollen *et. al.*, ed. *Indian University Press*, Allahabad, India, 1908-1958 (1967).

27. Nehrash AK. *J. Microbial*, Kiev. 23 (1961) 32.

28. Singh AK, Dikshit A & Dixit S.N. Beitr. Biol. Pflanzen, 58 (1983) 357-368.

29. Grover GS & Rao JT *Indian Drugs*, 15 (1978) 143.

30. Sharma GP, Jain NK & Garg BD. *Indian Drugs*, 16 (1978) 21.

31. Rao JT. *Ind. Jour. Pharm.*, 38 (1976) 53.

32. Pandey DK, Tripathi NN, Tripath RD & Dixit SN. *Angew Botanik*, 56 (1982) 259.

33. Dikshit A, Tripathi NN, Tripathi SC & Dixit SN. *Ann. Apl. Biol.* (100 suppl.) 3 (1979) 56.

34. Dubdy NK, Bhargav KS & Dixit SN. *Int. J. Trop. Plant Dis.*, 1 (1983) 177.

35. Eckert JW. Application and use of post harvest fungicides (Torgesson D.C. 1969, ed), Fungicides, *Academic Press*, New Delhi, 1 (1967) 288.

CHAPTER - 16

A BIOCONTROL FORMULATION FOR POST-EMERGENCE DAMPING-OFF OF *Moringa pterygosperma*

N.S.K. Harsh and B.M. Ojha

Forest Pathology Division, Tropical Forest Research Institute
P.O.-R.F.R.C., Jabalpur - 482021, M.P., India
Present Address : Medial Division, Directorate of Extention, Indian Council of Forestry Research & Education, P.O. New Forest, Dehradun-248006, India

ABSTRACT

Losses due to post-emergence damping-off disease of *Moringa pterygosperma* (*munga*, drum-sticks plant) caused by *Fusarium acuminatum* were estimated at nursery level in detail. It was found that there was a progressive reduction in seed germination, seedling growth, biomass, and survival with the increased level of infection. *Aspergillus terreus, Penicillum herquei and Trichoderma virens* were tested in laboratory and field conditions for their antagonistic potential against *Fusarium acuminatum.* Economic and feasible biocontrol formulations and delivery system of these antagonists have been prepared in powder form using bagasse for field application. The biocontrol formulations were found successful in controlling the disease and also compared with fungicidal (bavistin) 'treatments. Seed pelleting and soil mixing with *T. virens* formulation was found superior among the biocontrol formulations in reducing the disease incidence, i.e. 25.34% and 22.67%, respectively, which were at par with fungicidal seed treatment.

1. INTRODUCTION

Moringa pterygosperma Geartn. (= *M. oleifera* Lam.) the drum-sticks (*munga*) plant, is a deciduous, multipurpose, small-medium sized tree, found in many parts of India, which is known as a vegetable (tender fruits, flowers and leaves) and for its various uses. This tree is commonly grown in kitchen gardens and on the bunds of agriculture fields throughout India. It is a true multipurpose tree species because besides being used as vegetable, the leaves and twigs are of fodder value; wood is used for fuel and in making poles, pulp, paper, toys, floats and sandwiched material; has role in environmental conservation, and it possesses valued seed oil.

Due to its suitability in agroforestry, social forestry and wastelands afforestation programmes, there are increasing demands for its quality and disease-free nursery stock. Harsh *et al.* (1) have reported a post-emergence damping-off disease caused by *Fusarium acuminatum* Ell. & Ev. affecting nearly 41% of germinated seedlings. Thus it becomes imperative to control this particular disease in nurseries. Chemical fungicides released to environment and their persistence and accumulation over the years in the environment, especially in the soil and aquatic ecosystems and consequently their inherent hazards to plant and animal life and even to human beings, are the concerns advocated against their continued use. This leaves to the only alternative of biological control which offers an effective, safer and long term measure against the diseases. The present study is an attempt to develop a biocontrol formulation and delivery system for the control of post-emergence damping-off of *Moringa pterygosperma* caused by *Fusarium acuminatum.*

2. MATERIALS AND METHODS

The diseased seedling of *Moringa pterygosperma* and associated soil samples were collected during the survey of nurseries of Madhya Pradesh. Isolation and identification was made for pathogen and antagonists with standard procedures and principles.

Losses due to post-emergence damping-off disease in *M. pterygosperma* were estimated by conducting field experiments.

In laboratory, initial antagonistic tests were made between the test fungus and eight different soil fungi isolated using presumptive test (2) and dual culture method. The observations were made for both qualitative and quantitative effect. Qualitative evaluation was based on visual observations of the Petri plates showing growth interaction of the test fungus and antagonists. Whereas quantitative estimation was made with the help of a leaf area meter, using black paper discs of the size of the Petri plate, marked with the actual area covered by the test fungus, antagonists and inhibition zone and thereafter measuring the individual area. *Aspergillus terreus* Thom., *Penicillum herquei* Bainier & Sartory and *Trichoderma virens* showed effective antagonism against *F. acuminatum* and further studies were conducted. with them. The biocontrol formulations of *A. terreus* (At), *P. herquei* (Ph) and *T. virens* (Tv) were prepared in powder form using bagasse. For this the bagasse was collected from juice centres, chopped, surface washed, oven dried at 80°C for 12 h, powdered with Willey Mill grinder and sieved with 22 mesh screen. The bagasse powder (25 g) was packed in autoclavable polypropylene bags (23.5 x 17.5 cm), 50 ml water added, tightly sealed and autoclaved. After the bags cooled down to room temperature, 5 ml of spore suspension of antagonist spore suspension was inoculated through a hypodermic syringe and closed with cello tape at the inoculation point and incubated at room temperature 27± 2°C). The contents inside the bags were fully colonized by the inoculants in 15 days showing extensive sproulation, ready to use. The contents were taken out of the bags and dried in a laminar flow clean bench, packed in polythene bags and tightly sealed for storage. Mass culture of test fungus *Fusarium acuminatum* was prepared in carboxy methyl cellulose medium (3).

Colony forming units in the formulations were counted with the help of a haemocytometer slide.

The seeds of *Moringa pterygosperma* were collected, graded by hand and damaged, immature, dead, shrunken, discoloured, wrinkled and small sized seeds were discarded as recommended by Harsh and Gupta (4).

Experiments were conducted in field to control the disease, using biocontrol formulations and chemical fungicide bavistin (carbendazim) for comparison with nine treatments in three replications in randomized block design. In these treatments soil mixing and seed pelleting were given with biocontrol formulations and bavistin.

The observations regarding seed germination, symptoms of disease, growth behaviour trends, and disease incidence were recorded periodically. Seedling vigour index (SVI) and seedling vigour index in nursery (SVIN) was calculated by using formulae given by Abdul and Anderson (5). Statistical analyses were made to facilitate the results.

3. RESULTS

Results of antagonistic tests conducted with different isolates against test fungus *Fusarium acuminatum*, have been shown in Table 1 and 2. It is evident from Table 1 that isolate No. 12, 10 and 1 have higher percentage of covered area (in cm^2) of Petri plate in comparison to other isolates. Therefore, these isolates were screened for further tests (biculture test). The results of biculture test have been presented in Table 2 where isolate No. (88.20%) has better

Table 1. Antagonistic activity of some soil fungi against *Fusarium acuminatum* using presumptive tests

Sl. No.	Isolate No. and name	% area covered by test fungus (in cm^2)	% area covered by isolate (in cm2)
1	08 *Trichoderma harzianum*	2.24	19.56
2	09 *T. polysporum*	2.24	20.63
3	12 *T. virens*	2.24	28.45
4	18 *T. longibrachiatum*	2.24	19.98
5	02 *T. koningii*	2.54	18.63
6	06 *T. viride*	2.54	20.48
7	10 *Aspergillus terrus*	2.54	25.50
8	01 *Penicillium herquei*	2.54	28.39

Table 2. Antagonistic efficacy of selected soil fungi against *Fusarium acuminiatum*

Antagonist	% area of test fungus without antagonist (control) cm^2	% area of test fungus with antagonist in cm^2 (a)	% area of antagonist with test fungus in cm^2 (b)	% area of inhibition zone of antagonist in cm^2 (c)	% effectiveness of antagonist against test fungus (b+c-a)
Aspergillus terreus	52.51	10.55	80.39	12.25	82.09
Penicillium herquei	48.98	8.73	88.20	6.94	86.1
Trichoderma virens	47.32	6.60	83.13	11.28	87.81

percentage covered area followed by isolate No.12 (83.13%) and No. 10 (80.39%). The percentage of inhibition zone area was higher in isolate No. 10 followed by No. 12 and No.1. However, the area of test fungus was reduced maximum by isolate No. 12. The isolate No. 12 has been identified as *Trichoderma virens*, isolate No. 1 as *Penicillum herquel* and isolate No. 10 as *Aspergillus terreus*. The formulations have 2 - 3 x 10^9 c.f.u. per gram.

Effect of root pathogen (losses) on different parameters of *Moringa pterygosperma* has been exhibited in Table 3. It was clear that disease caused varied degree of losses at different levels, for example germination percentage, shoot height and root length have been reduced by 12%, 19.5% and 31%, respectively as compared to control (uninoculated seeds). Post-emergence damping-off disease incidence percentage was higher, i.e. 28.77% as compared with uninoculated control.

From Table 4, it is clear that all treatments (seed pelleting and soil mixing) made with biocontrol formulations and chemical fungicide bavistin (seed coating and soil drenching) were significantly superior to control (seeds treated with the test pathogen) for all the parameters except for shoot biomass where only soil drenching with bavistin and soil mixing with biocontrol formulation (Tv) were significantly superior to control. However, there were no significant difference between seed treatment with bavistin and treatments with Tv formulation. Soil drenching with chemical fungicide gave 72.93% gain in root length, followed by 47.085 with soil mixing with Tv formulation over control. The same trend was also noticed in fresh weight of roots. Gain in root length over control was more with seed pelleting with biocontrol formulation (Tv - 43.08%) than seed treatment with bavistin (30.00%). Gain in shoot height over control was maximum with soil drenching with bavistin (34.87%), followed by seed treatment with bavistin (19.84%) seed pelleting with At (19.33%), seed pelleting (17.78%) and soil mixing (16.94%) with Tv. The seedling vigour index (SVI) of the soil and seed treatments with bavistin and formulation Tv and soil mixing with

Table 3. Effect of root pathogen *Fusarium acuminatum* on different parameters of *Moringa pterygosperma*

Sl. N.	Parameters	Uninoculated control	Treated with pathogen	Per cent reduction in comparison to control
1	Germination (%)	56.67	48.67	12.00
2	Shoot height (cm)	82.70	66.50	19.50
3	Root length (cm)	23.50	16.00	31.00
4	Shoot wt. fresh (g)	28.55	19.69	31.10
5	Root wt. fresh (g)	33.35	26.42	20.00
6	Shoot wt. dry (g)	8.75	5.56	36.45
7	Root wt. dry (g)	2.96	1.72	41.18
8	Post emergence damping-off %	0	28.76	28.77*
9	Survival %	100	71.23	28.77
10	Not germinated seeds %	43.33	51.33	8.00*

* increase over control

Table 4. Observations on the effect of different treatments on growth parameters of *Moringa pterygosperma*

Treatment	Root length (cm)	% gain (over control)	Shoot height (cm)	% gain	Root fresh wt (g)	% gain	Shoot fresh wt (g)	% gain	Root dry wt (g)	% gain	Shoot dry wt (g)	% gain	SVI	SVIN
T1	25.00	0	19.30	0	3.90	0	2.97	0	0.31	0	0.25	0	2480.00	1949.20
T2	32.50	30.00	23.13	19.84	6.43	64.87	4.23	42.42	0.62	100.0	0.38	52.00	4598.37	4079.35
T3	35.77	43.08	22.73	17.78	6.27	60.77	3.70	24.58	0.61	96.78	0.34	36.00	4133.61	4055.80
T4	36.77	47.08	22.57	16.94	8.10	107.69	7.07	138.04	0.82	164.51	0.71	184.00	4272.48	3955.60
T5	28.30	13.20	19.93	3.26	4.97	27.43	3.87	30.30	0.39	25.80	0.35	40.00	3279.64	2893.80
T6	28.63	14.52	20.77	7.62	4.90	25.64	3.57	20.20	0.44	41.93	0.36	44.00	2832.11	2436.90
T7	30.53	22.12	23.03	19.33	5.43	39.23	4.73	59.25	0.51	64.51	0.46	84.00	4010.56	3499.07
T8	31.37	25.48	19.43	0.67	8.00	105.12	7.03	136.70	0.74	138.70	0.73	192.00	2962.33	2419.12
T9	43.23	72.93	26.03	34.87	10.06	157.94	7.37	148.15	1.18	280.64	0.75	200.00	5586.51	5263.76
SEM+	1.82		0.73		0.65		0.57		0.09		0.06		331.00	349.57
CD at 5%	15.93		2.40		2.12		1.87		0.29		0.21		1079.29	1136.61

T1 - Control (no treatment)
T2 - Seed dressing with Bavistin
T3 - Seed pelleting with Tv formulation
T4 - Soil mixing with Tv
T5 - Seed pelleting with Ph
T6 - Soil mixing with Ph
T7 - Seed pelleting with At
T8 - Soil mixing with At
T9 - Soil drenching with Bavistin

formulation of *Aspergillus* H was significantly superior to control. Highest SVI value achieved was with soil drenching with bavistin followed by its seed treatment, soil mixing and seed pelleting with Tv. The same trend was also found for SVIN except for seed pelleting with Tv which had a higher value than its soil mixing. Overall evaluation of the results of all the parameters indicates that treatments with biocontrol formulation with Tv are at par with that of bavistin and soil mixing has an edge over seed pelleting though there was no significant difference between the two application methods.

Table 5 shows that germination percentage in all treatments, except soil mix with formulations At and Ph, as significantly higher than control. Seed treatment and soil drenching with bavistin gave better performance (i.e. 82.67% and 80.00% respectively), followed by soil mixing (72%) and seed pelleting (70.67%) with Tv though there was no significant difference between the treatments. The same trend was also found in survival percentage. Post-emergence damping-off incidence percentage was reduced to maximum 2.67% by seed pelleting with Tv formulation followed by soil drenching with bavistin and soil mixing with Tv. Reduction in disease incidence was achieved by soil drenching and seed treatment with bavistin *(32% and 29.33%, respectively), followed by seed pelleting and soil mixing with Tv formulation (25.33% and 22.67%, respectively).

4. DISCUSSION

The suppression of growth of test pathogen was considered as an appropriate criterion for the efficacy of antagonists. Therefore, the antagonist occupying more area over medium in Petri plates were taken into consideration and on this basis *Aspergillus terreus*, *Penicullium herquei* and *Trichoderma virens* were selected for further studies. The reduction in the

Table 5. Effect of different treatments on post-emergence damping-off disease of *Moringa pterygosperma*

Treatment	Germination%	Survival	Post-emergence damping-off %	Pre-emergence daming-off %*	Non-viable seeds %	Disease %	Reduction in disease incidence %
T1	56.00	44.00	12.00	32.00	12.00	56.00	0
T2	82.67	73.33	9.33	2.67	14.67	26.67	29.33
T3	70.67	69.33	2.67	6.67	21.34	30.67	25.33
T4	72.00	66.67	4.00	9.67	19.67	33.33	22.67
T5	68.00	60.00	10.67	20.00	13.33	44.00	12.00
T6	57.33	49.33	8.00	26.67	16.00	50.67	5.33
T7	76.00	65.33	10.67	10.67	13.33	34.67	21.33
T8	58.67	49.33	9.33	26.67	14.67	50.67	5.33
T9	80.00	76.00	4.00			24.00	32.00
SEM+	3.33	3.83	1.14			3.88	
CD at 5%	10.85	12.50	3.72			12.65	

* % pre-emergence damping-off = % Disease for treatment - % disease for best treatment i.e. T9

** % Non-viable seeds = 100 - (% germination + % pre-emergence damping - off)

growth of test fungus, area of inhibition zone between antagonist and the test fungus, and area of growth of antagonist are the deciding factors for the efficacy of antagonists under study. The antagonistic efficacy of *T. virens* against *F. acuminatum* was maximum (87.81%) followed, by *P. herquei* (86.41%) and *A. terreus* (82.09%).

The formulation prepared on bagasse powder had 290 - 360 x 10^7 cfu per gram, which was much more than described by Jeyarajan *et. al.*, (6), i.e. 120 x 10^6 cfu/g on talc based formulation described by Shankar and Jeyarajan (7). The cfu in the present formulation was also much more than that obtained by fermenter biomass technology where it was reported to be 10^7 cfu/g (8).

The pathogen *F. acuminatum* accounts for low germination due to pre-emergence damping-off which was as much as 32% in the control (Table 5). The disease affects the overall growth of the seedlings as expressed by reduction in root-shoot length. McNew (9) reported that root diseases by *Fusarium* spp. cause destruction of seedlings or impairment of their growth. Ahmed *et. al.*, (10) while estimating the losses due to downy mildew disease in lucerne reported decrease in green fodder and dry matter. Severely diseased plants generally show subnormal development because the processes on which growth depends are impaired (11).

The fungicide bavistin was effective in controlling the disease plant growth when applied as seed treatment as well as soil drenching the results obtained with treatments of Tv formulation were at par with those of fungicidal seed treatment. The soil treatment showed an edge over seed treatment, as soil drenching with bavistin was superior to its seed treatment. Seed treatment with *Trichoderma* spp. have been found effective in controlling root diseases of agriculture crops (7,12,13). Harsh *et. al.*, (4) have reported the effectiveness of padding of *Trichoderma viride* infested wheat straw in controlling the post-emergence damping-off disease of

Moringa pterygosperma and found it at par with seed treatment with bavistin.

As an alternative to the chemical fungicides biocontrol agents are gaining importance in the management of plant pathogens in recent times (13). Biocontrol potential of *Trichoderma* spp. (8,14,15,6) and *Aspergillus* spp. (16) has been well established. Several commercial preparations of *Trichoderma* have appeared in European market (17). A talc based formulation of *T. viride* is produced by Tamil Nadu Agriculture University, Coimbatore, which is widely used by farmers against *Macrophomina phaseolina* root diseases in agriculture crops (6). In India Hoechst & Shering AgrEvo Ltd., Mumbai has brought a product Ecofit, which is a preparation of *T. viride* in talc powder for control of root rot, seedling rot, damping-off, collar rot and *Fusarium* wilt of agriculture crops. The attempts made by the authors to prepare a biocontrol formulation may be viewed in this commercial context. The formulation is developed on a biological waste which ensures its easier application, viability and effectiveness over longer period of storage. Churchill (18) has advocated the use of a medium having same osmotic concentrations and nutrients as the growth medium for prolonged storage of biocontrol agents in order to retain the viability. Besides microorganisms are known to loose their physiological characteristics when repeatedly subcultured on synthetic and semisynthetic media. The present formulation provides natural nutrients for the growth and survival of the fungi used. Bagasse powder formulation provides natural nutrients for the growth and survival of the fungi used. Bagasse powder contains a source of nutrients (carbohydrates) to initiate and support germination of biocontrol agents of fungal origin. The conidia of *Trichoderma* require an external source of nutrients for germination (19). Churchill (18) and Lisansky (20) have stated that a biocontrol formulation should possess characteristics such as ease of preparation, stability, adequate shelf life, abundant viable prapagules and low cost. The formulation prepared for *Aspergillus, Penicullium* and *Trichoderma* spp. on bagasse thus fulfills and the standard prerequisites of an effective biocontrol formulation.

ACKNOWLEDGEMENTS

We are thankful to Dr. R.B. Lal, IFS, the then Director of the Institute and to Dr. Jamaluddin, Head of Forest Pathology Division for kindly providing necessary facilities and encouragement for the work. The second author is grateful to Indian Council of forestry Research & Education, Dehradun for the financial assistance in the form of Junior Research Fellowship.

REFERENCES

1. Harsh, N.S.K. and B.N. Gupta. TFRI Information Series No.1 (1993) 31 p.
2. Broadbent, P., K.F. Baker and Y. Waterworth. Aust. J. Biol. Sci. 24 (1971) : 925-944.
3. Capellini, R.A. and J.N. Peterson. Mycologia 57 (1965) :962-966.
4. Harsh, N.S.K., S. Sharma, A.S. Bhandari and A.S.K. Asaiya. J. Trop. For. 10 (1994) : 241-248.
5. Abdul, A.A. and S.D. Anderson. Crop Science 3 (1973) :630-633.
6. Jeyarajan, R., G. Ramkrishnan, B. Rajamanickam and P. Sangeeta. Petria

1 (1991) : 43.

7. Shankar, P. and Jeyarajan, R. Indian Phytopath. 49 (1996) : 148-151.
8. Papavizas, G.C. Dann, M.T. and Ristaino, B.J. Phytopathology 74 (1984) : 1171-1175.
9. McNew, G.L. In : Plant Pathology Vol.2 (eds. J.G. Hrsfall and A.E. Diamond), (1960) pp 19-69, Academic Press, New York.
10. Ahmad, S.T., P.R. Sreenath and M.P. Gupta. Indian Phytoptah. 30(1977) : 466-468.
11. Tarr, S.A.J. Principles of plant pathology. Macmillan Publishers Ltd., Hong Kong (1972)
12. Raghuchander, T., R. Samiappan and G. Arjanan. Indian Phytopath. 46 (1993) : 379-387.
13. Ushamalini, C., K. Jayappa and Kousalya Gangadharan. Indian Phytopath. 50 (1997) : 504-507.
14. Elad, Y., Y. Zvieli and I, Chet. Crop Protection 5 (1986) :288-292.
15. Kehri, H.K. and S. Chandra. Indian Phytopath. 44 (1991) : 60-63.
16. Gogoi, R. and A.K. Roy. Indian Phytopath. 49 (1996) : 32-37.
17. Whips, J.M. Biocontrol Science and Technology 2 (1992) : 3-24.
18. Churchill, B.W. In : Biological control of weeds with plant pathogens, (eds. R. Charudattan and H.L. Walker (1982), pp. 139-156. John Willey and Sons, New York.
19. Danielson, R.M. and C.B. Davey. Soil Biol. Biochem. 5 (1973) : 517-524.
20. Lisansky, S.G. Production and commercialization of pathogens. In : Biological pest control (eds. N.W. Hossey and N. Scopes), (1985) pp. 210-218, Blandford Press Poole, England.

CHAPTER - 17

NEED FOR STRAIN IMPROVEMENT OF PLANT GROWTH PROMOTING FUNGI TO INCREASE GROWTH AND YIELD OF CROPS

Bharat Rai and T. Santavir Singh

Department of Botany, Banaras Hindu University, Varanasi - 221 005, India

ABSTRACT

Several plant growth promoting fungi (PGPF) have been recognised and tested in laboratory and glasshouse condition. *Trichoderma harzianum* BS-2, *Glioclatium and Trichoderma harzianum* BS-3 maximally enhanced the top length, pod numbers, total dry biomass and yield in case of chickpea, pea and mustard respectively than the control in natural soil. However, the PGPF exhibited more enhancement in terms of growth and yield in case of mustard as compared to chickpea and pea. These effective PGPF should be subjected to their strains improvement through biotechnolgical aids for their use in agriculture as biofertilizers.

1. INTRODUCTION

Recently, the beneficial effect of certain fungi in the rhizosphere independently or in association with roots as mycorrhizae in terms of plant growth promotion and biological control of plant disease has attracted renewed interest among researchers. Chang *et. al.*, (1) studied the potential of *Trichoderma harziamum* to induce increase growth of various floricultural and horticultural crops. Species of *Trichoderma* and *Rhizoctonia* have been shown to increase plant growth as well as to suppress plant pathogens by their antagonistic activities (2-6). *Trichoderma, Fusarium, Penicillum* and a *Mucor* sp. isolated from the rhizosphere of turfgrass and some cultivated crops were found to show the ability of both plant growth promotion as well as disease suppression (7). During the present study, several plant growth promoting fungi (PGPF) namely *Cladosporium cladosporioides, Curvularia lunata, Gliocladium virens, Fusarium moniliforme* and 5 isolates of *Trichoderma* spp. isolated from soil have been recognised and tested in laboratory and glasshouse conditions for their growth and yield promotion of chickpea, pea and mustard. Thus, there is a need to improve the strains of these PGPF for their effective use in agriculture as biofertilizer.

2. MATERIALS AND METHODS

Dried brown loam soil (pH 7.3) was collected from the campus of botanical Garden, B.H.U. (which was nutrient deficient) and was used for green house experiment. The soil was air dried at room temperature and ground and sieved through 2 mm pore size sieve. The PGPF mass cultured on wheat grains. Powder form of the wheat grain inocula of PGPF was well mixed with the soil at the rate of 1% (w/w). The PGPF inoculated soil was taken in polythene bags (2 Kg per bag) and the inoculated bags were kept at room temperature for 10 days. Appropriate amount of water was added to provide sufficient moisture for proper growth of the PGPF in the bags. Soil amended with unifested and autoclaved wheat grains was used as control. Seeds to be sown were surface sterilized with 0.1% NaOCI solution for 1 min and then washed 10 times with sterile water. Six healthy seeds of chickpea, pea and 12 seeds of mustard per bag were sown. Plant growth promotion in terms of top length, early flowering and fruiting, increase in yield and total dry biomass was recorded.

3. RESULTS AND DISCUSSION

All the PGPF showed pronounced stimulatory effects on growth and yield of the test crops (Table 1-3). On the other hand, the effective PGPF exhibited different degree of enhancement for different crops. The duration of plant growth promotion activities of PGPF was not same in all the crops. The PGPF showed a remarkable increase in terms of growth from the early stage of vegetative growth in case of mustard. Upto four weeks of sowing there was almost no difference in plant height in case of chickpea and pea but thereafter the PGPF exhibited considerable growth enhancement. Early flowering and fruiting were found in all the PGPF treated plants in case of chickpea and pea except mustard.

Almost all the PGPF, which enhanced the top length, also increased the number of pods, total dry biomass and yield of the plants. *Trichoderma harzianum* BS-2 increased the maximum top length to 33.4% (after 12 weeks), pod numbers to 145.9% total dry biomass to 114.25 and yield to 148.3% than the control in chickpea (Table 1). The minimum enhancement in yield was shown by *C. cladosporioides* and *C. lunata* (both 62.8%). *Gliocladium virens* increased the maximum top length, pod number, total dry biomass and yield (54.2%, 162.0%, 274.3% and 498.8% respectively) in pea. *Curvularia lunata* was less effective to enhance the yield than the other PGPF (219.7%) (Table 2). In case of mustard, all the PGPF showed more stimulatory effect in terms of yield in comparison to chickpea and pea plants (Table 3). The increase in yield by different PGPF ranged from 430.8% to 1161.5%. The maximum increase was exhibited by *T. harziamum* BS-3 and the minimum by *C. lunata*. Cowan (8) suggested that some growth promoting isolates of *Phialophora graminicola* increases the mineral nutrient uptake by plants in the same way as do mycorrhizal fungi. Dewan and Sivasitha-mparam (9-10) observed that the sterile red fungus invaded the inner root regions which helped plants in deriving nutrients from the soil and in protecting root against pathogens. The suppression of deleterious microorganisms by PGPF may be one of the mechanisms of plant growth promotion (7). Association of PGPF isolates with roots might help plants to derive mineral nutrients in the readily available form as has been shown in case of wheat (11). This might be closely related to the root colonization ability of PGPF isolates.

Table 1. Effect of soil amendment with PGPF inocula (1% w/w) on growth and yield of Chickpea in potted natural soil

Treated PGPF	Top length* after 12 weeks (cm)	Pod Nos.*	Total dry biomass* (g/plant)	Seed Nos.*
C. cladosporioides	37.02	3.00**	2.95	2.93
C. lunata	38.32	3.07**	2.78	2.93
F. moniliforme	39.71	3.80	3.61	3.47
G. virens	39.55	4.33	3.24	3.93
T. harzianum BS-2	40.59	4.60	4.54	4.47
T. harzianum BS-3	38.53	4.40	3.80	3.80
T. harzianum BS-1	37.64	3.80	3.52	3.33
T. viride BS-7	38.08	3.47	3.06	3.20
T. viride BS-8	38.65	3.40	2.95	3.13
Control	30.43	1.87	2.12	1.80

Note : The mean values in each column showed significant increase ($P<0.001$) except those followed by ** ($P<0.01$) when compared with their respective controls.
* Average of three replicates

Table 2. Effect of soil amendment with PGPF inocula (1% w/w) on growth and yield of Pea in potted natural soil

Treated PGPF	Top length* after 8 weeks (cm)	Pod Nos.*	Total dry biomass* (g/plant)	Seed Nos.*
C. cladosporioides	32.60	2.00**	2.82	6.75
C. lunata	33.38	2.08**	2.73	5.33
F. moniliforme	36.32	2.42	2.87	7.08
G. virens	39.70	2.83	4.08	10.00
T. harzianum BS-2	36.10	2.50	3.61	9.17
T. harzianum BS-3	36.27	2.25	3.16	8.33
T. harzianum BS-1	33.93	2.42	3.23	8.42
T. viride BS-7	33.38	2.25	3.14	6.25
T. viride BS-8	34.39	2.17	2.93	8.67
Control	25.75	1.08	1.09	1.67

Note : The mean values in each column showed significant increase ($P<0.001$) except those followed by ** ($P<0.01$) when compared with their respective controls.
* Average of three replicates

The production of NH_4-N and NO_2-N in soil was accelerated by amendment with PGPF infested barley grains (7). Analysis of soil samples amended with the PGPF, before and after growth of chickpea plants, revealed that the PGPF treated soil was richer in available phsophorus (P_2O_5), nitrogen (NH_3N) and potassium (K_2O) (12). The garden soil which was nutrient deficient supported the poor growth of

Table 3. Effect of soil amendment with PGPF inocula (1% w/w) on growth and yield of Mustard in potted natural soil

Treated PGPF	Top length* after 5 weeks (cm)	Pod Nos.*	Total dry biomass* (g/plant)	Seed Nos.*
C. cladosporioides	6.96	11.92	0.788	0.162
C. lunata	7.61	10.25	0.757	0.138
F. moniliforme	7.38	12.75	0.920	0.176
G. virens	9.24	16.25	1.058	0.212
T. harzianum BS-2	9.71	17.33	1.343	0.265
T. harzianum BS-3	10.06	18.17	1.204	0.328
T. harzianum BS-1	7.11	13.75	0.934	0.185
T. viride BS-7	8.59	15.83	0.982	0.207
T. viride BS-8	7.08	11.42	0.834	0148
Control	4.20	3.00	0.476	0.026

Note : The mean values in each column showed significant increase ($P<0.001$) when compared with their respective controls.

* Average of three replicates

control plants resulting in reduced yield. It suggests the increase of nutrient status of soil in the root region by the activity of the PGPF.

Singh (12) observed the increase in chlorophyll content in the said crops due to soil amendment with PGPF. The increase in chlorophyll content might be due to increase of nitrogen in soil by the activity of PGPF which is absorbed by the plant. Hence, another mechanism for plants' growth and yield promotion could be because of higher rate of photosynthesis by the presence of higher chlorophyll content. However, intensive researches are needed to observe the direct and indirect mechanisms of plant growth and yield promotion by PGPF. Thus there is need to improve strains of the well recognized PGPF through biotechnological aids in view of their roles in plant growth promotion and their yield so that they can be applied in agricultural fields as biofertilizers.

ACKNOWLEDGEMENTS

The authors are thankful to the Head, Department of Botany, B.H.U., Varanasi, for providing necessary facilities. The financial assistance from U.G.C., New Delhi to TSS is gratefully acknowledged.

REFERENCES

1. Chang YC, Chang YG & Baker R, *Plant Dis*, 70 (1986) 145.
2. Sneh, B, Zeidan M, Ichielevich-Auster M, Barash I & Koltin Y, *Can J Bot*, 64 (1986) 2372.
3. Windham MT, Elad Y & Baker R, *Phytopath*, 76 (1986) 518.
4. Baker R, *CRC Crit Rev Biotech*, 7 (1988) 97.
5. Baker R, In : *The Rhizosphere and Plant Growth*, (Keister DL & Cregan PG eds.). Kluwer Academic Publishers, Dordrecht, The Netherlands, (1991) 221.

6. Kleifeld O & Chet I, *Plant & Soil*, 144 (1992) 267.
7. Hyakumachi M, *Soil Micro*, 44 (1994) 53.
8. Cowan MC, *Plant & Soil*, 52 (1979) 1.
9. Dewan MM & Sivasithamparam K, *Mycol Res*, 93 (1989b) 161.
10. Dewan MM & Sivasithamparam K, *Mycol Res*, 94 (1990) 553.
11. Shivanna MB, Meera MS & Hyakumachi M, *Can J Microbiol*, 40 (1994) 637.
12. Singh TS, *Influence of non-pathogenic microfungi on growth and yield of crop plants*. Ph.D. Thesis, B.H.U., Varanasi, India, (1997).

CHAPTER – 18

INDUCTION OF CALLUS AND ORGANOGENESIS FROM CULTURED SHOOT APEX FOP THE MICROPROPAGATION OF *Cestrum nocturnum* LINN.

Manju Kumari, Manoj Kumar, J.D. Choudhary and L.N. Shukla

B.R.A. Bihar University, Muzaffarpur, Bihar, 843 001-India

ABSTRACT

Shoot apex explants of *Cestrum nocturnum Linn.* were cultured on MS medium variously supplemented with different concentrations of NAA, 2,4-D, KN, BA and CW either singly or in different combinations. Shoot apices were differentiated on the explants directly or through callus formation.Subculturing of stem calli resulted into differentiation of shoots. Differentiated shoots were rooted on MS medium with IAA (2- 3mg/1). The rooted plants were progressively acclimatized and transferred to pots and field conditions. The *in vitro* grown plants even flowered. Some of the plants showed improvement in leaf, flower size and in fragrance. The work shows the potentiality of micropropagation and improvement of the sp. through Tissue culture.

1. INTRODUCTION

Cestrum nocturnum Linn., belonging to the family solanaceae is considered pride of Indian gardens for its attractive, strongly sweet scented flowers which open only at night. Besides satisfying the aesthetic senses of human as ornamentals, these plants also show anti viral, fungitoxical and insecticidal properties (1). The leaf extracts induced resistance against tobacco mosaic virus in test plants. Powdered leaves of *C. nocturnum* gave complete control of eggs and newly hatched larvae of housefly (2), while petroleum ether extracts proved a protectant against pulse beetle in stored mung bean seeds (3).

Cestrum nocturnum is generally propagated either by seeds or by conventional nursery method of cuttings. In the first process, the plants may lose the genetic superiority and lack genetic uniformity while the second process is slow and requires lot of labour, space and money. Micropropagation, which is the plant propagation method by Tissue culture technique, can be cosnidered as an extention of the traditional methods of the plants propagation but with many advantages.

Micropropagation uses either *in vitro* produced axillary shoots for regenerating plants with genetic characteristic of original plants (4) or, regenerated plants through organogenesis and/or embryogenesis with original genotypes as well as somaclonal variants (5).

Shoot apex explants prove to be better due to possession of juvenile axillary buds and for their higher survival rates. Shoot apex culture of *Cestrum nocturnum* were taken up, to explore its potentiality for tissue culture response and micropropagation as there is hardly any report of successful tissue culture in genus *Cestrum* except some attempts (6,7).

2. MATERIALS AND METHODS

Pieces of 5mm long shoot apices with 2 to 3 leaves, taken from young branches were used as explants. They were washed in running tap water, rinsed in 1% cetavelone solution, again washed in running tap water, surface sterilized on 0.2% mercuric chloride solution for 5 minutes and washed several times with sterile distilled water. Surface sterilized explants were inoculated on Murashige and Skoog's (8) medium variously suplemented with different concentrations of NAA, 2,4-D, KN, BA & CW either singly or in different combinations.

The newly formed healthy calli were also subcultured on different media. The regenerated shoots were transferred to rooting medium for the development of root system. The rooted plantlets were transferred to field condition and grown to flowering. The characteristic of such plants were recorded and screened for somaclonal variants.

3. RESULTS

(a) Shoot apex culture

The cultured shoot apices swelled slighly to conspicuously in the 1st week of the culture. The swelling was followed by callus formation from the entire surface of the explants. Normally the callusing was so vigorous that except for 1 or 2 leaf tips rest of the explants was not recognisable. The callus in most of the cases was yellow, white compact one, while in some friable callus was also formed above the compact callus.

In *C. nocturnum* caulogenesis through callus formation was observed only in medium with 2 mg/l NAA, (Table 1 and Plate 1B). The direct regenetation of shoot-apices from cultured shoot apex was more frequent in case of *C. nocturnum* (Table 1). They were green and very healthy particularly in the media with Kinetin (Plate 1A). Medium with KN alone, showed better response for shoot differentiation as compared to medium with Auxin and Cytokinin both (Table 1).

(b) Shoot apex callus sub-culture

Shoot apex calli grown on MS medium with NAA and BA were subcultured on different media having different levels of BA or IAA only (Table 2). There was meagre or no callus and the caulogenesis increased in different concentration of BA (Table 2 and Plate 1C). However, increased level of BA in subcultured medium resulted in the suppression of caulogenesis.

(c) Rooting of differentiated shoots

The differentiated shoots developed directly through caulogenesis or through callus, were transferred to MS medium with different concentration of IAA (2,3 mg/l), where they developed well formed root systems. This response was found more with the former than with the latter medium (Plate 1D).

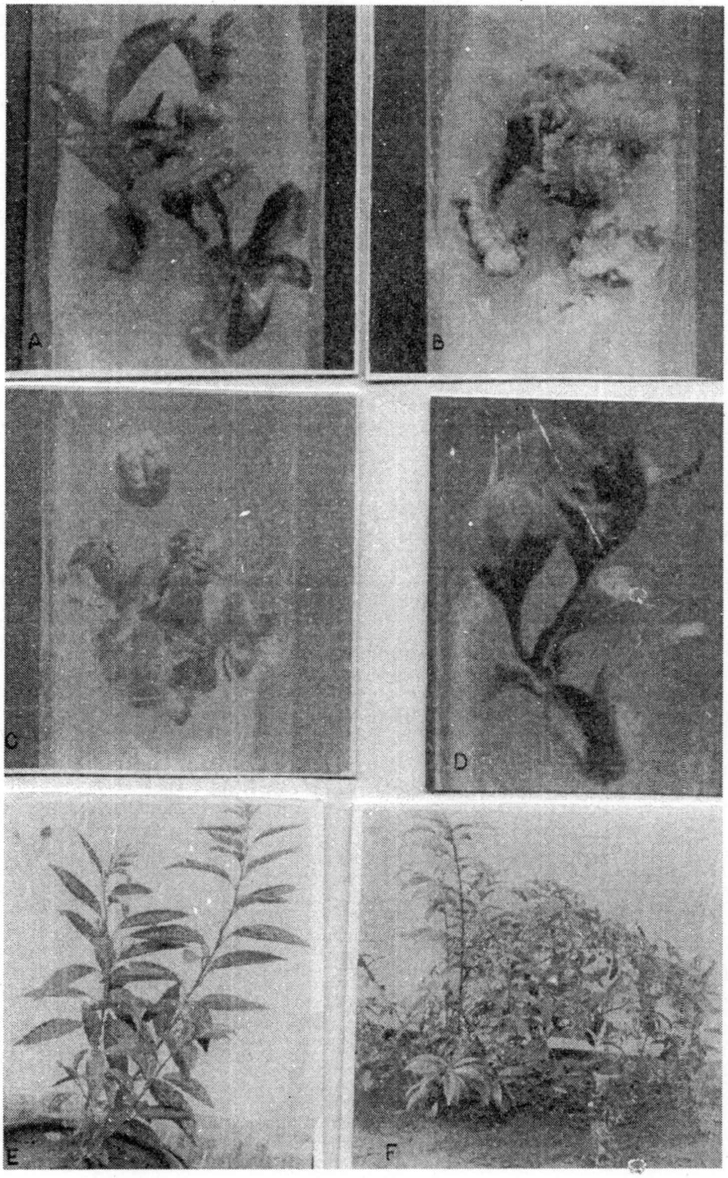

Plate 1. Shoot apex culture of *Cestrum nocturnum* Linn.
A. Cultured Shoot apices showing direct differentation of 4-6, Shoot apices on MS+KN (4mg/1)(*2.4), **B.** Cultured shoot apices showing comapct and friable callus formation mainly from the cut ends and differentation of shoot apices from meristem region on MS+NAA(2mg/1) (*1.88), **C.** Shoot apex callus, grown on MS+NAA (4mg/1) +BA(4mg/1) Subcultured on MS+BA(4mg/1) showing direct differentiation of many health green shoot apices (*2.76), **D**. Shoot apex, grown on MS+NAA(4mg/1)+BA(4mg/1) subcultured on MS+IAA (2mg/1) showing good growth and formation of 3 roots (82.36), **E.** Plantlet *Cestrum nocturnum* L in Pot (*0.35), **F.** *In vitro* plant of *Cestrum nocturnum* in field showing brisk flowering (*0.049).

Table 1. Shoot apex culture of *Cestrum nocturnum* L. on MS basal medium variously supplemented with different adjuvants.

S. No.	Media	% of culture showing Callus formation	Observation on callus			% of callus showing caulo-genesis	No. of shoots per culture	% of culture showing rhizo-genesis	No. of roots per culture
			Colour	Nature	Growth				
1.	MS+NAA (2mg/l)	91.66	Brown Yellow	compact triable	+++	83.33	4.5+	–	–
2.	MS+2,4-D (4mg/l)	25.00	Yellow white	compact	++	–	–	–	–
3.	MS+2,4-D (5mg/l)	54.54	Yellow white	compact	+++	–	–	–	–
4.	MS+KN (2mg/l)	50.00	White	compact	++	–	–	–	–
5.	MS+KN	–	–	–	–	72.72	4.75±0.66	–	–
6.	MS+KN (5mg/l)	–	–	–	–	90.90	2.1±0.7	–	–
7.	MS+NAA (4mg/l) +BA (4mg/l)	–	–	–	–	54.54	6.5±0.9	–	–
8.	MS+NAA (5mg/l) +BA (5mg/l)	–	–	–	–	45.45	2.4±0.48	–	–

-no response; + low callus growth; ++ moderate callus growth; +++ good callus growth.

Table 2. Shoot apex callus structure of *Cestrum nocturnum* L.

S. No.	Callusing media	Subcultu-ring media	Observation on new callus			% of callus showing caulo-genesis	No. of shoots per culture	% of culture showing rhizo-genesis	No. of roots per culture
			Colour	Nature	Growth				
1.	MS+NAA (4mg/l) +BA (4mg/l)	MS+BA (4mg/l)	–	–	–	83.33	5.1±0.94	–	–
2.	- Do -	MS+BA (2mg/l)	–	–	–	54.54	4.66± 1.10	–	–
3.	- Do -	MS+BA (1mg/l)	white	compact	++	41.66	3.00± 0.63	–	–
4.	- Do -	MS+IAA (2mg/l)	–	–	–	–	83.83	well developed roots on cultured shoot apices.	

- No response ; ++ Moderate callus growth.

(d) Acclimatization of plantlets

The plantlets with well developed root systems were transferred to the pot with sterile soil mixture. Initially, they required high humidity conditions, which was provided by covering the pots with polythene bags, which reduced the humidity and acclimatized the plants for the field transfer. The plants were further acclimatized by transferring them in to natural light condition out side the culture room. The frequency of the plantlets survived during acclimatization was about 20%.

(e) Field Transfer and Somaclonal variants

The acclimatized plants were transferred to bigger pots or directly in the soil. About 8% of the plantlets survived in final field transfer. The plants were grown to full height and even flowered (Plate 1E & F).

An analysis of the characteristics of tissue cultured plants, has shown that most of the characters were like the original in-vivo grown plants, but some characters like the size of the leaf, size of the flower bud, size of floral parts and even fragrance showed some increases.

4. DISCUSSION

The most widely used and commercially successful application of plant biotechnology has been the rapid and large scale clonal multiplication of plants, through micropropagation (9). Plants can be regenerated *in vitro* from cultured explants by the formation of somatic embryos or adventitious buds either through callus formation or directly on the explants. This process has been used for mass production of many ornamental plants and thus can be extended to *Cestrum nocturnum* also which is being propagated through stem cuttings and seeds (10). The analysis of plants that have demonstrated the potential for clonal multiplication through the culture, has shown that virtually every organ or tissue can be used as a source of explants. The most commonly used medium in micropropagation is MS and its modification and the same has been used for shoot apex culture of *Cestrum nocturnum.* The MS medium was supplemented with auxins namely NAA which is more stable and generally more effective and 2,4-D which is most active (11). Addition of cytokinin is generally required for initiation of culture and shoot multiplication. The Cytokinins used were BA, which is most widely used and KN, which has been shown to be essential for tissue culture of most of the Solanaceous plants.

The cultured shoot apices of *Cestrum nocturnum* formed massive normally compact calli, from which after few days caulogenesis occurred. Direct regeneration of shoots was more frequent in *C. nocturnum.* Such observations from cultured shoot apex, were also found in *Cynara scolymno* (12) and *Pharbitis nil* (13). Direct differentiation of shoots buds from cultured shoot apex and their use in micropropagation has been found in *Glycine hispida* (14).

The shoot apex callus when subcultured, showed more differentiation of shoots and roots. The increased differentiation might have resulted with change or decrease in phytohormone level in subculture medium. The shoots differentiated through caulogenesis, lacked roots and were transferred to rooting medium either without hormones or with Auxins only. Among the Auxins, NAA and IBA are the most effective for induction of roots and shoots (15). However in *C. nocturunum,* the IAA at 2-3mg/1, level induced well developed root systems in the differentiated shoots. Commercial exploitation of the micropropagation has been limited by the failure in the field transfer of test tube plants (16). Dessication and wilting are the two major causes of failure. In *C. nocturunm* few plants were able to survive in pot and field conditions after congressive acclimatization. These plants grew like normal plants and flowered, showing that micropropagation is feasible in this important plant species. Some improvement *in vitro* grown plants, prove that the plant can even be improved through exploitation of somaclonal variation.

REFERENCES

1. Roy Choudhary, R, *Acta Botanica, India.* (1980).
2. Ahmed SM, Chander H & Pereira J, International pest control, 23 (1981) 170.
3. Chander H & Ahmed SM, *J Food Sci and Tech* 19 (1982) 50.
4. Hussey G *In : Plant Biotechnology Manteil SH & Smith H (eds.) Cambridge University Press, Cambridge (1983) 111.*
5. Larkin PJ & Scrowcoft WR, *Theor Appl Genet* 60 (1981) 197.
6. Klimaszewska K, *Acta Agrobotanica* 34 (1981) 5.
7. Weissenbergh M, Madz A, Levy A & Wasserman RH, *Planta medica* 54 (1988) 63.
8. Murashige T & Skooque F *Physiologia Pl* 15 (1962) 473.
9. Debergh, P & Zimmerman RH (eds.), *Micropropagation Technology and Application Kluwer, Dordecht.* (1990)
10. Siddique MA, *J Hort* 30 (1973) 440.
11. Hussey G, In Plant cell culture Technology Yeoman MM (ed) Narosa publishing Hosue New Delhi (1987) 29.
12. Debergh, P & Zimmerman RH (eds.), *Pl* 53 (1981) 181.
13. Bapat VA & Ray PS, *Pl Sci Lett* 10 (1977) 327.
14. Sharma, VK & Kothari SL, *J Indian Bot Soc* 70 (1991) 431.
15. Ancora G, Bell-Donni ML & Cuozzo L, *Scient Hort* 14 (1981) 207.
16. Broom OC & Zimmerman RH, *Hort Sci* 13 (1978) 151.

CHAPTER - 19

CLONING OF NITRATE REDUCTASE GENE (*nia*D) OF *Aspergillus niger* AND IT'S USE FOR TRANSFORMATION OF *Beauveria bassiana*

J.R. Kinghorn[1], *Shiela E. Unkles*[2], *S.S. Sandhu*[3], *R.C. Rajak*[3] *and G. Venkateswerlu*[4]

1. Molecular Genetics Unit Sir Harold Mitchell Building, Greenside place, St. Andrews Fife, KY16 9TH, U.K.,
2. Dept. of Microbiology, Monash University, Clayton-Victoria- 3160, Australia.
3. Dept. of Biological Science, R.D. University, Jabalpur - 482 001 (India).
4. Dept. of Biochemistry, Osmania University, Hyderabad (India)

ABSTRACT

An heterologous transformation system for the entomopathogenic fungus *Beauveria bassiana* has been developed based on the use of the *Aspergillus niger* nitrate reductase gene (*nia*D). *B. bassiana nia*D$^-$ *stable* mutants were selected by treatment of protoplast with Ethane Methane Sulphonate and regenerated on chlorate medium. The cloned gene is capable of transforming *B. bassiana* at a frequency of 5.8 to 20 transformants per µg of DNA. Mitotic stability studies demonstrated that while some transformants were as stable as the wild type (wt), others were markedly less so.

1. INTRODUCTION

The Deuteromycete fungus *Beauveria bassiana* (Balsamo) Vuill. parasitizes many insect species and are in use as mycoinsecticide against several pests of agricultural crops and human health (1,2). However, the potential for rational strain improvement is limited by ignorance of the molecular and biochemical basis of pathogenicity and by absence of a sexual cycle in these fungi. However, the successful application of gene cloning technology to these entomopathogenic fungi relies on an efficient transformation system. We have recently identified *nia*D mutants in *Beauveria bassiana*, and present here the conditions for the isolation of stable nitrate reductase mutants in *B. bassiana* and their transformation with pUC*nia*D and pSTA8 plasmids carrying the *Aspergillus niger* wild type *nia*D gene.

2. MATERIALS AND METHODS

(a) Strains, media and plasmids

Escherichia coli strain DH5α (F$^-$, endAl, hsd17, $SupE_{44}$, thi-1, recA1, gyrA96, relA1, r^-_K, m^+_K) was used for construction, propagation and amplification of hybrid plasmid. *Beauveria bassiana* was used for selection of recipient mutants. Standard media for cultivation of bacteria and fungi was used as described by (3, 4) respectively.

(b) Nitrate assimilation pathway and the isolation of *nia*D mutants of *B. bassiana*

The nitrate assimilation pathway in filamentous fungi is shown in Fig.1. Mutants defective in the genes required for nitrate assimilation can be selected in *B. bassiana* on the basis of EMS mutagenesis subjected to chlorate resistance (5). Such mutants canbe assigned by growth tests to one of a number of gene loci (Table 1). Of particular interest in this communication are *nia*D mutants, which fails to grow on nitrate, but grow as wild type on other sole nitrogen source. Nitrate assimilation defective mutants of *B. bassiana* were obtained by treating the protoplasts with 15 μl EMS (d=1.17 *g* Sigma) in 1 ml of stabilizing medium (0.02 M phosphate buffer pH 7.2 containing 0.6 M KC1 and 2mM $MgC1_2$) at 28°C ± 1°C for 1 hr. These protoplasts were then washed twice with 10 ml of stabilizing medium by centrifuging at 1000rpm for 10 minutes, diluted and then spread on plates containing a regeneration medium (2% Sucrose; 1% Peptone; 0.5% NaCl; 3% yeast extract and 2% agar-agar; pH 7.0). Conidia from surviving colonies were harvested and suspended in tween-80 aqueous suspension (50 μl/100 ml distilled water). Dilution of this suspension has plated again on a minimal medium (1% glucose; 0.1% K_2HPO_4; 0.5% $MgSO_4$; 0.001% $FeSO_4$, $7H_2O$; 0.005% EDTA disodium salt, pH 6.5) containing 10mM glutamate as sole source of nitrogen and 470 mM chlorate and incubated at 28°C ± 1°C for five days.

(c) *Beauveria bassiana*

Mycelia grown at 28°C ± 1°C for 48hrs were collected and washed with 0.6 M $MgSO_4$ and 2% mercaptoethanol and 0.1% tween 80 aqueous solutions respectively. The mycelial pellet has resuspended in a sterile 150 ml flask containing 5 ml cold osmotic medium (1.2 M $MgSO_4$, 10mM NaOP(pH 7.0) and the pH was adjusted to 5.8 with 0.2M Na_2HPO_4). Novozyme- 324 was added to a concentration of 4 mg/ml and mixture was incubated at 30°C for 2 hrs. Protoplast were washed 3 times in 1.2 M sorbitol; 50 mM $CaCl_2$; 10 mM Tris pH 7.5 and finally resuspended in the above osmotic medium at a concentration of 2 x 10^7 protoplast in 1 ml.

To 1ml of protoplast suspension 10 μl of plasmid DNA was added and was incubated at room temperature for 20 min. Four volumes of 50% PEG (w/v); 50 mM $CaCl_2$; 10 mM Tris pH 7.5 was added and allowed to stand at room temperature for 10 min. The PEG solution was diluted with 20 ml of 1.2 M Sorbitol. Protoplast harvested by gentle centrifugation at 2000 rpm for 7 min. and resuspended in 1-2 ml of 1.2 M Sorbitol. -The suspension was plated on transformation selection medium (minima) medium containing 10 mM nitrate as sole source of nitrogen) and incubated at 30°C for 5 days. Control sets

$$NO_3^- \xrightarrow[\text{Permease}]{\text{crnA}} NO_3^- \xrightarrow[\text{nitrate reductase}]{\textit{nia}\text{D}} NO_2^- \xrightarrow[\text{nitrite reductase}]{\text{niiA}} NH_4^+$$

Fig. 1. The catalytic conversion of nitrate to ammonium ions. Extracellular nitrate is transported into the cell by a permease encoded *crnA* gene, converted to nitrite by the action of nitrate reductase encoded *nia*D gene, and finally converted to ammonium ions by nitrite reductase, the *nii*A gene product.

were treated as above but without addition of DNA. Protoplast viability was tested by spreading protoplast on non-selective complete medium.

(d) Construction of *Aspergillus niger* gene library

Chromosomal DNA for library construction was isolated from nuclei essentially by the method of Garber and Yoden (10) and Unkles *et al.* (8). Following partial *Sau*3A digestion, fragments of 6 - 12 Kb were isolated from a 1% agarose gel by centrifugation through nitrocellulose (11) and ligated to *Bam* H1 digested pUC8 to yield a library. DNA isolated from each pool was digested with *Eco*R1 or *Bam* H1, subjected to 0.8% agarose gel electrophoresis followed by southern blotting and hybridization to a 2.7 Kb Xba I fragment of pSTA8. This gragment containing the 5 coding region of *A. niger* *nia*D gene (9).

(e) Mitotic stability of transformations

Conidia of transformants grown on minimal medium containing 10 mM $NaNO_3$ were harvested and suspension containing 10% viable spores per ml was diluted as appropriate in saline Tween (0.9% NaCl w/v; 0.01% Tween 80 v/v). Dilution in triplicate were spread evenly on either complete medium or minimal medium containing 10 mM glutamate and 30 mM chlorate. After 8 days growth at 28°C ± 1°C the percentage of colonies diverting to chlorate resistance was calculated.

(f) Detection and analysis of proteins expressed from cloned gene

Analysis of proteins expressed from cloned gene was done by the method Sambrook *et al.* (6) on SDS-Polyacrylamide Gel Electrophoresis and stained with silver nitrate. SDS-Polyacrylamide gel was photographed on UV-visible Transillumin-ator Gel Documentation System.

(g) Plasmid DNA

Plasmids pUC*nia*D and pSTA-8 were isolated from *E. coli* DH5 α by lysis in alkali method (6) and purified by gene cleane kit (Bangalore, Genei) checked on 1.5% Agarose gel for purity.

(h) Genetic manipulation

Transformants were isolated on solid medium containing nitrate as sole source of nitrogen and grown in 100 ml cultures without selective pressure. DNA was extracted from mycelium by Nucleon DNA isolation kit (Amersham Life Science, U.K.). DNA samples were digested with restriction enzymes that were digested and used according to the manufacturer's directions. Electrophoresis and hybridisation of DNA was performed according to standard procedure (6) and plasmid DNA was also purified according to standard procedure (6), and was nick translated with ^{32}p-thymine triphosphate (9).

3. RESULTS AND DISCUSSION

(a) Isolation of *nia*D mutants of *Beauveria bassiana*

It was observed that the *nia*D mutants obtained by EMS treated protoplast and regenerated on chlorate medium were more stable than the spontaneous *nia*D mutants. Their reversion frequency was less than one in 10^7 viable conidia, whereas comparatively very high reversion frequencies were obtained, 100 in 10^7 viable conidia whereas comparatively very high frequencies were obtained, 100 in 10^7 viable conidia of spontaneous mutants of *B. bassiana* (Table 1).

It was found that EMS treated mutants arose at a frequency of 10 and 8 in 10^6 viable conidia of *B. bassiana* on chlorate medium. These mutants were isolated and further purified on a minimal medium containing chlorate and glutamate. The ability of these mutants to grow on nitrate, nitrite, ammonium, hypoxanthine, proline and glutamate as sole source of nitrogen was assessed (Table 2). Of these 18 mutants under test, two of *B. bassiana* were phenotype indicative of *nia*D⁻ and one *crn*A⁻ of *B. bassiana* was noticed.

Generally *nia*D mutants of fungi are selected by spontaneous mutations on appropriate minimal medium supplemented with various concentrations of $KC10_3$ and a nitrogen source (7,8,9). But in case of entomopathogenic fungus *B. bassiana* it has been observed that *nia*D mutants isolated simply by spontaneous mutation on chlorate were not stable (Table 1). Therefore, a methodology has been evolved to isolate stable *nia*D mutants of these fungi by treating the protoplasts with EMS (5).

(b) Transformation experiments

In the typical experiment approximately 10^7 protoplasts were mixed with 10 - 15 μg of plasmid DNA following the protocol of (9). Transformation of *B. bassiana* with circular pSTA-8 resulted 58 transformants per μg DNA with *nia*D, as recipient, whereas 176 transformants of *B. bassiana nia*D, were obtained by pUC *nia*D transformation (Table 3).

(c) Mitotic stability of transformants

Previous studies on transformants stability in fungi have examined the reversion frequency on non-selective medium. Such studies suffers with the disadvantage that often only small number of cells are phenotypically screened for presence or absence of incoming gene. Furthermore, normally a few transformed cells are studied. The reason may be the screening of cells derived from transformants can be very laborious, especially with auxotrophic markers. As nitrate non-utilizing strains can be selected directly by growth on chlorate, a much larger sample can be analysed and therefore allows a more precise investigations of the stability of gene introduced by transformation.

Table 4 shows the frequency of reversion to chlorate resistance of 4 *nia*D transformants of *B. bassiana* compared with the mutation frequency of wild type *B. bassiana*. *B. bassiana* transformed with pUC*nia*D showed frequencies almost similar to wild type, whereas pSTA-8 showed notably instability at 150 and above. These figures represent the lowest instability that could be expected, since the transformants were maintained under selective conditions. However, no direct correlation could be made between stability and location of incoming *nia*D gene in *B. bassiana*.

(d) Analysis of proteins pattern expressed by transformants

On analysis of protein pattern expressed by wild type, *nia*D mutants and transformants of *B. bassiana*, it was found that an additional prominent protein band of approximately 85 Kd was observed in the *nia*D mutants, persisting even after the transformation of mutant with plasmid pSTA-8 having *nia*D gene of *Aspergillus niger* suggesting that during mutation by EMS treatment for chlorate resistance these might have developed some alternative pathways and in addition, persistance of this 85 Kb protein band in the transformants also suggest that the mutant were stable and therefore are suitable for transformation experiments.

(e) Molecular analysis of the transformants

Total cellular DNA of transformants was isolated and southern transfers of undigested DNA were hybridised with pUC *nia*D as a probe. The wild type *Beauveria*

Table 1. Showing reversion frequencies of mutants of *B. bassiana* to wild type growth*

Name of the fungus				
	Spontaneous**	EMS	Spontaneous	EMS
B. bassiana	4	10	>100	<1

* Reversion frequencies in 10^6 viable conidia.
** Methods of Malairdair *et al.*, (1989) was followed.

Table 2. *B. bassiana* mutants defective in nitrate assimilation

Gene* mutation	Chlorate# resistance	Utilization of sole nitrogen sources						Summary**
		A	B	C	D	E	F	
*nia*D⁻	R	–	+	+	+	+	+	Nitrate reductase structural gene
*crn*A⁻	R	+	+	+	+	+	+	Encode for nitrate uptake in young cells

* = The minus superscript denotes loss of functional mutations.
** = Summary of the role of these genes
\# = R denotes resistance to Chlorate
Symbol (+) denotes wild type level of growth and (–) denotes poor growth. A = nitrate; B = nitrite; C = ammonium; D = hypoxanthine; E = prolien; F = glutamate.

Table 3. Transformation frequencies of plasmid pSTA-8 and pUC *nia*D.

Fungus mutation	Mutant allele	Plasmid	Conformation	Maximum number of transformants per 10 μg of plasmid DNA obtained
B. bassiana	$niaD_1$	pSTA-8	Circular	58
B. bassiana	$niaD_1$	pUC*nia*D	linear	176

Table 4. Mitotic stability of transformants

Fungal strain	Transformed with	Reversion rate of mutation frequency (x 10^{-6})*	
		Exp. 1	Exp. 2
*B. bassiana nia*D⁺ (wild type)	—	<3	<2
*B. bassiana nia*D⁺ 1	pSTA-8	196	153
*B. bassiana nia*D⁺ 1	pUC*nia*D	<5	<8

(*) Conidia were taken from five days cultures grown in minimal medium containing 10 mM NaNO3 as sole nitrogen source. Ten fold dilutions were made in saline-Tween. Viable count was obtained by plating on complete medium. All counts were as a result of three individual dillution and plating. Experiments one and two are independent.

bassiana strains contained no sequence which hybridise to the vector at this stringency. In all undigested transformant DNAs hybridization occurred only in the genomic DNA bands suggesting that *nia*D gene intergrated into chromosomal DNA and did not replicated autonomously.

We report the successful isolation of stable *nia*D mutants of *B. bassiana* and their transformation with *nia*D gene of *Aspergillus niger*. The entomopathogenic fungus used in this study are not genetically well characterised. Despite this drawback, we were able to transform if by nitrate reductase gene of *A. niger*. We are therefore able to conclude that this transformation system may be applied to any entomopathogenic fungus even those in which it is often difficult to screen for auxotrophic mutants. This system of isolation of *nia*D stable mutants and transformation is simple and rapid, and believe that it should prove useful in fungal species in which a genetic transformation system is not yet established.

ACKNOWLEDGEMENT

We gratefully acknowledge the financial support by D.B.T. New Delhi, UGC New Delhi and MPCST Bhopal in the form of Biotechnology National Associateship and research projects respectively to Dr. S.S. Sandhu.

REFERENCES

1. Sandhu S S, Rajak R C & Agarwal G P, *Biocontrol Science* and *Technology*, 3 1993 (a) 47.
2. Sandhu S S, Rajak R C & Agarwal G P, *Anals of Forestry*, 1(2) 1993(b) 136.
3. Maniatis T, Fritsch E F & Sambrook J, 1982, Molecular cloning Laboratory Manual. Cold Spring Harbor Laboratory N. Y. 1982.
4. Cove D J, *Biochem Biophys Acta, 113 1966 51.*
5. Sandhu S S & Venkateswerlu G, *Fungal Genetic Newsletter USA*, 44 1997 50.
6. Sambrock J, Fritsch E F & Maniatis T, Molecular cloning. A Laboratory Manual, Cold Spring Harbor Laboratory, Cold Spring Harbor, N.Y. 1989.
7. Daboussi M J, Djeballi A, Gerlinger C, Blaieau P L, Bouvier I, Cassan M, Lebrun M H, Parisot D & Brygoo Y, *Current Genetics*, 15 1989 453.
8. Johnstone I L, Mac Cabe P C, Greaves P, Cole G E, Brow M A D, Gurr S R, Unkles S E, Clutterbuck A J, Kinghorn J R & Innis M A 1990, *Gene*, 90 1990 181.
9. Unkles S E, Campbell E I, Carrez D, Grieve C, Contreras R, Fiers W, Van den Hondel C A M J J & Kinghorn J R, *Gene* 78 1989 157.
10. Garber R C & Yoder O L, *Anals Biochem*, 135 1983 416.
11. Zhu J, Kampenaers W, Van der Straeten D, Contreras L & Fiers W, *Bio/Technology*, 3 1985 1014.

CHAPTER - 20

MICROBIAL QUALITY OF OSMO-DEHYDRATED BANANA

P. Heartwin Amala Dhas, K. Thangavel, Z. John Kennady and R. Kailappan

Senior Research Fellow, Assistant Professor, Assistant Professor and Professor & Head
Department of Agricultural Processing, College of Agricultural Engineering, Tamil Nadu Agricultural University, Coimbatore.

ABSTRACT

Nendran and Red Banana slices were immersed in sugar syrup at different concentrations for partial dehydration by osmosis. KMS and citric acid of 0.1 per cent, each were added to the syrup to enhance the shelf-life of the product. The slices after osmosis were dried in air or under vacuum. Water loss and solid gain occurring during osmosis were analysed. Osmosed as well as osmovac dried banana slices were stored in polyethylene bags of 86.35 μm thickness and evaluated for water activity and microbial load.

1. INTRODUCTION

Banana is one of the most important sustaining fruits of our country. It occupies 12.36 per cent of the total area under fruit crops and 23.97 per cent of fruit production. About 7.9 million tonnes of banana is produced in India, offering considerable scope for export. One major blockade to the export of banana is its limited shelf-life. Dehydration is one of the easiest methods of enhancing the shelf life and is practiced from earliest times. A dried product apart from increased shelf-life, offers the advantage of less weight, which has the potential for savings in the cost of packaging, handling and transporting the product.

Fruits can be partially dehydrated by osmosis since they contain sugars and solutes in dilute solution and their cellular periphery acts as an effective semi-permeable membrane. Dehydration or removal of water by osmosis takes place due to the difference in osmotic potential between fruit pieces and surrounding hypertonic solution. Osmosis is conducted at relatively lower temperatures, reducing the severity of heat treatment. The dehydrated product therefore, retains the desirable qualities of natural fruits to maximum extent.

To produce a stable dehydrated product that could be stored for very long time, the slices after osmosis should be dried in a dryer. The advantage over other forms of drying is that the sugar which penetrated into the fruit during osmosis prevents enzymatic browning and helps in the retention of colour after drying.

2. MATERIALS AND METHODS

(a) Osmosis

Ripened Nendran and Red Banana slices of 5, 7.5 and 10mm thickness were dipped in 40, 50 and 60°Brix sugar syrups maintaining a fruit to syrup ratio of 1:4 by weight. In each treatment, 50g of slices were soaked and osmosis was carried out at RT, 50 and 75°C. KMS and citric acid of 0.1 per cent each were added to the syrup. Osmosis was done for 10, 5 and 2.5 h at RT, 50 and 75°C, respectively. At the end of osmosis, the slices were taken out, blotted to remove adhering syrup and weighed to find out the weight reduction. Samples of 5g were taken and kept in the vacuum oven at 70°C and 100mm Hg for the determination of solid gain. Water loss was computed by adding weight reduction and corresponding solid gain. The slices after osmosis were air dried at 65°C or vacuum dried at 70°C and 160 mm Hg. Only osmosed slices found to be best based on water loss and organoleptic qualities were dried. Slices partially dehydrated by osmosis (intermediate moisture banana) and osmovac dried slices were stored in polyethylene bags of 86.35 μm thickness for 30 days. The observations were average values of three replications.

(b) Water activity

Water activity determines the shelf-life of any product. Water activity of vacuum dried and intermediate moisture banana slices were determined using desiccator technique. Samples weighing 5g were taken in small cups and exposed to relative humidity ranging from 0 to 99 per cent. Different relative humidities were maintained inside these desiccators using saturated salt solutions. The samples were allowed to equilibrate for 7 days and their weights were noted for determination of moisture content. The Brunauer-Emmet-Teller (BET) isotherms were drawn by plotting moisture content of the sample against equilibrium relative humidity or water activity. From the sorption curve, the water activity of the sample was found (1). the experiments were repeated three times and the mean values were calculated.

(c) Microbial quality

The bacterial, yeast and fungal count in the stored banana were analysed by serial dilution and plate technique in periodical intervals of 15 days. Bacteria, fungi and yeast were isolated using nutrient agar, Martin's rose bengal agar and yeast extract malt extract agar media, respectively (2,3,4). The microbial load of the samples was tested as per the standard procedure (5). The bacterial count was taken after 2 days while fungal and yeast counts were taken after 4 days (6).

3. RESULTS AND DISCUSSION

(a) Water loss and solid gain

The initial moisture content of Nendran and Red Banana was found to be 2.03 and 2.84 kg H_2O Kg^{-1} DM. Water loss after osmosis ranged from 24.7 to 53.1 per cent in Nendran and from 26.43 to 57.9 per cent in Red Banana (Table 1). Water loss was maximum for 5 mm slices subjected to osmosis in 60° Brix syrup held at 75°C. For 10 mm slices soaked in 40° Brix at RT, water loss recorded the lowest value (7).

Table 1. Water loss in Nendran and Red Banana

Thickness	Temperature	Nendran			Red BAnana		
		Concentration ^{O} brix					
		40	50	60	40	50	60
5	RT	31.36	38.40	44.70	39.83	47.43	52.93
	50	36.36	46.03	52.20	41.70	49.16	54.63
	75	40.70	49.70	53.10	44.26	50.90	57.90
7.5	RT	28.33	33.26	39.66	32.16	38.66	46.36
	50	35.20	35.40	41.40	34.66	41.93	49.66
	75	35.23	40.90	43.00	36.06	42.06	50.60
10	RT	24.70	31.13	37.90	26.43	32.30	39.53
	50	29.53	34.06	38.43	30.60	38.83	43.73
	75	31.36	39.53	42.06	31.93	38.86	44.26

Table 2. Solid gain in Nendran and Red Banana

Thickness	Temperature	Nendran			Red Banana		
		Concentration ^{O} brix					
		40	50	60	40	50	60
5	RT	5.23	6.10	7.43	7.56	8.16	8.56
	50	6.60	8.53	9.30	9.33	9.66	10.50
	75	10.50	11.36	11.80	13.93	14.43	15.50
7.5	RT	4.66	5.76	6.36	6.63	7.16	7.63
	50	6.86	7.56	8.43	8.43	9.13	9.66
	75	9.66	10.53	11.50	10.53	11.36	12.26
10	RT	4.16	5.80	6.43	5.80	6.36	6.86
	50	6.23	7.46	8.36	7.33	9.06	9.60
	75	8.90	10.30	11.46	9.90	10.50	11.26

Solid gain observed for Nendran and Red Banana varied from 4.16 to 11.8 per cent and 5.8 to 15.50 per cent, respectively. Similar to water loss, solid gain was maximum when 5mm slices were dipped in 60° Brix at 75°C and minimum for 10 mm slices osmosed in 40° Brix at RT (Table 2). Slices osmosed in syrups maintained at 50°C were found to be the best based on water loss and organoleptic qualities. Hence, studies on drying and microbial qualities were carried out for those treatments only.

(b) Drying

The drying curves obtained for Nendran and Red Banana showed drying to occur in the falling rate period. Untreated slices (fresh) needed 8 h of drying to bring the moisture content below 0.2 kg H_2O, Kg^{-1} DM while the osmosed slices required 4 to 7 h only to attain the same moisture content. Vacuum drying took 1 to 2 hours more compared to air drying. The final moisture content after drying fell in the range of 0.05 to 0.2 kg H_2O Kg^{-1} DM.

Table 3. Water activity of Nendran and Red Banana after osmosis

Thickness	Temperature	Syrup concentration	Nendran	Red Banana
5	RT	40	0.935	0.940
		50	0.930	0.930
		60	0.920	0.925
	50	40	0.915	0.930
		50	0.910	0.920
		60	0.905	0.915
	75	40	0.910	0.920
		50	0.910	0.915
		60	0.910	0.910
7.5	RT	40	0.955	0.960
		50	0.950	0.950
		60	0.950	0.955
	50	40	0.940	0.945
		50	0.930	0.935
		60	0.920	0.920
	75	40	0.930	0.935
		50	0.920	0.925
		60	0.920	0.920
10	RT	40	0.960	0.960
		50	0.950	0.955
		60	0.945	0.950
	50	40	0.950	0.955
		50	0.945	0.945
		60	0.940	0.940
	75	40	0.935	0.935
		50	0.930	0.930
		60	0.925	0.925

Table 4. Water activity of osmovac dried banana

Thickness	Temperature	Syrup concentration	Nendran	Red Banana
Untreated	—		0.16	0.30
5	50	40	0.21	0.18
	50	50	0.23	0.24
	50	60	0.26	0.38
7.5	50	40	0.15	0.18
	50	50	0.15	0.22
	50	60	0.15	0.26
10	50	40	0.21	0.21
	50	50	0.23	0.26
	50	60	0.27	0.32

Table 5. Microbial counts of intermediate moisture Nendran after 15 days

Thickness	Temperature	Syrup concentration	Fungi x 10^3	Yeast x 10^4	Bacteria x 10^6
5	RT	40	9	18	17
		50	9	20	14
		60	10	16	13
	50	40	9	10	5
		50	8	11	4
		60	8	9	4
	75	40	6	8	4
		50	4	6	4
		60	5	7	3
7.5	RT	40	10	19	15
		50	10	27	17
		60	8	21	8
	50	40	10	12	6
		50	8	10	5
		60	7	10	5
	75	40	5	7	5
		50	4	7	5
		60	5	9	4
10	RT	40	12	20	22
		50	10	25	21
		60	12	21	17
	50	40	11	9	7
		50	10	8	8
		60	7	9	4
	75	40	9	7	10
		50	8	7	5
		60	5	8	5

Vacuum drying yielded a superior quality product and hence osmovac dried samples were selected for storage study.

(c) Water activity

The water activity values of fresh Nendran and Red Banana were observed to be 0.985 and 0.988 respectively. Due to osmosis, the water activity of the slices decreased and lied in the range of 0.905 to 0.960 (Table 3). The decrease in water activity was caused by dewatering of the cells and penetration of sugar into the slices (8). Water activity was minimum for 5 mm slices osmosed in 60° Brix syrup at 75°C. The reduction in water activity was related to the temperature of osmosis, concentration of syrup and thickness of slice.

The water activity of osmo-vacuum dried slices lied between 0.20 and 0.40 depending upon the residual moisture content in slices after drying (Table 4). The water activity decreased substantially due to the loss of water by vacuum drying.

Table 6. Microbial counts of intermediate moisture Red Banana after 15 days

Thickness	Temperature	Syrup concentration	Fungi x 10^3	Yeast x 10^4	Bacteria x 10^6
5	RT	40	17	25	21
		50	15	21	26
		60	16	12	16
	50	40	13	11	7
		50	8	13	6
		60	12	10	6
	75	40	11	12	6
		50	11	9	7
		60	8	9	6
7.5	RT	40	15	18	19
		50	18	18	19
		60	20	12	16
	50	40	17	16	9
		50	14	12	9
		60	11	9	9
	75	40	12	13	8
		50	11	15	6
		60	9	8	6
10	RT	40	24	29	26
		50	22	28	25
		60	23	28	22
	50	40	13	15	11
		50	12	14	10
		60	12	11	10
	75	40	13	14	9
		50	10	12	7
		60	11	12	8

(d) Microbial quality

The total bacterial, yeast and fungal count in the intermediate and osmovac dried slices were analysed after 15 and 30 days of storage. Observations showed that the bacterial, yeast and fungal count in one gram of intermediate moisture Nendran stored for 15 days ranged from 3 x 10^6 to 22 x 10^6, 6 x 10^4 to 27 x 10^4 and 4 x 10^3 to 12 x 10^3 respectively (Table 5). Corresponding counts in Red Banana were 6 x 10^6 to 26 x 10^6, 2 x 10^4 to 29 x 10^4 and 8 x 10^3 to 24 x 10^3 (Table 6). The microbial population was well above the limit of 50,000 total count per gram permissible for dehydrated foods indicating that the slices were not fit for consumption after 15 days (9). In order to be shelf-stable, the water activity should be below 0.88 when all microbial activities subside or cease completely. This proves the need for air/vacuum drying of slices after osmosis. The microbial population was high in intermediate moisture banana because the water activities were above 0.90. Therefore, there existed a chance for microorganisms to grow. The microbial load depended on the water activity of the slices. At higher water activities, more water is available for the microorganisms to act and hence the population is high. The microbial load in the slices osmosed at 75°C were relatively less because of the high level of sugar infusion into the slices during osmosis. The microbial growth continued during storage and all the samples showed high level of contamination after 30 days. There was not much difference in the Microbial population among samples.

The osmovacuum dried slices wee literally free from microorganisms at the dilutions tested. The non-availability of water for microbial activity at lower water activities hindered the growth of microorganisms. Osmovac dried slices were, therefore, microbiologically sound for consumption.

REFERENCES

1. Karel M, Fennema OR & Lund DB, Principles of Food Science Part II, Marcel Dekker Inc, (1975) 237.
2. Allen DN, Experiments in soil bacteriology (IIed), Burgees Publishing Co, (1953) 127.
3. Martin JP. Soil Sci, 69 (1950) 215.
4. Phaff HJ, Isolation of biotechnological organisms for nature, McGraw Hill Publishing Co, (1990) 53.
5. Anonymous, Manual of microbiological methods, American Phytopathological Society, (1967) 1.
6. Subba Rao NS, soil microorganisms and plant growth (II ed.), Oxford and IBH Publishing Co, (1986) 290.
7. Ponting JD, Watters GG, Forrey RR, Jackson R & Stanley WL, J Fd Technol, 20 (1966) 125.
8. Mazza G. J Fd Technol, 18 (1983) 113.
9. Scott WJ, Adv in Fd Res, 7 (1957) 83.

CHAPTER - 21

NATURAL INCIDENCE OF TOXIGENIC *Aspergillus flavus* LINK STRAINS ON THE SURFACE OF PREHARVEST JUJUBE FRUITS (*Ziziphus mauritiana* LAMK.)

Yash Paul Singh and Geeta Sumbali

Mycology and Plant Pathology Laboratory, Department of Biosciences, University of Jammu, Jammu - 180 006, India.

ABSTRACT

Among the various fungal species comprising surface mycoflora of Jujube fruits, *Aspergillus flavus* was consistently recorded during the entire growth period. Fifty isolates of *A. flavus* recovered from the preharvest fruits were observed to cause extensive postharvest rot of mature Jujube when inoculated artificially. As this species is a known producer of aflatoxins all the fifty isolates were screened for their aflatoxigenic potential *in vivo.* 54% of the tested isolates wer recorded positive for different aflatoxins (B_1, B_2, G_1 and G_2) and their level ranged from 31 to 2874 ppb which is very high as compared to the tolerance level of 20 ppb fixed by WHO. This study demonstrates for the first time that mature Jujube fruits are a favourable substrate for infection and aflatoxin production by *A. flavus* strains.

1. INTRODUCTION

Mature fruits are susceptible to attack by a variety of microfungi to which they are resistant during their development on the plant. Majority of such rot producing fungi result in enormous wastage of fruits even before reaching the consumers [1]. Much of the inoculum responsible for these losses may get associated with the surface of the fruit right from the orchard [2] or may be carried along with the packing leaves, straw and baskets [3] or may originate within the enclosure of the storage houses and fruit shops [4].

Propagules of pathogenic fungi found abundantly on the fruit surface during their maturation in the orchards have been implicated in postharvest spoilage of some fruits [5-8]. In addition to the loss of food calories and nutritive elements, production of toxic metabolites (mycotoxins) by fruit rot fungi is emerging as a major health problem especially in the developing countries [9]. Among the known mycotoxins, aflatoxins have attracted global attention mainly because of their toxicological characteristics [10] and the fact that they are leading natural contaminants found in most of the harvested agricultural commodities consumed by man [11-13]. It has been established that the regions of the world with the highest incidence of liver cancer correspond to the areas where climatic conditions favour aflatoxin contamination [14].

Aflatoxin contamination in fresh fruits have received comparatively little attention [15 - 17]. Among the common Indian fruits, Jujube (*Ziziphus mauritiana* Lamk.) also referred as the poor man's fruit is grown extensively in the arid and semiarid regions [18] for its nutritive value. These fruits are quite resistant to fungal attack before harvest, but once harvested the susceptibility increases as they matures [19]. Taking all these views into consideration, the present investigation was designed to find out whether *A. flavus* strains constituting the surface mycoflora of preharvest Jujube fruits have the potential of producing infection and aflatoxins in the fruit during postharvest period.

2. MATERIALS AND METHODS

(a) Isolation of *A. flavus* strains associated with fruit surface during pre-harvest stage

Surface washing technique was employed to isolate *A. flavus* from fruit samples obtained during the entire growth period. In this method, whole fruit (25g) was put in sterilized distilled water and subjected to horizontal shaking for 30 minutes. The liquid was then centrifuged at 3000 rpm for 20 minutes and the residue was made into a homogenous suspension. This suspension was poured in sterilized petriplates (1ml/plate) and modified Czapeks agar medium supplemented with Rose Bengal (0.20g/1) and Streptomycin sulphate (0.06g/1) was used for plating. Petriplates thus prepared were incubated at 28°C for 6 days. In this way 50 isolates of *A. flavus* were selected randomly from the samples.

(b) Inoculation of fruits

Mature fruits of Jujube were inoculated individually with the spores from fresh cultures of the selected *A. flavus* isolates by Pin prick method [20] and incubated at 28°C for 10 days.

(c) Extraction and estimation of aflatoxins from diseased fruit

Natural production of aflatoxins was estimated by extracting 10g of fruit pulp adjacent to the inoculated region, following the method of Thomas *et al.* [21]. Chloroform extracts thus obtained were evaporated to dryness. The residue was dissolved in 1ml of chloroform for thin layer chromatography. Qualitative separation of aflatoxins (B_1, B_2, G_1, G_2) was performed on activated TLC plates using toluene : iso-amyl alcohol : methanol (90 : 30 : 2 v/v) as the solvent system [22]. Identity of the aflatoxins was further confirmed chemically by spraying 25% H_2SO_4 which changed the blue fluorescent spots to yellow [23]. Finally, the aflatoxins were quantified by spectrophometric method [24].

3. RESULTS AND DISCUSSION

A total of fifty isolates of *A. flavus* were collected from the surface of Jujube fruits during the growing season, and all could equally cause extensive spoilage of mature Jujube when artificially inoculated (Table 1), proving their opportunistic pathogenic

Table 1. Production of aflatoxins in mature jujube fruits by *Aspergillus flavus* isolates recovered from fruit surface during growth period (before harvest).

Number of *A. flavus* isolates screened = 50
Percentage of pathogenic isolates = 100%
Number of toxigenic *A. flavus* isolates = 27
Percentage of toxigenic isolates = 54%

Toxigenic *A. flavus* isolates	Amount of aflatoxin produced (ppb)			
	B_1	B_2	G_1	G_2
AF – 11	–	–	–	93
AF – 13	2811	–	–	–
AF – 14	343	–	–	–
AF – 17	–	–	–	93
AF – 18	–	–	–	343
AF – 19	–	–	–	437
AF – 20	–	–	–	437
AF – 21	–	–	–	62
AF – 22	–	–	–	62
AF – 23	–	–	–	31
AF – 27	–	–	343	–
AF – 28	–	93	–	–
AF – 29	–	593	–	–
AF – 30	–	62	–	–
AF – 31	–	93	–	–
AF – 32	–	343	–	–
AF – 33	–	343	–	–
AF – 34	–	–	374	62
AF – 35	–	–	624	–
AF – 36	–	–	156	499
AF – 42	–	–	–	93
AF – 44	–	–	–	343
AF – 45	–	–	–	2874
AF – 46	–	–	–	343
AF – 48	–	–	–	562
AF – 49	–	–	–	156
AF – 50	–	–	–	93

– Not detected

behaviour. Thin layer chromatographic analysis of diseased tissue extracts revealed that about 54% of these isolates were aflatoxigenic (Table 1). There was no correlation between the pathogenic behaviour and aflatoxin producing potential of these isolates. Among the confirmed toxigenic isolates, 2 produced AfB_1 (343-2811 ppb), 6 produced AfB_2 (62-593 ppb), 4 produced AfG_1 (156-624 ppb) and 17 produced AfG_2 (31-2874 ppb). Higher production of aflatoxins in Jujube fruits may be attributed to its high ascorbic acid content which is known to be good source for aflatoxin production [25]. It was noted that *A. flavus* isolates even differed in their capacity to interconvert the aflatoxins, a behaviour described earlier also [26].

This analysis indicated that Jujube fruit is a suitable substrate for aflatoxin elaboration. Atleast two isolates could produce very high concentration of aflatoxin B_1, the most toxic member of the group whose tolerance level as fixed by the World Health Organisation [27] is 20 ppb. Under the prevention of food adulteration rules, Government of India has also imposed maximum permissible level of 30 ppb for aflatoxin in all foods. Nearly sixty six other countries have also introduced legislation on the maximum permissible levels of aflatoxin ranging from zero to 50 ppb in different food commodities of human consumption [28].

These results also suggest that since Jujube fruits contain plenty of water and other solutes, *A. flavus* strains may be readily utilising it and thus persist throughout the growing season. Besides, *A. flavus* spores have a tendency to become air borne [4] and are even endowed with long period of viability which enables them to remain on the fruit surface for the entire preharvest period. Being an opportunist, it easily invades the wounded and the undamaged mature fruits while some strains produce variable amount of aflatoxins. This association of toxigenic *A. flavus* strains with Jujube fruits can pose a health problem for those people who consume such molded fruits due to ignorance or duress. However, magnitude of aflatoxin contamination may vary with various preharvest and postharvest factors.

ACKNOWLEDGEMENT

The authors wish to thank Head, Department of Biosciences, University of Jammu, Jammu India for providing necessary laboratory facilities during the investigation.

REFERENCES

1. Salunkhe D K & Desai B B, Post - harvest biotechnology of fruits. Boca Raton, FL : CRC. 1984.
2. Meredith D S, Nature, 196, 1961 : 555.
3. Panduranjan A G & Suryanarayanan T S, J Econ Tax Bot, 7, 1985 : 309.
4. Sumbali G & Badyal K, Indian Phytopath, 44, 1991 : 214.
5. Dennis C & Mountford J, Ann Appl Biol, 79, 1975 : 141.
6. Kainsa R L, Singh J P & Gupta, O P, Haryana Agric Univ J Res, 8, 1978 : 86.
7. Singh J P & Kainsa R L, Indian Phytopath, 36, 1983 : 72.
8. Mclean M A & Sutton J C, Can J Bot, 70, 1992 : 846.
9. Rodricks J V, Adv Chem Ser No 149, Am Chem Soc, 1976 : 409.
10. IARC, International Agency for Research on Cancer, Lyon, Franch. Monograph, 1993 : 56.
11. Payne G A, Curr Rev Pl Sci, 10, 1992 : 423.
12. Ahmad S K J Stored Prod Res, 29, 1993 : 33.
13. Miller J D, J Stored Prod Res, 31, 1995 : J Stored Prod Res, 31, 1995 : 1.
14. Casteel S W & Braun W F, In : Encyclopedia of Microbiology, [Lederberg, J. (ed). Academic Press, London (1992).
15. Sinha K K & Singh A, Nat Acad Sci Lett, India, 7, 1982 : 213.

16. Madhukar J & Reddy S M, Geobios New Reports, 11, 1992 : 188.
17. Doster M A, Michalides T J & Morgan D P, Plant Dis, 80, 1996 : 484.
18. Yamdagni R, 1986 Ber. In : Fruits of India, tropical and sub- tropical. (Bose T K ed. Naya Prakashan, Calcutta India (1986).
19. Singh Y & Sumbali G, J Ind Bot Soc, 76, 1997 : 127.
20. Tomkins R G & Trout S A, J Pomol Hort Sci, 9, 1931 : 257.
21. Thomas F, Eppley R M & Trucksess M W, J Assoc Off Anal Chem, 58, 1975 : 114.
22. Reddy T V, Viswanathan L & Venkittasubrammanian T A, 1970. Analyt Biochem, 38, 1970 : 568.
23. Stack M E & Pohland A E, J Assoc Off Analyt chem, 58, 1975 : 110.
24. Nabney J & Nesbitt B E, Analyst London, 90, 1965 : 155.
25. Detroy R W, Lillehoj E B & Ciegler A, In : Microbial toxins, (Ciegler A, Kadis S & Ajl S J (eds). Academic Press, New York, 1971.
26. Heathcote J G & Hibbert J R, Aflatoxins - chemical and biological aspects. Elsevir Scientific Company, Amsterdam, 1978.
27. WHO, Environmental health criteria. II. Mycotoxins. World Health Organisation, Geneva, 1979.
28. Sinha A K, Mycotoxin-a health hazard Yojna, 39, 1995 : 15.

CHAPTER – 22

ALGALIZATION STUDIES ON CHICKPEA (*Cicer arietinum* L)

S.B. Angadi Jagannath, Umapati Dengi and Eshwarlal Sedamakar

Phycology Laboratory, Department of Botany,
Gulbarga University, Gulbarga - 585 106

ABSTRACT

The effect of BGA as potent biofertilizer on chickpea solely and also its combined effect with cowdung and chemical fertilizers (NPK) was investigated. Algalization enhances the overall growth of chickpea. It is found to enhance all the morphological and biochemical characters such as proteins, carbohydrates, total nitrogen content, net grain yield and biomass of the chickpea. The wilting of chickpea can also be controlled by using BGA.

1. INTRODUCTION

The chemical fertilizers provide required amount of nitrogen to plants. The Blue Green Algae (BGA) are the part of largest contributors to the process of biological nitrogen fixation **[1]**. Many studies have been made by using algal biofertilizer on various important crop plants. In the present investigation an attempt was made to know the effect of algalization on chickpea which is an important pulse crop and contains large amount of proteins, fats, carbohydrates, minerals, like Phosphorous, Iron, Calcium and Vitamins. Further popularization was made to the farmers of this area for utilization of algal biofertilizers by field demonstrations.

2. MATERIALS AND METHODS

Multistrain BGA viz; *Nostoc* spp., *Anabaena* spp., *Tolypothrix* spp., and *Scytonema* spp., were isolated and cultured in the laboratory. The unialgal cultures were subjected to mass cultivation in cultivation devices. Samples of all these algae were taken and mixed to make multistrains. The mixed inoculum was then transferred to small polythene tanks (4 x 2 size) containing 1:50 litres of De's modified Beneck's medium **[2]** and sterilized tap water. The polythene tanks were covered by hard transparent polythene sheets. Profused growth of multistrain BGA was attained between 15-20 days. This mass cultivated BGA was used for algalization experiments in the fields.

Plan of field work

The work was broadly divided into five experimental study. Accordingly five different plots were made each of size 10' x 4'.

1	Control field	No inoculum
2	Field-I	500 g of multistrain BGA alone
3	Field-II	250 g of multistrain BGA and 250 g of cowdung
4	Field-III	250 g of multistrain BGA and 50 g Chemical fertilizers (NPK)
5	Field-IV	100 g of NPK only

The method of inoculation followed was of Singh and Singh **[3]** but it was slightly modified by adding 10 mg of sodium molybdate to BGA for good establishment **[4]**.

The observations were made on various parameters such as leaf length, plant height, length of the pods and number of pods per plant. Further the studies were also made on chemical estimations such as, proteins **[5]**, carbohydrates (Anthrone reagent method) and total nitrogen content **[6]**.

3. RESULTS

The field inoculated multistrain BGA integrated with chemical fertilizers NPK had shown overall effect on growth, development and morphological variations of chickpea as compared to using only BGA, BGA integrated with cowdung and only NPK (Plate-1 and Table-1).

Early flowering and early maturation was observed in field-III (BGA+NPK), Similar observation were also made by many workers. **[1, 7-12]**. The effect of algalization on net yield, biomass and economics of algalization is given in the table-2 and graph-4. It was observed that field-III (BGA+NPK) has given yield and biomass as compared to control and other three fields [I, II and IV]. It was also observed that the net grain yield was three fold over control and two fold over field I, II and IV.

The observations made on the total proteins, carbohydrates and nitrogen content of the chickpea grains are presented in the table-3 and graph-1. It was noticed that field-III (BGA+NPK) enhanced all the biochemical parameters over control as well as field- I, II and IV.

Table 1. The effect of algalization on various morphological characters of chickpea (*Cicer arietinum* L.)

S.N.	Mode of experiment conducted	No. of plants per row	Total Plants/ field (Plot)	Length of the leaf (in cms)	Height of the Plant (in cms)	Length of the Pod (in cms)	No. of Pods/ Plants
1	Control	75	300	0.8	31.2	1.5	24
2	Field-I Treated with only BGA	75	300	1.3	41.3	1.9	42
3	Field-II 50% BGA + 50% cowdung	75	300	1.1	38.8	1.7	38
4	Field-III 50% BGA + 50% NPK	75	300	1.6	54.6	2.4	56
5	Field-IV Treated with only NPK	75	300	1.4	46.2	2.2	52

Plate -1. The display of various photographs showing the 5 experimental fields — 1. Control : without inoculum, 2. Field-I : Treated with 500 g of BGA; 3. Field-II : Treated with 250g of BGA + 250 g cowdung; 4. Field-III : treated with 250 g of BGA + 50g chemical fertilizer NPK; 5. Field IV : Treated with 100 g of chemical fertilizer NPK only; 6. Photographs showing all the 5 fields; 7. Photographs showing the pods of all the 5 fields.

Table 2. The Effect of algalization on net grain yield, net biomass and economics of algalization

S.N.	Mode of expt. conducted	Grain yield/ acre (in kgs)	Expendt. spent/ acre (in Rs.)	Expendt. spent for mainte-nance (in Rs.)	Gross gain income (in Rs.)	Net income (in Rs.)	Net gain (in Rs.)	Net biomass produced per acre (in Kgs)
1	Control	307	—	15	2763	2748	——	169.82
2	Field - I Treated with only BGA	806	87	15	7254	7152	(1) 4404 over control (2) 0693 over field II	756.64
3	Field - II 50% BGA + 50% cowdung	729	43	15	6561	6459	(1) 3711 over control	732.28
4	Field - III 50% BGA + 50% NPK	1075	390	15	9675	9573	(1) 6825 over control (2) 2421 over Field-I (3) 3114 over Field-II (4) 0693 over Field-IV	1016.77
5	Field - IV Treated with NPK only	998	696	15	8982	8880	(1) 6132 over control (2) 1728 over Field-I (3) 2421 over Field-II	963.51

Table 3. Effect of algalization on Protein, carbohydrates and Nitrogen content of chickpea

Sl.No.	Mode of expt. conducted	Total Protein content (mg/1 gm)	Total Carbohydrate content (mg/1 gm)	Total Nitrogen content (mg/1 gm)
1	Control	191.0	515	7.6
2	Field-I Treated with only BGA	208.5	553	9.8
3	Field-II 50% BGA + 50% cowdung	195.6	537	8.3
4	Field-III 50% BGA + 50% NPK	223.8	628	12.8
5	Field-IV Treated with only NPK	214.3	592	11.9

4. DISCUSSION

It was observed that algalization has beneficial effect on chickpea crop either by multistrain BGA or integrated with other fertilizers. Algalization has found to enhance the height of the plants and number of pods per plant. Among various algalization experiments (control, BGA, BGA+Cowdung, BGA+NPK and NPK), the field integrated with BGA+NPK showed maximum beneficial effect. Similar observations were also made by many investigators on some crop plants **[1, 7, 11-14]**.

It was also observed that field-I treated with only BGA and field-II BGA+Cowdung enhanced the grain yield two fold over the control while the maximum benefit of grain yield was obtained in field-III treated with BGA+NPK as compared to other treatments. However, field-IV (Only NPK) better than the control, field I & field II. Observations were also made by many investigators on other crop plants and they obtained 30-36% increased rice yield by algalization **[3, 10, 15, 16]**.

Further in the present study, it was observed that there was a considerable

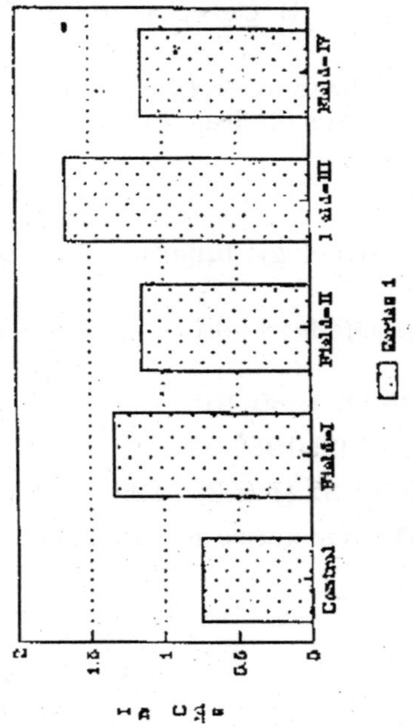

Graph - 1
Comparative Length of Leaf of Five Plots

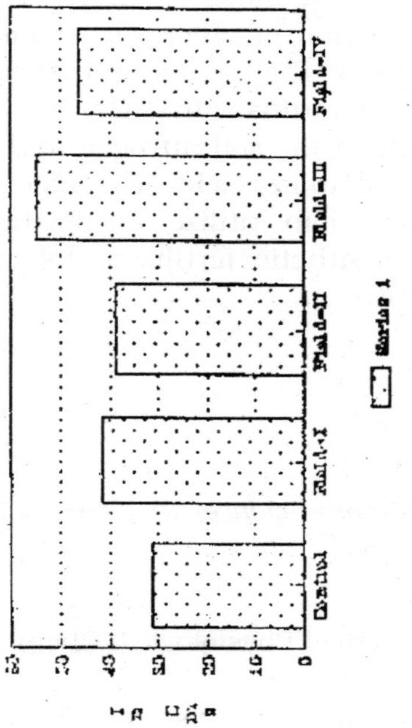

Graph - 2
Comparative Height of Plants of Five Plots

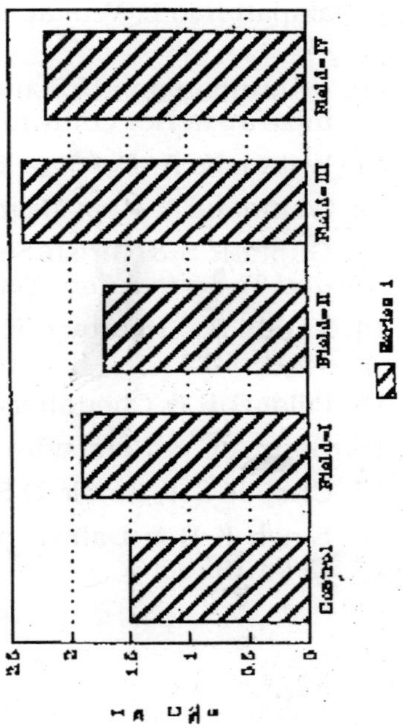

Graph - 3
Comparative Length of Pod of Five Plots

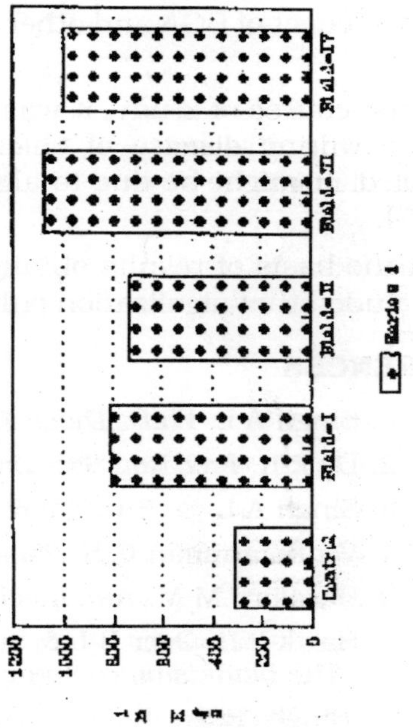

Graph - 4
Comparative Grain Yield of Five Plots

Graph 1-4

increase in the biochemical composition of the chickpea due to algalization as well as combined effect of BGA and other fertilizers (Table 3).

In the course of study, it was observed that the wilting disease of chickpea was controlled. It might be due to algalization. **[17- 18]**.

On the basis of results obtained it can be concluded that algalization enhances all the morphological characters, net grain yield, biomass and biochemical composition of chickpea. Hence it can be suggested that the marginal farmers can use the multistrain BGA to increase the fertility of the soil nitrogen and get at least 2-3 fold more yield. The medium or rich farmers can utilize algalization coupled with synthetic fertilizers for much more benefit.

REFERENCES

1. Singh A L. *Ph.D. Thesis* CRRI, New Delhi 1984.
2. De P K, *Proc Roy Soc London*, 1939, 127-132.
3. Singh A L, & Singh P K, *Proc Ind Acad Sci (Plant Science)* (2) 1986 147.
4. Venkataraman G S. *Phykos*, 5 1982 p 164.
5. Bardford M M. *Ann Biochem*, 72 1976 248.
6. Hawk P B, Oser B L & Summerson W H. Practical Physiological Chemistry. *The blankiston company*. USA 1948.
7. Singh R N. Role of Blue-Green algae in nitrogen economy of Indian agriculture. *ICAR* New Delhi 1961 p 175.
8. Aiyer R S, Salaudean S & Venkataraman G S. *Ind J Aroric* Kerala 1972 3 p. 100-104.
9. Jalapati Rao L, Venkataraman A & Sunder Rao W B, *Curr Sci*, 46 (2) : 1997 50.
10. Subramanian R. Some observations on utilization of Blue-Green algae mixture in rice cultivation in India. *Mad Univ Pres* Madras, India 1972 281.
11. Watanabe A, *Bot Maz*, 69 (820-821) 1956 530.
12. Watanabe A, *Appl Microbiol*, 8 (2) 1962 85.
13. Gupta R S & Girish Seth. Perspectives in Phycology. Rajarao V.V. *Today and Tomorrow Pub.* New Delhi 1990 401.
14. Nanda B, Tripathy S K & Padhi S, *World J. Of Micro and Biotech* Vol. 7 (6) 1991 622.
15. Pillai G K & Choudhary D B, *Fert News*, 25 (2) 1980 40.
16. Gour A C & Ramendra Singh, *Fert. News*, 27 (2) 1982 87.
17. Chopde S T, Kotale O S & Khan Y D, *Phykos*. 32 (1 & 2) : 1993 115.
18. Singh K B & Dahiya B S, Breeding for wilt resistance in Chickpea. *New Delhi* 1973 13.

CHAPTER - 23

MANAGEMENT OF MANGO LEAF LITTER USING MICROORGANISMS

P.D. Desai and K.J. Shah

Department of Microbiology, Bhavan's College, Andheri (W), Mumbai - 400 058

ABSTRACT

Three set of laboratory scale experiments : Mango leaves with cowdung slurry, Mango leaves with cowdung slurry and consortium of cellulase and amylase producing microorganisms and Mango leaves with addition of cowdung slurry, consortium of cellulase and amylase producing microorganisms and earthworm (*Eudrilus euginea*) were set up to study the enhancement of the process of composting mango leaf litter. It was observed that addition of consortium of microorganisms and earthworms brought rapid conversion (35 days) of the mango leaf litter in to compost as compared to the one with only cowdung slurry (50-55 days).

1. INTRODUCTION

During the last few decades, India has drastically progressed in the field of agriculture. Agricultural production has increased in this country because of newer techniques, high yilding varieties of cash crops and use of biofertilizers, biopesticides etc. All these developments in the agriculture have gradually led to increasing problem of pollution. Normal practices such as burning of leaf litter is one of the causes leading to pollution. With the growth in agriculture produce particularly mango orchards, the present study was taken up with a view to rapidly recycle leaf litter nutrients which are normally burnt releasing precious organic carbon as CO_2 in the atmosphere. Mango leaf litter has total nitrogen 0.99%, organic carbon 14%, calcium 0.582%, magnesium 0.181%, available phosphous 0.011% and total phosphorus 0.2%, thus providing a nutritive medium for growth of the microorganisms. Soil microflora bring about decomposition of leaf litter at different rates depending upon the nature of the substrates, the characteristics of the environment etc.

In this study, different types of

microorganisms having amylase and cellulase activities were selected and used as consortium to enhance the process of decomposition of mango leaf litter. Also, the effect of consortium of these micro-organisms along with earthworm (*Eudrilus euginea*) was studied.

2. MATERIALS AND METHODS

(a) Screening of cellulase and amylase producing microorganisms

Different groups of microorganisms such as fungi, bacteria and actinomycetes which possess the ability to degrade cellulose or starch were isolated by enrichment technique using samples of soil, compost and manure collected from various places around Mumbai.

(b) Culture Media

Nutrient agar, McCarty medium, Rose Bengal medium having 1% CMC/filter paper/soluble starch were used for isolating bacteria, actinomycetes and fungi respectively.

(c) Cellulase and Amylase activity

For primary screening of cellulase producing microorganisms, the isolates were spot inoculated on the above mentioned agar media and the cellulase activity was detected by flooding the CMC plates with 1% Cetyl trimethyl ammonium bromide (CTAB) (5) or 1% Congo red solution (6). The amylase activity was detected by flooding starch agar plates with Lugol's iodine.

The cellulose and starch degraders were selected by measuring the zone size and calculating the selection ratio as per given below (3).

$$SR = \frac{\text{diameter of zone of clearance in mm}}{\text{diameter of zone of growth of the culture in mm}}$$

For secondary screening, the rate of degradation of filter paper, CMC and starch by the cell free broth cultures of the selected microorganisms were measured using DNSA reagent to determine the concentration of reducing sugars released (4). One E.U., is defined as the quantity of enzyme releasing 1 micromol of reducing groups (maltose or glucose equivalents) from the substrate per minute under assay condition (1,2).

(d) Composting of mango leaf litter

Composting of mango leaf litter was carried out on a laboratory scale using the selected amylase and cellulase producing microorganisms and earthworms. Following set of lab scale experiments were conducted with a view to enhance the process of composting mango leaf litter.

SET-1 : 5 kg Mango leaves with addition of 2 lit cowdung slurry.

SET-2 : 5 kg Mango leaves with addition of 1 lit of cowdung slurry and 1 lit of consortium of cellulase and amylase producing microorganisms.

SET-3 : 5 kg Mango leaves, with addition of 1 lit of cowdung slurry and 1 lit of consortium of cellulase and amylase producers and earthworms.

The cowdung slurry was prepared using fresh cowdung each time, mixing 1 kg of cowdung with 1 lit of water.

Out of the 7 cellulase producers 3 were bacterial, 2 fungal and 2 actinomycete cultures. Among the 7 amylase producers 3 were bacterial, 2 fungal and 2 actino-mycete isolates. Consortium of micro-organisms was prepared using equal volumes of 48 h broth cultures of each isolate. A box of vermicasting was obtained from Bio-Genik Sys., Vile-Parle, Mumbai. It was processed for 15 days to allow growth of earthworms.

3. RESULTS AND DISCUSSION

(a) Selection of microorganisms

From the primary screening, 48 cellulase producing isolates were obtained out of which 28 were bacteria, 14 fungi and 6 actinomycetes. Fifty three amylase producing isolates included 32 bacteria, 15 fungi and 6 were actinomycetes. Each of the amylase producers detected on starch agar plate was grown in starch broth and the cellfree broth was used to measure amylase activity. Similarly, each of the cellulase producers detected on the CMC agar plates was grown in CMC broth and filter paper medium and the cellfree broth was used to measure cellulase activity.

Tables 1 and 2 show the isolates having high level of amylase and cellulase activity selected for preparing the consortium. Isolate numbers BM 52, BM 53, BM 54, BM 55, BM 56, and BM 57 were observed to have both cellulase and amylase activities and hence were selected to minimize the cost and labour of preparing the consortium.

(b) Composting of mango leaf litter

Large drums were used to carry out the three sets of composting mango leaf litter. The cowdung slurry or the consortium/water was sprinkled on alternate days and were mixed well every 5 days. The vermiculture contents of the box were added to the third set after 7 days of initial composting. Addition of cowdung slurry/consortium of microorganism/water was continued till the softening and black colouration of leaf tissues was observed. This was followed by sprinkling only water (500-700 ml/drum) to keep the contents moist. After completion of composting period (earthworms were removed from the set 3 for reuse) the contents were dried by spreading on flat ground and mascerated to obtain the final product for use as fertilizer. Table 3 shows that the addition of consortium of selected microorganisms and earthworms bring about rapid conversion of the mango leaf litter in to compost (35 days) as compared to the traditional method of using only cowdung. Thus the organic matter in the mango leaf litter is mineralized through detritus food chain i.e. dead leaves - fungi - bacteria - earthworms and the resulting nutrients in the form of fertilizer can be taken up by the plants.

Table 1. Amylase Activity of the Isolates

Isolates No.	Type of organism	Selection ratio	Amylase activity
BM 52	Bacteria	4.2	16.11
BM 53	Bacteria	3.9	19.20
BM 54	Actinomycete	4.5	15.45
BM 55	Actinomycete	4.2	16.11
BM 56	Fungi	3.6	14.11
BM 57	Fungi	4.2	12.44

Table 2. Cellulase Activity of the Isolates

Isolates No.	Type of organism	Selection ratio	Amylase activity
BM 52	Bacteria	3.2	14.91
BM 53	Bacteria	2.9	15.00
BM 54	Actinomycete	3.2	13.23
BM 55	Actinomycete	3.5	13.12
BM 56	Fungi	2.8	11.14
BM 57	Fungi	2.9	12.10

Table 3. Effect of microorganisms on the composting time for Mango Leaf Litter

Set No.	Process of Composting	Total time required for Composting
SET-1	mangoleaves+ cowdung slurry	50-55 days
SET-2	Mangoleaves+ cowdung slurry + consortium of microorganisms	45 days
SET-3	Mangoleaves+ cowdung slurry + consortium of microorganisms+ earthworm	34 days

REFERENCES

1. Gonzalez Siso, MI, Lang E, Carreno-Gomez, B, Becerra, M, Espinar, M & Blanco Mendez, J, *Process Biochemistry*, 32 (1997) 32 211-216.
2. Jayaraman, J, In : Laboratory manual in Biochemistry. Newsare International Ltd., New Delhi (1996).
3. Khandeparker, VG, & Dhala, SA, *Ind J of Microbiol*, 13 (1973) 73.
4. Miller, GL, *Analytical Chemistry* 31 (1959) 426.
5. Hankin, L. and S.L. Anagnostakis, *J. Gen. Microbiol.* 98 (1977), 109-115.
6. Teather, RM & Wood, *Applied Environ Microbiol* 43 (1982) 43 777.

CHAPTER - 24

EFFECT OF *Azotobacter* INOCULATION AND NITROGEN LEVELS ON GROWTH AND YIELD OF RADISH (*Raphanus sativus* L.)

O. Basavaraju, A.R. Madhava Rao, and T.H. Shankarappa

Department of Agricultural Microbiology, University of Agricultural Sciences, G.K.V.K. Campus, Bangalore - 560 065

ABSTRACT

Investigation was conducted to know the effect of *Azotobacter* inoculation on germination and seedling development of radish by using germination paper test under controlled condition. Inoculation of *Azotobacter chroococcum* strain C_2 significantly increased the germination percentage by 9.33 percent, radical length by 90.47 per cent and plumule length by 54.37 percent over uninoculated control.

Further a pot culture experiment was carried out to know the effect of *Azotobacter* inoculation on growth and yield of radish along with different levels of nitrogen employing completely randomized factorial design under green house condition. Inoculation of radish seeds with *Azotobacter* showed increase in height of the plant, number of leaves, leaf area, root girth, root length, fresh weight and dry weights of root and leaf and root nitrogen contents over uninoculated control. Combination of 75 percent recommended dose of nitrogen per ha^{-1} and *Azotobacter* inoculation found to be advantageous by way of reducing the dependency on nitrogenous fertilizers and at the same time maintaining good yield prospects.

1. INTRODUCTION

Radish — *Raphanus sativa* L. an important vegetable crop cultivated all over India is known for its fleshy edible root, which is eaten raw with salt or as salad or cooked in vegetable curry. Radish posses many medicinal values like, it contains vitamin A and C, prevents constipation, increases appetite. It is considered to be good food for patients suffering from piles lever trouble enlarged splean and Jaundice.

The average yield of the Radish in India is low owing to poor fertility status, poor economic conditions of the farmers, which restrict the use of costly nitrogenous fertilizers. The biological nitrogen fixation mediated through *Azotobacter* in vegetables is already an established fact (4, 11). Besides nitrogen fixation the *Azotobacter chroococcum* a free living diazotraph, known to produce growth promoting substances like vitamins, auxins, gibberellins and cytokinins and antifungal substances (6), which favors its use as biofertilizer. Therefore inoculation of suitable *Azotobacter* strain is an important factor to minimize the expenditure on chemical fertilizers and to maximize the production, with this in view the present investigation was undertaken to study the effect of *Azotobacter* inoculation with different levels of nitrogen on growth and yield of Radish.

2. MATERIALS AND METHODS

Two *Azotobacter* strains namely *Azotobacter chroococcum* strain C_1 and strain C_2 isolated from rhizosphere soils of Radish were used to study the inoculation effect on germination and vigour of radish var. Japanese white under controlled condition at Department of Agricultural Microbiology, University of Agricultural Sciences, GKVP Bangalore.

Three replications of hundred seed lots were soaked for four hours in seven days old cultures separately and made up into paper dolls with moist, sterilized germination paper and placed in germinator at 25± 1°C and 85 percent relative humidity. Uninoculated control was provided by using same number of seeds which were soaked in Waksman No. 77 medium for four hours. (The Waksman No. 77 medium had the following composition : $CaCO_3$ - 5.0 g., mannitol - 10.0 g., K_2HPO_4 - 0.5 g., $Mgso_4$ - 0.2 g., Nacl - 0.2 g., $Fecl_3$ - Traces, $Mnso_4$ - Traces and distilled water one litre). Observations were recorded on 3rd day and vigour test was done based on length of the seedlings. (1).

Vigour index = shoot length + root length x germination percentage

Based on seedling vigour test the selected strain of *Azotobacter chroococcum* C_2 was multiplied in Waksman No. 77 liquid medium and stored in sterilized lignite for further studies.

The pot culture experiment conducted under green house condition at the above said Department during 1995 from June to August for a period of 45 days using red loamy soil collected from Agricultural farm at GKVK. The soil had pH of 6.55, electrical conductivity 1.70 m mhos per cm., organic carbon 0.04% and *Azotobacter* population of $15x10^2$ cells per gram dry soil.

The experiment was laid out in factorial randomised design with 4 levels of nitrogen i.e. 0, 50, 75 and 100% of recommended nitrogen ha^{-1} together with *Azotobacter* inoculation and without inoculation respectively making 8 treatments with 3 replications.

Fertilizer Nitrogen as per the treatment and full dose of P and K were applied at the time of sowing. The recommended dose of N P K is 75 : 37 : 37 ha^{-1}.

Radish seeds (C.V. Japanese white) inoculated by mixing the seeds with lignite based inoculum slurry, shade dried and sowed in the pots (22cm x 20cm making 5 kg of soil thoroughly mixed with sand and FYM at the rate of 2:1:1) and one seedling was maintained per pot. Plant growth parameters such as height of the plant (cm), number of leaves, leaf area (cm^2), root girth (cm), root length (cm), fresh weight (g) and dry weight (g) of root and percent nitrogen content of leaf and root was recorded at harvest. Total nitrogen content of leaf and root was estimated by microkjeldahl method (3). The amount of nitrogen (percent) in sample was calculated by using the formula :

$$\text{Percent N} = \frac{\text{Titre value x Normality of acid x 0.014}}{\text{weight of the sample. x 100}}$$

3. RESULTS AND DISCUSSION

Inoculation of *Azotobacter chroococcum* strain namely C_1 and C_2 significantly increased the germination percentage by 4.33 and 9.33 percent, radical length by 63.36 and 90.47 percent and plumule length by 38.08 and 54.37 percent respectively over uninoculated control in radish. Again *A. chroococcum* strain C_2 found to be more efficient than C_1 with respect to above parameters. (Table 1). Similar kinds of results were reported (8) in cereals and vegetables may be attributed to the production of physiologically active substances like IAA, IBA, Gibberellins, vitamins and aminoacids (6), which might have stimulated the germination and vigour indix of radish.

The individual treatment supplied with 100% nitrogen, 75% N and 50% N performed better than the control which received no nitrogen (Table No. 2 and 3). Application of 100% N showed maximum plant height (29.83 cm), No. of leaves (10.17), leaf area (802 cm^2), root length (19.17), root girth (4.03 cm), fresh weight (236.50 g), dry weight (9.90 g) of root and total nitrogen contents of leaf (1.93%) and root (1.95%) over other nitrogen levels. The significant differences in the above mentioned parameters on application of nitrogen from zero to 100 percent could be due to better photosynthesis and respiration by the plants (2) and better response of plants to nitrogen uptake during the early stages of growth (5, 9) as nitrogen is one of the important factors causing wider yield differences in crop plants, while the increase in nitrogen contents of leaf and root is attributed to increased concentrations of soil nitrogen through addition of nitrogen which might have contributed towards accumulation of more nitrogen in the plants.

Inoculation of *Azotobacter* strain alone showed no significant differences with respect to plant height (26.58 cm), no. of leaves (9.23) and root growth (3.32 cm) while showed leaf area (754.91 cm^2), significant differences with respect to root length (18.58 cm), fresh weight of root (215.75 g), dry weight of root (8.78g) and total per cent nitrogen content of leaf (1.65%) and root (1.67) over uninoculated control. Observations on Radish (7) similates the present observations regarding increase in leaf area, dry matter production and leaf and root nitrogen contents as a result of *Azotobacter* inoculation. Here as nitrogen is capable of improving radish growth so also is the case with *Azotobacter* inoculation may be attributed to better plant response to uptake of fixed atmospheric nitrogen, influence of growth promoting substances

Table 1. Effect of *Azotobacter chroococcum* cultures on germination of seeds and seedling vigour of Radish

Treatment	Germination (%)	Radicle length (cm)	Plumule length (cm)	Vigour index
Control	83.00	2.73	2.63	444.88
Azotobacter chroococcum (C_1)	87.33 (4.33)	4.46 (63.36)	3.63 (38.08)	706.49
Azotobacter chroococcum (C_2)	92.33 (9.33)	5.20 (90.47)	4.06 (54.37)	854.97
F test	*	*	*	
C.D. (P = 0.05)	4.26	0.57	0.67	

Note : Figures in parantheses indicates per cent increase over control.
Vigour index on seedling length basis.

Table 2. Effect of levels of Nitrogen and *Azotobacter* inoculation on Radish plant height, number of leaves, leaf area, root girth and root length at harvest (45 days)

	Treatments	Plant height (cm)	No. of leaves (cm)	Leaf area (cm2)	Root girth (cm)	Root length (cm)
I. INTERACTION EFFECT						
1	Uninoculated Control	21.33	7.66	518.33	2.57	15.67
2	*Azotobacter* inoculation	22.00	8.33	706.00	2.63	16.66
3	Recommended Nitrogen at 50%	24.00	8.33	696.66	2.90	17.67
4	Recommended Nitrogen at 50% + *Azotobacter* inoculation	25.33	9.33	698.66	2.96	18.33
5	Recommended Nitrogen at 75%	27.00	9.66	707.67	3.40	19.00
6	Recommended Nitrogen at 75% + *Azotobacter* inoculation	28.66	9.67	777.00	3.63	18.33
7	Recommended Nitrogen at 100%	29.33	10.00	766.00	4.0	19.33
8	Recommended Nitrogen at 100% + *Azotobacter* inoculation	30.33	10.00	838.00	4.06	20.33
	F. Test	NS	NS	NS	NS	NS
	SEM ±	0.56	0.31	31.87	0.09	0.55
	CD (P = 0.05)	–	–	–	–	–
II. INDIVIDUAL EFFECT						
(a) Nitrogen levells						
	0%	21.66	8.00	612	2.60	16.16
	50%	24.67	8.83	697.66	2.93	18.00
	75%	27.83	9.66	742.33	3.52	18.66
	100%	29.83	10.17	802.00	4.03	19.83
	F. Test	*	*	*	*	*
	SEM ±	0.40	0.22	22.53	0.07	0.39
	CD (P = 0.05)	1.19	0.66	67.56	0.21	1.17
(b) *Azotobacter* inoculation						
	Uninoculated	25.41	9.00	572.16	3.21	17.75
	Inoculated	26.58	9.33	754.91	3.32	18.58
	F. Test	NS	NS	*	NS	*
	SEM ±	0.28	0.15	15.93	0.05	0.27
	CD (P = 0.05)	–	–	47.77	–	0.82

* Significant
NS : Non significant
All values are means of 3 replications
Recommended fertilizer dosage : 75:37:37 kg NPK ha^{-1}

Table 3. Effect of levels of Nitrogen and *Azotobacter* inoculation on fresh weight and dry weight, total percent nitrogen content of leaf and root of Radish at harvest (45 days)

	Treatments	Leaf Nitrogen (%)	Root Nitrogen (%)	Fresh weight (g)	Dry weight (g)
I. INTERACTION EFFECT					
1	Uninoculated Control	1.14	1.16	143.30	6.04
2	*Azotobacter* inoculation	1.28	1.31	191.67	7.36
3	Recommended Nitrogen at 50%	1.43	1.44	198.33	8.14
4	Recommended Nitrogen at 50% + *Azotobacter* inoculation	1.52	1.58	206.66	8.26
5	Recommended Nitrogen at 75%	1.50	1.61	216.67	8.36
6	Recommended Nitrogen at 75% + *Azotobacter* inoculation	1.81	1.83	26.67	9.33
7	Recommended Nitrogen at 100%	1.89	1.91	235.00	9.63
8	Recommended Nitrogen at 100% + *Azotobacter* inoculation	1.97	1.99	238.00	10.19
	F. Test	NS	NS	*	*
	SEM ±	0.03	0.02	7.03	0.15
	CD (P = 0.05)	—	—	21.09	0.45
II. INDIVIDUAL EFFECT					
(a) Nitrogen levells					
	0%	1.22	1.23	167.50	6.69
	50%	1.48	1.51	202.50	8.20
	75%	1.71	1.72	221.6	8.84
	100%	1.93	1.95	236.50	9.90
	F. Test	*	*	*	*
	SEM ±	0.02	0.02	4.97	0.10
	CD (P = 0.05)	0.07	0.06	10.54	0.31
(b) *Azotobacter* inoculation					
	Uninoculated	1.52	1.53	198.33	8.04
	Inoculated	1.65	1.67	215.75	8.78
	F. Test	*	*	*	*
	SEM ±	0.01	0.01	3.51	0.07
	CD (P = 0.05)	0.050	0.042	21.09	0.02

* Significant
NS : Non significant
All values are means of 3 replications
Recommended fertilizer dosage : 75:37:37 kg NPK ha^{-1}

and antifugal substances produced by the bacterium, which might have enhanced the plant growth and inhibited the disease occurrence.

Inoculation of *Azotobacter* along with 100% recommended dose of nitrogen gave superior yield prospects (10) than the *Azotobacter* inoculation with 75% N and 50% N levels, where the interaction effect of *Azotobacter* and 100% recommended dose of nitrogen showed maximum plant height (30.33 cm), number of leaves (10.00), leaf area (838.00 cm_2), root length (20.33 cm), root girth (4.0 cm), fresh weight of root (238.0 g), dry weight of root (10.19 g) and total nitrogen contents of leaf (1.97%) and root (1.99%).

It is evident from the present study that

use of *Azotobacter* as a biofertilizer along with nitrogen may solve the problem of dependency on costlier inorganic nitrogenous fertilizers to some extent which are short in supply, besides maintaining the soil health and productivity.

REFERENCES

1. Abdul Baki A A & Anderson J D. Crop Sci 13 1973 630.
2. Hodges T & Kenemasu E Y. Agron. J 69 1977 674.
3. Jackson M L. Soil chemical analysis, printice Hall of India private Ltd. 1973 New Delhi.
4. Mehrotra C L & Lehri L K. J. Indian Soc Soil Sci 19 (3) 1971 243-248.
5. Mehta C P, Lal S & Singh H. Haryana J Hort Sci 16 (3-4) 1987 287-293.
6. Mishustin E N & Shilnikova V K. *In soil biology Reviews of research* UNESCO 1969 65.
7. Neito K F & Frankenberger Jr. W T. Pl soil 127 (2) 1990 147.
8. Rajkumar K & Lakshmanan M, Indian J. Microbial, 30 (2) 1990 221-224.
9. Rastogi K B, Sharma P P & Korla B N, veg. Sci, 14 (2) 1987 105.
10. Sharma M L, Namdeo R N & Mishra U R, Indian J Agron, 33 (3) 1987 204-207.
11. Subba Rao N S, Soil microorganisms an plant growth. Mohan primlani, oxford and IBH publishing Co. 1978 p. 254-255.

CHAPTER – 25

APPLICABILITY OF FLY - ASH, PHOSPHATE SOLUBILIZING MICROBES AND MYCORRHIZAE FOR SUSTAINABLE PRODUCTIVITY OF BANANA

NV Phirke, SB Chincholkar, KR Yadav and RM Kothari*

School of Life Sciences, North Maharashtra University, Jalgaon- 425 001
*Biotechnology Division, Thapar Corporate R & D Center, Patiala- 147 001

ABSTRACT

Research in the last two decades indicated a possibility of using fly-ash to amend agricultural soil for ramified root system, as also a source of micro-nutrients as well as some macro-nutrients such as Ca and phosphate. To take advantage of these inherent attributes of fly-ash, it has been utilised for the higher productivity of banana, at lesser cost to be more remunerative to the farmers. To optimise the use of phosphate from fly-ash and reduce dependence on imported phosphatic fertilisers, the use of phosphate solubilizing microbes is developed. For this purpose, attempts have been made to screen the most efficient phosphate solubilizing microbes from the elite banana rhizosphere. Their nutritional requirements are optimised on two litre scale fermentation for farm scale trials. Besides optimising sustained availability of solubilized phosphate, the application of mycorrhizae has been explored to minimise the chances of entrance of toxic elements from fly-ash (Cd, Hg, As, Pb and Cr) in the food chain. Thus, a natural waste has been explored to serve as a resource of nutrients through the catalytic roles of phosphate solubilizers and mycorrhizae in eco-friendly and sustainable manner.

1. INTRODUCTION

India's total production of banana is 13.2 million tons, of which Jalgaon district alone has been contributing 2.08 million tons (about 15.8%). Through personal experience, progressive farmers have cultivated high yielding varieties of banana such as *Basrai, Shrimanti, Laxmi, Hari saal*, varieties fetched from Karnataka and Tamilnadu and recently tissue culture-derived Grand Nain, Williams variety from Israel. In spite of application of recommended doses of NPK fertilisers/urea, abundant irrigation (20-25 times) from sowing to harvesting, these varieties have not significantly enhanced banana productivity. In fact, a stage has reached where high cost of tissue culture derived plantlets, drip irrigation network, increased cost of NPK fertilisers, increased cost of labour and soaring electricity bill have led the farmers to perceive banana cultivation less remunerative compared to other cash crops. However, any reduction in acreage under banana cultivation has the danger of reduced availability of banana to common man and higher cost to economically sound class. In view of these facts, it is considered highly desirable to reduce the cost of cultivation without compromising the quality or yield of banana so as to restore acceptable profit margin to the farmers. This article outlines a strategy to achieve these objectives.

2. MATERIALS AND METHODS

(a) Banana suckers

Dwarf variety of *Shrimanti* was procured locally, assuming that less biomass due to dwarfness will afford availability of more nutrients, water and surplus energy for higher yield of banana.

(b) Fertilisers

It has been decided to reduce the input of NPK fertiliser by 25-50% by substituting with freely available fly ash as a source of Ca, PO_4, and micronutrients. The fly ash was procured from nearby Bhusawal Thermal Power Station, Deep Nagar at mere transportation cost.

(c) Elemental analysis

While nitrogen was analysed by Kjeldahl method (1), total and available phosphorus as per John (2) and Olsen's method respectively and potassium by flame photometry. Ingredients of fly ash were estimated by atomic absorption spectrophotometry.

(d) Microbial inputs

To provide sustained release of Ca and PO_4, application of the most efficient phosphate solubilizing fungus, tentatively identified as *Aspergillus* sp. has been used. It was economically cultivated at 28°C for 48 hours for optimal biomass as per Ramamurthy *et al.* (3)

(e) To arrest the entry of toxic ingredients of fly ash (As, Hg, Pb, Cd, Cr etc.) in banana pseudostem, leaves and fruits, multipurpose ectomycorrhiza *Tricholoma imbricatum* has been used. It was grown at 25°C for 96 hours as above to get optimal biomass.

(f) Soil conditioner

It was prepared from wild grass by solid state fermentation using mixed consortium of microbes at ambient temperature, 55 % moisture in about 8-10 weeks essentially as per Ramamurthy *et al.* (4). It was used to serve as a source of organic carbon for the growth and sustenance of *Aspergillus* sp. and *Tricholoma imbricatum*.

(g) Experimental design

To establish a cause and effect relationship in a statistically meaningful manner, a randomised block experimental design by computerised programme in triplicate is summarised in Table 1. It was implemented on 1.3 acre farm having loamy sandy type of soil.

(h) Experimental set-up

In each block, there were 4 rows, each with 5 plants, thus totally 20 plants per block, with the plant to plant and row to row distance of 150 cm.

(i) Plant care

Rhizome transplantation and post-plantation care was taken as per Patil (5). This involved drip irrigation (M/s. Jain Irrigation Systems, Jalgaon), and application of recommended dose of fertiliser (6).

(j) Parameters to be monitored

Percentage survival of banana plants, height, girth and chlorophyll contents were estimated at one month's interval. The utilisation of fly ash was monitored by estimating Ca and PO_4. The entry, if any, of toxic ingredients of fly ash was monitored by atomic absorption spectrophotometer.

3. RESULTS AND DISCUSSION

An attempt has been made to devise an integrated strategy for the higher yield of banana by adding biotech inputs so that not only the technology be cost-effective, but it be eco-friendly also for sustainability. For accomplishing this, a stepwise systematic effort has been made to optimise each input, by varying only one input at a time and keeping other inputs constant so that a cause and effect relationship could be established, as also an optimised protocol of banana production evolved. We have succeeded in this strategy earlier (7).

The dwarf Shrimanti variety of banana was accorded prime importance, almost on the same line as hybrid jowar (dwarf) was a success to replace traditionally tall variety of jowar. The rationale behind sowing the dwarf variety was to use minimal necessary nutrients towards biomass (pseudostem and leaves) development and divert the major quantum of nutrients for optimal production of banana. Thus, nutrient quality/size of banana was not compromised.

Table 1. Randomised block design for banana plantation

Sr. No.	Treatments	Coding	Description
1	T_1	$S_oF_oM_o$	No inputs
2	T_2	$S_oF_oM_1$	Only microbes
3	T_3	$S_oF_1M_o$	Only fly ash 2.5 t/ha
4	T_4	$S_oF_1M_1$	Fly ash 2.5 t/ha + Microbes
5	T_5	$S_oF_2M_o$	Only fly ash 5 t/ha
6	T_6	$S_oF_2M_1$	Fly ash 5 t/ha + Microbes
7	T_7	$S_1F_oM_o$	Only soil conditioner 5 t/ha
8	T_8	$S_1F_oM_1$	Soil conditioner 5 t/ha + Microbes
9	T_9	$S_1F_1M_o$	Soil conditioner 5 t/ha + Fly ash 2.5 t/ha
10	T_{10}	$S_1F_1M_1$	Soil conditioner 5 t/ha + Fly ash 2.5 t/ha + Microbes
11	T_{11}	$S_1F_2M_o$	Soil conditioner 5 t/ha + Fly ash 5 t/ha
12	T_{12}	$S_1F_2M_1$	Soil conditioner 5 t/ha + Fly ash 5 t/ha + Microbes
13	T_{13}	$S_2F_oM_o$	Only soil conditioner 10 t/ha
14	T_{14}	$S_2F_oM_1$	Soil conditioner 10 t/ha + Microbes
15	T_{15}	$S_2F_1M_o$	Soil conditioner 10 t/ha + Fly ash 2.5 t/ha
16	T_{16}	$S_2F_1M_1$	Soil conditioner 10 t/ha + Fly ash 2.5 t/ha + Microbes
17	T_{17}	$S_2F_2M_o$	Soil conditioner 10 t/ha + Fly ash 5 t/ha
18	T_{18}	$S_2F_2M_1$	Soil conditioner 10 t/ha + Fly ash 5 t/ha + Microbes

To ensure the sustained release of organic and chemical nutrients for healthy growth of banana plant and fruit, the strategy comprised of application of soil conditioner, varying doses of chemical fertilisers, supplementary doses of fly ash and biofertilizer mix, should any shortfall in nutrients arise at any stage of experimental trials (Table 1). It was anticipated that soil conditioner will not only help in absorbing transplantation shock to the banana suckers, but will also help in sustained release of organic carbon, chemical nutrients, moisture and serve as a matrix for the sustained growth of beneficial microbes (4). The addition of varying amounts of fly ash was purposeful to determine the feasibility of reduced requirement of synthetic fertilisers which are cost-prohibitive, as also capital-, energy-, and pollution-intensive. Their partial replacement by fly ash available at merely transportation cost was an effort to reduce input cost to the farmers, but also to devise a meaningful application of a thermal power station waste which has the potential to provide requisite amount of Ca and PO_4 for banana growth. Its application was thus a step in the direction of good agro-practice (GAP), to eliminate waste at the point of generation. Incidentally, fly ash supplementation has imparted porosity and water holding capacity to the soil, besides micronutrients, thereby enabling banana plant to procure adequate nutrition from soil with minimal energy input.

The fear of toxic ingredients of fly ash entering in the food chain has always prohibited its agricultural application. However, our experience on large scale application of fly ash in *Eucalyptus* plantation has conclusively shown that such fears are misplaced (8). In the first place, most of the ingredients are insoluble by virtue of being oxides. Secondly, some of these elements have existed naturally in the soils without causing any harmful effects on the human system by consumption of produce raised in such soils. And yet, as a matter of abundant precaution, application of mycorrhizae was experimented. Literature is replete indicating that mycorrhizae have the capacity to imbibe/encapsulate toxic elements, thereby preventing their entry in the food chain. Another reason for the application of mycorrhizae was to create an in-built mechanism by virtue of its multifaceted characteristics, to ward of any contingency during the 15 months long harvest period of banana. Thus, the above strategy provided requisite nutrition at minimal cost, pollution and long term effects on ecosystem.

Due to indiscriminate use of NPK fertilisers in the last two decades in a greed to produce more banana, the fertility of soil was stagnated and infact large quantum of phosphate was immobilised in the soil as revealed by analysis. This provided optimism to reduce NPK inputs and instead utilise inherently available PO_4 by solubilizing it with the help of efficient *Aspergillus* sp. Thus, at an incremental cost, not only the cost of nutrients was reduced, but the soil fertility is hoped to be improved by reducing the imbalance in its nutrients.

Out of ignorance, the farmers always resorted to flood irrigation, with least concern for (i) consumption of electricity for pumping of water, (ii) gradually reducing water tables, (iii) increasing salinity in the soil and (iv) stagnated fertility and yet enhanced banana yield was not realised. In our opinion, flood irrigation placed root system under stress by denying it constant access to air for respiration and energy (9). To alleviate an inadvertent mistake, we have opted for drip irrigation for sustained availability of moisture and economy in wastage of water and electricity, without compromising long term effects on soil fertility.

The experimental design is in its place; it would mean additional twelve months to show tangible results. However, the preliminary observations are indicating right direction of strategy. The objectives of cost-reduction, eco-friendly approach, sustainability and cost-reduction seem to have been achieved. Whether these are complemented with the increased yield of banana remains to be seen.

REFERENCES

1. Jayaraman J, In : Laboratory Manual in Biochemistry, Wiley Eastern Ltd; New Delhi (1981) 74-75.
2. John MK, Soil Science 109 (1970) 214-220.
3. Ramamurthy V, Upadhyay CM, Nehete PN, Shah NK, Shah DN, Sharma RK, Thakar SP, Shankar V & Kothari RM, Hindustan Antibiot. Bull. 35 (1993) 43-76.
4. Ramamurthy V, Sharma RK, Yadav KR, Kaur J, Dev Vrat & Kothari RM, Biodegradation, 7 (1995) 121-127.
5. Patil AV, In : Kelibagh, Continental Publications, Pune-411 030 (1998).
6. Patil B & Kale P, In : Alpamoli bahugunee - Keli, Extension press, MPKV, Rahuri (1989).
7. Shah DN, Shah VD, Nehete PN & Kothari RM, Biotechnol Letts, 8 (1980) 103-106.
8. Kothari RM, Sharma RK, Goyal D & Ramamurthy V, DST sponsored fly ash mission programme on Eucalyptus plantation (1997).
9. Sharma RK, Ramamurthy V & Kothari RM, In : Bioresource Technology for Sustainable Agriculture (Ed Kanniyan, S) Associated Publishing Co., New Delhi (1999) 82-108.

CHAPTER – 26

ACUTE TOXICITY OF SELECTED INSECTICIDES TO THE INFECTIVE STAGE OF *Steinernema* SP. NEWFOUNDLAND STRAIN NO. 1 (L1C)

Worthman, MG[1] and Finney-Crawley, J.R.[2]

[1]*Environment Canada, P.O. Box 5037, St. John's Newfoundland, A1C 5V3, Canada.*
[2]*Department of Biology, Memorial University of Newfoundland, St John's Newfoundland, A1B 3X9, Canada.*

ABSTRACT

Infective juveniles of a locally isolated steinernematic nematode, *Steinernema* sp. Newfoundland strain No. 1 (L1C) were killed following 24h of exposure to 1.0 mg/ml chlordance and methoxychlor. Nematodes exposed for 168 hr to ≤ 0.1 mg/ml of these 2 chemicals showed no toxic effect. Diazinon at 1.0 mg/ml showed toxicity to the nematode over the time course of the experiment. No toxic effects were observed after exposure to ≤0.1 mg/ml diazinon. Carbaryl exerted a toxic effect at concentrations of 1.0 mg/ml and 0.1 mg/ml. Dimilin, Matacil and Malathion were not toxic at ≤1.0 mg/ml. Efficacy of surviving infective juveniles to the Greater wax moth, *Galleria mellonella* was unaffected. Nematode migration in soil was not impaired by the organochlorine insecticide chlordane. Diazinon reduced migration of nematodes at ≥ 0.1 mg/ml. There appear to be variations in sensitivity of steinernematid species and strains to different insecticides.

1. INTRODUCTION

Unimodal pest control strategies have been successful in reducing pest populations below economic injury thresholds. Insecticide use generally been successful but has come under scrutiny from elevated public awareness and concern over adverse effects on the ecosystem, as well as target organism resistance and increasingly higher costs of manufacturing selective insecticides. The integration of entomogenous nematodes with an insecticide program may make it possible to lower the volume of insecticide used. This reduces the levels of insecticide in the environment and the deleterious effects on the ecosystem, while not compromising agricultural and forest yields. This study examined the compatibility of several insecticides with a locally isolated 'cold tolerant' steinernematid with potential for use against forest and agricultural insect pests in Canada (1,2).

2. MATERIALS AND METHODS

Steinernema sp. Newfoundland strain No. 1 (L1C) (3-5) was propagated by passage through *Galleria mellonella* at 15°C (6). Infective nematodes were stored in water contained in tissue culture flasks at 4°C. All juveniles were less than 1 month old and still retained the second larval stage cuticle when utilized in experiments. Seven insecticides were chosen to encompass the major categories of pesticides. These included 2 organochlorine insecticides : Chlordane and Methoxychlor, 2 organophosphates : Malathion and Diazinon, 2 carbamates : Carbaryl (Sevin[R]). The Aminocarb (Matacil[R]) and an insect growth regulator : Diflubenzuron (Dimilin[R]). The organochlorine and organophosphate insecticides as well as carbaryl were commercial formulations from Chipman Inc. Limited. Carbaryl was a water soluble miscible liquid while the other formulations were emulsifiable concentrates. Dimilin and Matacil (1.8D OSC (T2)) were technical grade material. Dimilin was a powder with 99% active ingredient. Matacil was formulated in oil (Ultramar stove oil) with 19.9 ± 0.5% active ingredient. The commercial formulations were used as stock solutions with the manufacturers guarantee used for insecticide concentration content. The technical grade Matacil was used as a stock solution. A stock solution of diflubenzuron was obtained by dilution with reagent grade dimethylsufoxide (DMSO) from BDH chemicals. Insecticide stock solutions had the following concentrations : chlordane, 400 mg/ml; methoxychlor, 288 mg/ml, malathion, 500 mg/ml; diazinon, 62.5 mg/ml; carbaryl, 150 mg/ml, Matacil 175.8 mg/ml; Dimilin, 100 mg/ml.

(a) Behaviour effects due to Direct Exposure to Insecticides

Glass petri dishes (60 x 15 mm) were filled with 9.0 ml of appropriately concentrated insecticide. A stock suspension of infective juvenile nematodes was made to a final concentration of 1000 per ml of water. Final insecticide concentrations were 1.0, 0.1, 0.01 and 0.001 mg/ml following the addition of 1.0 ml of stock nematode suspension. Untreated and solvent-treated nematode suspensions (1% v/v) served as controls. Behaviour of juvenile nematodes (212), was assessed after exposure to insecticides for 24, 72, 120 and 168 h. To assess nematode response, each petri dish was swirled several times to make an even suspension of nematodes. A pipette was used to remove 0.5 ml of nematode suspension to a clean petri dish and the response of 25 nematodes was assessed by mechanical stimulation with a probe. Tests were conducted at 15°C and each insecticide concentration was replicated 3 times with 25 nematodes per replicate. Untreated controls were also examined.

(b) Infectivity effects due to direct exposure to Insecticides

Insecticide stock solutions were diluted in 1.5 ml distilled water such that the addition of 1.0 ml of stock infective juvenile solution (1000 ml) yielded finall insecticide concentrations of 1.0, 0.1, 0.01, and 0.001 mg/ml. Both insecticide and nematode suspensions were added together in a glass test tube (16 x 125 mm) and incubated at 15°C for the incubation period. At 24, 72, 120 and 168 h, 0.62 ml of suspension, containing 250 nematodes, were pipetted onto filter paper at the bottom of a petri dish (60 x 15 cm). Five *G. mellonella* larvae were placed on the filter paper and exposed to insecticide-treated nematodes for 48 h at 15°C. After incubation, the larvae were washed with distilled water and placed on fresh, moistened filter paper. Five to 6 days after initial exposure, alive and dead larvae were enumerated. Dead larvae were dissected and observed for presence of nematodes. Infectivity tests were conducted with all insecticides based on behaviour data. Larvae exposed to diazinon, chlordane and methoxychlor were dissected and adult nematode(s) counted. The mean number of adult nematodes per infected *G. mellonella* larva was referred to as the Infection Index (6). Each insecticide treatment was replicated 3 times at each concentration and time exposure, with 5 insects per replicate. Baseline Infection Index was determined with untreated nematodes. Infection Index data were arcsine transformed and subjected to analysis of variance and Student-Newman -Keuls (SNK) test procedure. Nematode scores after dissection were subjected to analysis of variance and SNK test procedure (7)

(c) Migration of Nematodes in Soil

Five 2 cm (deep) by 5 cm (diameter) plastic rings were fastened together with adhesive tape to form a 10 cm vertical cylinder. Ten late instar *G. mellonella* larvae were secured at the base of the cylinder with several layers of cheesecloth and elastic bands. Pro-Mix^R, presterilized soil was sifted through a Canadian standard sieve to give soil with particle size <85 μm. Soil was dried overnight at 50°C, then mixed with distilled water in a 3:1 ratio until homogeneous. Cylinders were filled with 90-100 g of soil mixture and the soil was allowed to settle by gentle tapping of the cylinder against the work bench. Cylinders were covered with aluminum foil to prevent evaporation and held verticdally at 15°C overnight. Columns with *G. mellonella* at the base served as control.

(d) Column Inoculation

Each insecticide was mixed with distilled water before addition to dry soil. Chlordane and diazinon were tested at 1.0, 0.1, 0.01 and 0.001 mg/ml distilled water. The experiments were replicated 5 times, columns without insects served as controls. Stock solutions of L1C were adjusted to yield a final concentration of nematodes of 3000 per 0.5 - 1.0 ml. Aluminum foil was removed from the top of the cylinder, a small amount of soil cleared to the perimeter, and 3000 nematodes inoculated into the soil. The soil was gently pushed back into place and the foil cover replaced. Cylinders were incubated vertically at 15°C for 48h.

(e) Enumeration of Migrated Nematodes

Adhesive tape was carefully removed from the top cylinder ring and a sheet of paper inserted between layers. The initial layer of soil (0-2 cm) was removed and placed on cheesecloth which was then fastened with an elastic band. The soil was placed in a Baermann funnel and left overnight at 17°C (8, 9). A sufficient volume of distilled water was added to cover the soil. The remaining 4 layers of 2 centrimetres each were recovered in a similar manner. Approximately 20-25 ml of water was collected from each funnel and all nematodes counted. Additionally, *G. mellonella* recovered from each cylinder were placed in a petri dish on moistened filter paper. After 5 to 7 days, these larvae were dissected and adult nematodes counted. Experiments were replicated 10 times. Data were square-root transformed by analysis of variance and SNK test procedure (7).

3. RESULTS AND DISCUSSION

The condition of nematodes after

exposure to insecticides was categorically divided into 3 groups; alive, affected and dead. Affected or inactive nematodes typically had a j-shaped posture and responded weakly to mechanical stimulation with a blunt, metal probe. However, for LIC juvenile nematodes, this state occurred infrequently over all time periods and insecticide concentrations. Dead nematodes had a straight posture and did not respond to mechanical stimulation. The organochlorine insecticides, chlordane and methoxychlor were toxic to juvenile L1C nematodes. Mortality at 24h was above 50% for both chemicals at 1.0 mg/ml and rose above 75% at 168h

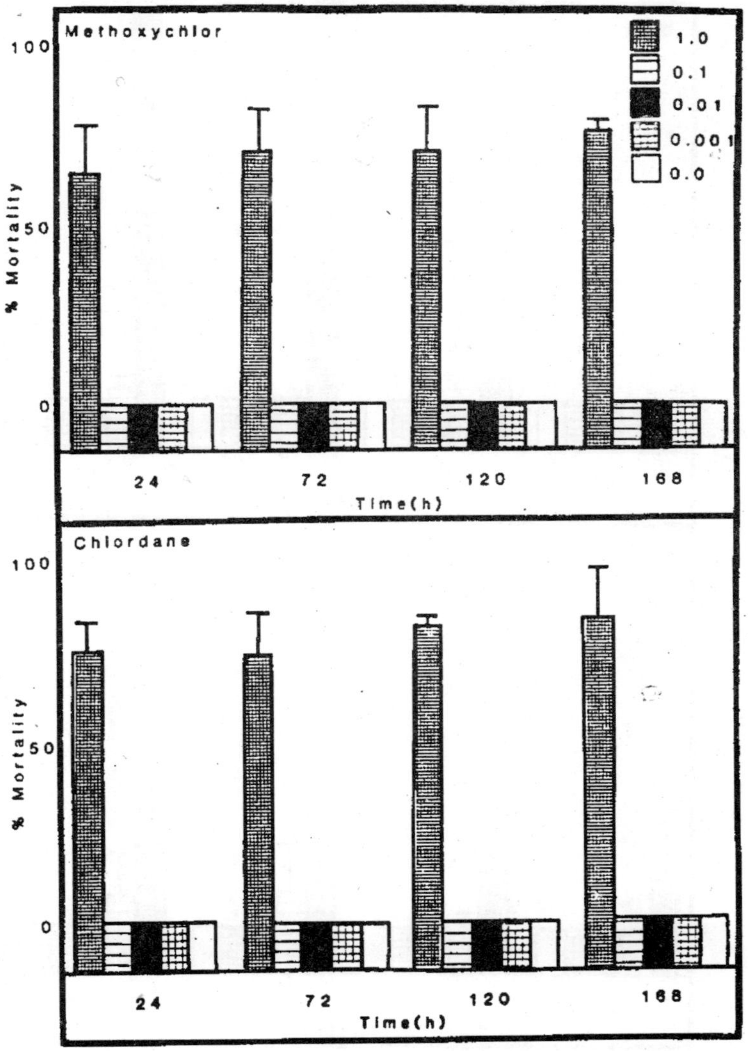

Figure 1. Percent mortality of *Steinernema* sp. Newfoundland strain No. 1 (L1C) juvenile nematodes after exposure to the organochlorine insecticides, chlordane and methoxychlor. Insecticide concentrations were as indicated, mg/ml. Sample statistics are mean ± standard deviation. Three replicates with 25 nematodes per replicate.

for both chlordane (85.3%) and methoxychlor (77.3%) but there was no toxic effect at concentrations ≤ 0.1 mg/ml (Fig. 1). The organophosphate insecticide, diazinon, showed toxicity after 72, 120 and 168 h, there was no mortality at 24 h but mortality increased to 45.3% by 168 h (Fig. 2). Carbaryl, treatments at 1.0 mg/ml resulted in 13.3% mortality at 24h and 22.7% by 168h. As well, there was some mortality (8.0%) for 0.1 mg/ml at 120 and 168h (Fig. 2). Although mortality was low, it was significantly higher than the lower carbaryl concentrations and controls where no toxicity was observed. Dimilin, matacil and malathion were not toxic to L1C

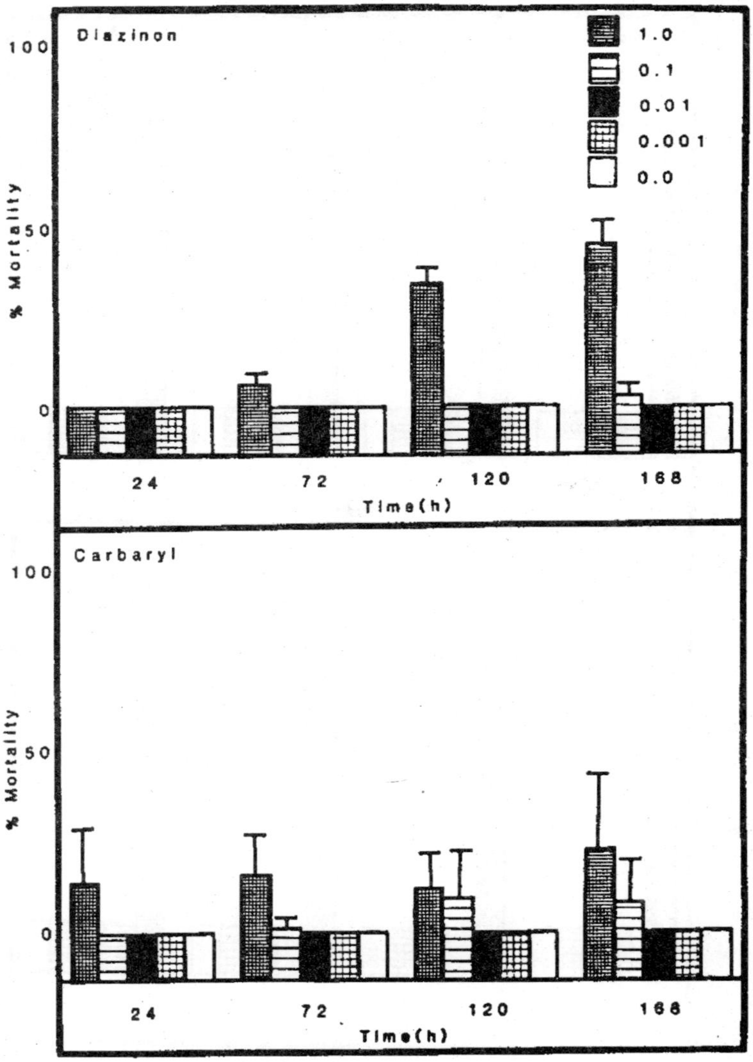

Figure 2. Percent mortality of *Steinernema* sp. Newfoundland strain No. 1 (L1C) after exposure to the organophosphate diazinon and carbamate insecticide, carbaryl. Insecticide concentrations were as indicated, mg/ml. Sample statistics are mean ± standard deviation. Three replicates with 25 nematodes per replicate.

juvenile nematodes at ≤ 0.1 mg/ml. Thus, the toxic effects of certain insecticides were usually at concentrations of 1.0 mg/ml for nematodes.

(a) Infectivity

Infection of *G. mellonella* larvae by L1C juveniles was significantly lower ($P<0.05$) for nematodes treated with chlordane and methoxychlor. Comparison of Fig. 2 with Table 1 illustrates the relationship between mortality and infection success. Diazinon and carbaryl increased mortality at 1.0 mg/ml (Fig. 2), but did not significantly affect the capacity of juveniles to infect *G. mellonella* (Table 1). Chlordane showed no sublethal effects to L1C nematodes at concentrations ≤ 0.1 mg/ml. Likewise, methoxychlor did not show significant

Table 1. Recovery of *Steinernema* sp. Newfoundland strain No. 1 (L1C) after exposure to insecticides and subsequent infection in *G. mellonella.**

% *G. mellonella* with nematodes after exposure (h) to insecticides**

Concentration (mg/ml)	*Organochlorines*							
	Chlordane				Methoxychlor			
	24	72	120	168	24	72	120	168
0.0	100	100	100	100	100	100	100	100
0.001	100	100	100	100	100	100	100	100
0.01	100	100	100	100	100	100	100	100
0.1	100	100	100	100	100	100	100	100
1.0	8* (11.5)	0*	13* (23.1)	0*	0* (11.5)	0*	0*	0*

	Organophosphates			
	Malathion	Diazinon		
	24-168h	24h	72-120h	168h
0.0	100	100	100	100
0.001	100	100	100	100
0.01	100	100	100	100
0.1	100	100	100	100
1.0	100	93 (11.5)	100	100 (34.6)

	Carbamates	*Insect Growth Regulator*	
	Carbaryl 24-168 h	Matacil 24-168 h	Dimilin 24-168 h
0.0	100	100	100
0.001	100	100	100
0.01	100	100	100
0.1	100	100	100
1.0	100	100	100
0.0 (S.C.)***			100

* Combined data of 3 replicates with 5 larvae per replicate. Standard deviations are bracketed. Standard deviations were ± 0.0 unless indicated

** Significantly different ($P<0.05$) from control by SNK test

*** Solvent - treated Control (1% v/v).

effects on Infection Index at concentrations 0.1 mg/ml (Table 2). Both these chemicals were toxic to L1C infective juveniles at 1.0 mg/ml (Fig. 1) and as a result, the Infection Index was significantly lower because most nematodes were killed by the chemical.

(b) Soil Migration

The presence of 10 *G. mellonella* larvae at the base of a soil column significantly (P<0.05) affected the degree of migration of L1C infective juveniles. The percentage of inoculated nematodes reaching the 8-10 cm depth in the cylinder was significantly different (P<0.05) in cylinders containing *G. mellonella* (2.7%) from those in cylinders without *G. mellonella* larvae (0.1%) (Table 3). Although the magnitude of migration in the middle layers (2-8 cm) of the cylinders was not significantly different, there was a trend towards more downward migration of L1C in cylinders with *G. mellonella* larvae. Recovery rates of nematodes between cylinders with and without *G. mellonella* were not significantly different (Table 4). During this series of experiments, nematodes were recovered from soil by the Baermann funnel method. Recovery in this study was usually 30-40% of the original inoculum and therefore, values were converted to a proportion of the original inoculum. It was assumed that recovery efficiency throughout individual layers of soil from a single column was equal. Chlordane had no significant (P<0.05)

Table 2. Recovery of *Steinernema* sp. Newfoundland strain No. 1 (L1C) from *G. mellonella* by dissection after exposure to selected insecticides. Sample statistics are mean ± SD.*

Insecticide Conc. (mg/ml)	Mean No. Nematodes per larva**			
	24h	72h	102h	168h
Methoxychlor				
1.0	1.0 (1) aA	0.0 (0) aA	0.0 (0) aA	0.0 (0) aA
0.1	6.9 ± 1.9 (15) aB	7.2 ± 2.2 (15) aB	7.4 ± 2.3 (15) aA	8.1 ± 3.1 (15) aB
0.01	8.5 ± 3.2 (15) aB	7.7 ± 2.2 (15) abB	6.1 ± 1.6 (15) bB	8.7 ± 3.3 (15) aB
0.001	6.1 ± 2.2 (15) aB	6.0 ± 2.0 (15) aB	6.4 ± 3.2 (15) aB	9.3 ± 3.2 (15) bB
0.0	6.9 ± 3.2 (15) aB	7.3 ± 2.1 (15) aB	6.9 ± 2.9 (15) aB	8.6 ± 2.3 (15) aB
Diazinon				
1.0	4.3 ± 2.2 (14) aA	7.1 ± 2.6 (15) bA	6.3 ± 2.7 (15) bA	3.0 ± 2.1 (12) aA
0.1	5.2 ± 1.4 (15) aAB	7.3 ± 2.5 (15) bA	7.3 ± 2.1 (15) bA	6.6 ± 1.2 (15) bB
0.01	4.9 ± 2.1 (15) aAB	7.3 ± 1.9 (15) bA	6.5 ± 3.0 (15) abA	8.2 ± 3.1 (15) bBC
0.001	5.9 ± 1.7 (15) aAB	8.1 ± 1.8 (15) bA	6.2 ± 2.5 (15) aA	6.3 ± 2.0 (15) aB
0.0	6.9 ± 3.2 (15) aB	7.3 ± 2.1 (15) aA	6.9 ± 2.9 (15) aA	8.6 ± 2.3 (15) aC
Chlordane				
1.0	1.0 (1) aA	0.0 (0) aA	1.0 ± 0.0 (2) aA	0.0 (0) aA
0.1	8.0 ± 2.1 (15) aB	7.1 ± 2.5 (15) aB	7.4 ± 1.3 (15) aB	8.9 ± 2.7 (15) aB
0.01	7.3 ± 2.6 (15) aB	7.4 ± 1.4 (15) aB	6.1 ± 1.7 (15) aB	7.9 ± 3.4 (15) aB
0.001	7.3 ± 2.7 (15) aB	7.1 ± 2.7 (15) aB	6.3 ± 2.3 (15) aB	8.5 ± 3.2 (15) aB
0.0	6.9 ± 3.2 (15) aB	7.3 ± 2.1 (15) aB	6.9 ± 2.9 (15) aB	8.6 ± 2.3 (15) aB

* Sample size is bracketed.

** Means followed by different lowercase letters in a row and by different uppercase letters in a column are significantly different (P < 0.05) by SNK test.

Table 3. Vertical distribution of *Steinernema* sp. Newfoundland strain No. 1 (L1C) infective juveniles in the presence and absence of 10 *G. mellonella* larvae located at the base of a 10 cm soil column 40h after placement of 3000 nematodes in the top soil layer

	Mean No. Juveniles ± SD (n = 10)*	
Depath (cm)	P*	A*
0-2	1807.8 ± 333.0 Aa	2431.7 ± 107.2 Ab
2-4	736.1 ± 240.9 Ba	485.7 ± 166.1 Ba
4-6	258.0 ± 116.8 Ca	68.0 ± 45.3 Ca
6-8	117.3 ± 52.4 Da	12.1 ± 10.3 Da
8-10	80.8 ± 48.9 Da	2.5 ± 3.1 Db

*Means followed by different uppercase letters in a column and different lowercase letters in a row represent significant differences ($P < 0.05$) by SNK test.
P* = *G. mellonella* larvae present at 8-10 cm depth
A* = *G. mellonella* larvae absent

Table 4. Recovery of infective juvenile nematodes 48h after inoculation of 3000 nematodes in cylinders with and without 10 *G. mellonella* larvae located at the base.

	Mean % Recovery ± SD (n = 5)*	
Species	P*	A*
L1C	37.4 ± 6.5 a⁺	35.5 ± 13.3 a

*Means followed by different lowercase letters represent significant differences ($P < 0.05$) by Student T-test.
P* = *G. mellonella* larvae present at 8-10 cm depth.
A* = *G. mellonella* larvae absent.

Table 5. Vertical distribution of *Steinernema* sp. Newfoundland strain No. 1 (L1C) infective juveniles in chlordane-treated soil columns 48 hours after inoculation of 3000 nematodes in top soil layer

	Mean No. Juveniles ± SD (n = 5)*									
	Concentration (mg/ml)									
Depth (cm)	0.0		0.001		0.01		0.1		1.0	
0-2	1812.2 (±293.7)	Aa	1916.4 (±249.0)	Aa	2129.8 (±218.2)	Aa	2175.8 (±260.1)	Aa	2169.4 (±250.5)	Aa
2-4	728.6 (±235.0)	Ba	603.9 (±44.6)	Ba	521.8 (±123.5)	Ba	495.0 (±110.6)	Ba	489.2 (±86.3)	Ba
4-6	281.4 (±134.7)	Ca	212.4 (±91.0)	Ca	175.4 (±49.8)	Ca	166.6 (±69.3)	Ca	161.0 (±67.7)	Ca
6-8	105.4 (±41.6)	Da	160.0 (±90.0)	Ca	97.4 (±31.8)	Da	81.8 (±29.3)	Ca	86.8 (±38.6)	Ca
8-10	72.2 (±42.4)	Da	108.2 (±58.9)	Ca	75.6 (±40.4)	Da	80.8 (±76.0)	Ca	93.6 (±100.1)	Ca

*Means followed by different lowercase letters in a row and by different uppercase letters in a column are significantly different ($P < 0.05$) by SNK test. Standard deviation is bracketed.

effect on migration of L1C over the range of concentrations tested (Table 5). Diazinon did not significantly alter migration at concentrations ≤ 0.01 mg/ml, however, a larger proportion of nematodes remained in the initial (0-2 cm) layer at 1.0 mg/ml and differences in downward migration could be detected in all layers. At 0.1 mg/ml,

Table 6. Vertical distribution of *Steinernema* sp. Newfoundland strain No. 1 (L1C) infective juveniles in diazinon-treated soil columns 48 hours after inoculation of 3000 nematodes in top soil layer.

	Mean No. Juveniles ± SD (n = 5)*									
	Concentration (mg/ml)									
Depth (cm)	0.0		0.001		0.01		0.1		1.0	
0-2	1812.2 (±293.7)	Aa	1953.0 (±196.4)	Aa	1764.4 (±405.5)	Aa	2244.0 (±238.5)	Aa	2725.0 (±149.0)	Aa
2-4	728.6 (±235.0)	Ba	667.0 (±58.8)	Ba	653.4 (±102.3)	Ba	562.0 (±127.5)	Ba	237.2 (±135.9)	Bb
4-6	281.4 (±134.7)	Ca	205.6 (±60.8)	Cab	286.8 (±119.1)	CDab	122.0 (±79.3)	Cb	35.0 (±32.2)	Cc
6-8	105.4 (±41.6)	Dab	116.6 (±85.5)	Dab	191.6 (±149.9)	DEa	49.2 (±43.0)	Db	2.8 (±6.3)	Cc
8-10	72.2 (±42.4)	Da	57.8 (±24.3)	Da	103.8 (±71.2)	Ea	22.8 (±29.1)	Db	0.0 (±0.0)	Cb

*Means followed by different lowercase letters in a row and by different uppercase letters in a column are significantly different (P < 0.05) by SNK test. Standard deviation is bracketed.

Table 7. Recovery of infective juveniles 48h after inoculation of 3000 nematodes in insecticide treated cylinders

	Mean % Recovery ± SD (n = 5)*		
	Insecticide (mg/ml)		
Species	Concentration	Chlordane	Diazinon
	1.0	23.0 ± 3.9	25.1 ± 6.0
	0.1	31.1 ± 2.8	38.6 ± 6.6
	0.01	37.6 ± 12.0	37.8 ± 8.3
L1C	0.001	32.8 ± 5.0	38.1 ± 15.5
	0.0	38.5 ± 5.5	38.5 ± 5.5

*Means followed by an asterisk are significantly different (P < 0.05) from control by SNK test.

migrational differences were significant at the 4-6 cm level (Table 6) although recovery was unaffected. The migration rate of nematodes was not affected even at 1.0 mg/ml chlordane but recovery was significantly different from control at this concentration (Table 7). In the presence of diazinon >0.1 mg/ml the migration of the nematodes was clearly altered although their recovery was unaffected (Table 6 and 7).

Immersion of juveniles in aqueous suspensions of insecticide served to substantiate the protective effect of the double cuticle in the free-living infective stage of the steinernematid nematodes. While some insecticide concentrations caused mortality, exposure to lower concentrations resulted in decreased mortality and paralysis. Several organophosphate, such as monocrotophos, phosalone and fenitrothion have been found to be directly toxic to *S. dutkyi* at 0.5$\overline{>}$ mg/ml, while vanidothion was not toxic (10) and phenamiphos was toxic to *S. feltiae* at 0.5 mg/ml (11). In this study, when infective juveniles of L1C were directly exposed to diazinon at 1.0 mg/ml, substantial (*ca*, 50%) mortality occurred after 120h of exposure. Thus, there appear to be variations in sensitivity of steinernematid species and strains to different organophosphate insecticides and this should be considered in any integrated pest management program. Carbamate insecticides, dioxacarb and oxamyl were toxic to *S. feltiae* (LC_{50} or 0.1 mg/ml), but there was no mortality at concentrations below 1.0 mg/ml(12) and there was little direct mortality after exposure to 1.0 mg/ml methomyl after 168h of exposure (11). During this study, the carbamates, carbaryl and Metacil (aminocarb) were assayed and while Matacil had no observable toxic effect on the nematode, carbaryl caused low (25%) mortality of L1C at 1.0 mg/ml (1000 ppm), a finding in general agreement with previous reports. The Matacil assayed was formulated in oil and thus did not mix well with the water suspension of nematodes; as a result contact was limited and likely affected these results. The strain of *Steinernema* examined in this study resembles *S. feltiae* in its tolerance to the insect growth regulator, Dimilin (11). In separate studies (11, 12), methoxychlor at 1.0 mg/ml, was not toxic to *S. feltiae*. However, L1C juvenile nematodes were found to be sensitive to 1.0 mg/ml of methoxychlor and chlordane. Differences noted for methoxychlor may again be attributed to species or strain differences, alternatively, commercially formulated methoxychlor was used in the present study, as opposed to the technical product used by previous researchers, this may account for the difference in results.

Nematodes in the affected state, which was characterized by slow movement and delayed response to probing, infrequently exceeded 5% of the sampled population after exposure to insecticides, this affected condition, termed 'sublethal effect', was consistently observed after exposure of *S. feltiae* junveniles to the organophosphates, methomyl (0.1 mg/ml) and phenamiphos (0.01 mg/ml) (11). The lethal or sublethal affects were not observed for juveniles exposed to the organochlorine, methoxychlor (11). Nematodes in this condition progress to death unless the insecticide is removed by washing (13), subsequently nematode infectivity would be restored (11). The reason why nematodes were not observed in the sublethal state, prior to mortality after exposure to other insecticides, is difficult to explain. In some cases, the time course from alive to sublethal to a lethal effect may be short if the insecticide is particularly toxic (e.g. 1.0 mg/ml chlordane at 24h).

The paralysis and eventiual death of L1C infective juveniles occurred after exposure to 1.0 mg/ml of both organochlorine insecticides, chlordane and methoxychlor. The organophosphate, diazinon was also toxic at 1.0 mg/ml. It was likely that the cuticle provided some protection which eventually broke down over time. This may be due to rates of penetration, metabolism and detoxification of the chemical by the nematodes. No toxic effects were observed for dimilin, malathion and matacil at non-lethal concentrations.

Recovery of adult nematodes from moribund *G. mellonnella* larvae, after infection with juveniles previously exposed to insecticide, was unaffected for L1C after exposure to dimilin, malathion and both carbamates. This result was not unexpected as significant toxicity was not observed at any of the tested concentrations. However, diazinon, methoxychlor and chlordane exerted observable toxic effects of 1.0 mg/ml in previous experiments so that in addition to examining the effect of these insecticides on infection success, the infection Index was also assessed. Infection Index dropped concomitantly as lethal effect increased, yet it was unclear if this decrease in Infectivity Index was due to sublethal effects on remaining nematodes or if the value was simply a reflection of the decline in the number of living nematodes.

The lethal effect of methoxychlor and chlordane on L1C at 1.0 mg/ml resulted in no infection of *G. mellonella*. The Infection Index for L1C was unaffected at 0.1 mg/ml. The Infection Index data displayed a high level of variability among replicates of the same treatment group. Some infection Index values were statistically different from controls but clearly, restraint must be shown in interpreting such results.

Infective juvenile neamtodes were attracted to insect larvae, as seen by significant migration of L1C nematodes in soil towards *G. mellonella* larvae located 10 cm below their inoculation point. The kairomone or chemical responsible for eliciting favorable nematodes response is unknown; carbon dioxide, heat and fecal excrement have been implicated by other researchers (14, 16). Nematodes response to host stimuli appears to be rapid. Carbon dioxide moves rapidly through air spaced in a soil matrix and movement of *S. feltiae* has been shown to be impaired in poorly aerated, overly saturated soil (17). Soil columns used in this experiment were loosely packed and carbon dioxide probably contributed to nematode migration. Soil moisture, texture and motility of the nematode have also been shown to be important factors in neamtodes movement in soil (18). The organochlorine insecticid, chlordane, and the organophosphate insecticide, diazinon, were consistently the most toxic compounds of the 7 insecticides investigated, both were toxic to infective juveniles after immersion in aqueous suspension. Chlordane did not significantly impair migration of L1C in soil treated with 1.0 mg/ml. Diazinon significantly impaired migration of L1C at 0.1 and 1.0 mg/ml. There was no direct toxicity at 0.1 mg/ml diazinon as noted previously. Therefore, it was possible that at this non lethal concentration, chemosensory response was impaired (15).

It is likely that the physiochemical interactions of insecticides with soil components affected the bioavailability of the insecticides to the migrating nematodes. Nematodes migrate in the water film of the soil pores and thus, the water is the site of insecticide-nematode contact. Chlordane is highly hydrophobic and as such, tighty absorbs to soil components, whereas diazinon is water soluble and thus, remains in the water fraction. Diazinon is an anticholinesterase and can impair nematode motility, dispersion and attraction to host stimuli by disrupting neuromuscular function. Studies with plant parasitic nematodes have implicated the somatic musculature as target sites of organophosphate insecticides (20, 22), the observations in the study are consistent with the proposed mechanism of insecticide action. Although migration of L1C was not impaired by chlordane, nematode recovery from soil was significantly lower at 1.0 mg/ml for L1C. This was due to a toxic effect exerted on the nematodes when water was added to the soil to regain nematodes in the recovery process. In suspension, the infective juveniles were directly exposed to insecticide. Chlordane, methoxychlor and diazinon were toxic at 1.0 mg/ml. Migration in soil was shown to be impaired by 0.1 and 1.0 mg/ml of diazinon. Lower concentrations of chlordane, methoxychlor and diazinon and all tested levels of carbaryl, matacil, dimilin and malathion, did not impair infectivity of the larvae.

The commercially formulated chlordane used in this study was recommended for use against soil and lawn insects, although no longer licensed for commercial application in Canada it is still used in other parts of the world. The rate of application, as recommended by the manufacturer, varied from 2-8 mg/ml, depending on the insect target and spray area. L1C nematodes were sensitive to direct immersion in 1.0 mg/ml chlordane. Methoxychlor was also directly toxic to both juvenile nematodes at 1.0 mg/ml. This chemical is often used as an insecticide on fruit trees; dosage ranges from 2.2-3.8 mg/ml. The organophosphate, malathion, is widely used to control pests of vegetables and trees at a recommended dosage of 0.5-1.5 mg/ml. In the current study, malathion was not directly toxic to either nematode at the tested concentrations. Diazinon is applied in dosages from 0.625-1.0 mg/ml to control fruit and garden insect pests and was found to be directly toxic to L1C at 1.0 mg/ml. Carbaryl exposure resulted in low mortality at 1.0 and 0.1 mg/ml. It is applied on ornamental plants and domestic vegetable gardens for insect pest control. Depending on application area, the recommended dosage of carbaryl is 0.75-1.5 mg/ml. Both matacil and dimilin have been used in foliar sprays to control insect pests in timber producing areas and on fruit trees. Traditionally, these chemicals are applied in spray programs in terms of an amount of chemical per given area. Both insecticides were not directly toxic to L1C juveniles at 1.0 mg/ml. Therefore, combined use of nematodes with the organochlorine insecticides methoxychlor and chlordane is not recommended, as spray programs use both chemicals at concentrations above juvenile nematode tolerance. Nematodes were not compatible with diazinon at the upper levels of recommended usage but malathion was compatible at levels often used in the field. Carbaryl was not compatible with juvenile nematodes at field application levels.

Insects have a higher sensitivity than nematodes to anticholinesterase insecticides and this factor is important in integrated programs (23). Nematodes are attracted to, infect and reproduce in moribund insect larvae which have been killed by insecticide (5). The organophosphate and carbamate insecticides usually degrade within 2 weeks and this time factor can be important in integrated pest management. If L1C nematodes are to be used in integrated programs along with insecticides, compatibility could be achieved by altering the use pattern of the insecticides. Insecticide application rate could be lowered to have minimal effect on the nematode, but still affect the insect. Notwithstanding, nematodes can be applied to soil after a sufficient times has elapsed for insecticide degradation. The combination of insecticide and entomogenous nematodes could result in additive mortality of the pest species and result in the application of lowered amounts of insecticide required for insect control.

ACKNOWLEDGEMENTS

Funded by Canadian Forestry Service and School of Graduate Studies, Memorial University of Newfoundland.

REFERENCES

1. Eidt D C, Zervos S & Finney-Crawley J R, Can Ent 127 (1995), 439-442.
2. Eidt D C, Zervos S, Pye A E & Finney-Crawley J R, Can Ent 127 (1995), 431-438.
3. Finney J R, CANUSA Newsletter, 34 (1984), 5.
4. Finney-Crawley, J R Canada, Nematol, 17(1985) : 496.
5. Hara, A H & Kaya H K, J Econ Entomol, 76 (1983b): 423-426.
6. Dutky S R, Thompson J V & Cantwell G E, J Insect Pathol 6 (1964) : 417-422.
7. Sokal R R, & Rohlf F J, 1969. Biometry. The principles and practice of

Statistics in Biological Research. W.H. Freeman and Co., San Francisco.

8. Goodell P B, In : Nematodes in Soil Ecosystem. (Ed. Freckman D W), Univ. Texas Press, Austin (1982) 178-198.
9. Hooper D J, 1986. In Laboratory methods for Work with Plant and soil Nematodes. Southey J F, Ministry of Agriculture, Fisheries and Food. Her Majesty's Stationary Office, London, (1986).
10. Prakasa Rao P S, Das & Padhi G, Indian J Agric Sci 45(1975) : 275-277.
11. Hara A H & Kaya, Env Entomol 12 (1983a) : 496-510.
12. Fedorko A, Kamionek M, Kozlowska J, & Mianowska E, Pol Ecol Stud, 3 (1977a) : 79-88.
13. Keetch, D P, Namatoligica, 20 (1974): 107-118.
14. Schmidt, J & All JN, Env Entomol 8 (1979) : 55-61.
15. Gaugler R, LeBeck L, Nakagaki B, & Boush G M, Evn Entomol, 9(1980) : 649-652.
16. Byers J A & Poinar Jr G O, Behaviour 79 (1982) : 1-10.
17. Molyneaux A S, & Bedding R A, Nematologica, 30 (1984) : 358-365.
18. Schroeder, W J & Beavers J B, J Nematol 15 (1987) : 173-182.
19. Marban-Mendoza, N & Viglierchio D R, J Nematol 12 (1980b): 114-118.
20. Croll N A. Can J Zool. 53 (1974) : 894-903.
21. Greco N & Thomason I J, J Nematol, 12 (1980) : 91- 96.
22. Marban-Mendoza, N & Viglierchio D R, J Nematol 12 (1980a) : 102 - 114.
23. Pree D J, Townshend J L & Archibald D E, J. Nematol 19(1987) : 188-193.

SECTION – 2

INDUSTRIAL BIOTECHNOLOGY

CHAPTER - 27

S.W.O.T. ANALYSIS OF COMMERCIAL BIOTECHNOLOGY — THE INDIAN CONTEXT

Dr. Abhay Deshpande

J.K. Drugs and Pharmaceuticals Ltd., Faridabad.

1. INTRODUCTION

I intend to present before you, a very succinct account of commercial biotechnology in form of a S.W.O.T., i.e. Strengths, Weaknesses, Opportunities and Threats. I wish to offer a very gross analysis, and with the time constrains, I may miss out on a few points, which I am sure shall be covered in the subsequent technical sessions. Three major scientific revolutions were witnessed by our generation, the first one being harnessing Nuclear energy, the second being Computers while the third and the most latest being the Biotechnology. Mr. Bill Gates, CEO Microsoft, has already predicted that the next century belongs to the Biotechnology. We have all witnessed since the famous patent of Cohen and Boyer a flurry of activity in the exploitation of the technology of genetic engineering, four expression of heterologous proteins to the birth of Dolly. The mankind is looking forward to the solutions of the burning problems in human health and well being like AIDs, Cancer, Genetic dis-order oriented diseases, Food supply, and Environment to be solved by using the biotechnological means.

Let us question ourselves :

* Are all the hopes placed on our fraternity realistic ?
* Is the science of Biotechnology likely to stand-up to the expectations of our society ?

Let us try to find some answers to these questions, and possibly present a realistic picture of Biotechnology before the society. A few days ago, Prof. M.G.K. Menon, the distinguished scientist, defined the Modern Biotechnology, as the science of making using Biology. I would like to add a few words to this and define for Commercial

Biotechnology as, the Technology which involves commercial exploitation of the biosynthetic capabilities of the organisms. The life forms in various forms, from viruses to the most advanced human beings offer a very large diversity of the biosynthetic capabilities. The bio-synthetic capabilities range from simple low molecular weight molecules to large macro-molecules.

2. STRENGTH

Before getting into the analysis, let us answer a simple question to ourselves and justify "Why Biotechnology?" The answers to this questions would portray the "STRENGTHS" of commercial biotechnology.

(a) Biotechnology is natural

We are all fond of this fad of naturality. It is a corollary to the concept of the products being mad-made. Honestly, we are going to make the products which are made by human intelligence of exploitation of knowledge, but using the natural Biosynthetic capabilities.

(b) Biotechnology is eco-friendly

Another in-thing of the day is eco-friendliness of the technologies. Most of the man made technologies generate wastes of different degrees of environmental pollutions. The biotech based technologies are known to be relatively environment friendly. There exists relatively good knowledge base to protect the environment from possible biotechnology oriented environmental hazards. For example consider the extent of pollutants generated by the synthetic method for a synthetic indigo dye and natural one. It is probably one tenth in terms of biodegradability and environmental half life of the pollutants.

(c) Biotechnology based processes consume natural renewable resources

The biotechnology based processes are aimed at exploitation and maximization of the specific biosynthetic capability of the organisms. The required growth media therefore are always natural based natural renewable resources. e.g. sugars and proteins. It thus consumes minimum of non-renewable energy and most of the wastes and biomass can be recycled for energy generation.

(d) Biotechnology is not hazardous, unless we engineer with unnatural genes

The biotechnology based processes are the reactions which are typically the "NTP", normal temperature and pressure reactions. They do not involve any explosive reactants. The Biotechnological processes are therefore, extremely non- hazardous.

(e) Biotechnological processes are less energy intensive

All the fermentation/biotransformation reactions are carried out (with exception of DAO in 7-ACA production) at NTP. There is neither a need to heat/pressure to be provided for the reactions, nor does it lead to generation of unsermountable amounts of heat. The only energy consumption avenue is oxygen transfer and steam for maintenance of sterile envelope. The advances in technology of mass transfer have led to reduction in power consumption to much lower levels as compared to the era of sixties and seventies.

(f) The synthetic capabilities of Mankind cannot synthesise the range of Molecules which are generated efficiently by the Organisms

The biological systems bring about the synthesis and organisation of a wide range of chemicals, groups, side chains as well as polymers. It is well known that, the synthetic capabilities at the present level of human knowledge base, cannot commercially synthesise the molecules like sugars, citric acid etc.

(g) Availability of trained manpower

Thanks to DBT efforts, there is no paucity of the trained manpower in the respective specialisations of Biotechnology, all over India. India is thus a goldmine of the trained but under utilised technical manpower.

3. WEAKNESSES

I have been singing all the praises for our fraternity, and it would be unfair, if we do not balance this portrayal with a "Weakness Analysis".

(a) Biotechnology is extremely hi-tech. the expertise is more dependent on individual skills than equipment

The success with the modern molecular biology and genetic engineering techniques is extremely dependent on highly skilled scientific manpower, using highly specialised equipment. The rapid dissemination and deployment of the biotechnological processes, thus gets hampered. The development of a clone containing multiple copies of a gene specific for production of a foreign gene, with full expression capabilities, is a job of a specialised team of molecular biologists and biochemists. The downstream processing of purification of the products like interferons and alike is a major technological challenge. The knowledge/ experience of running such a system successfully resides with only those with the specialised skills. The reproducibility of results in biotechnology, is thus a function of the skills of an individual.

(b) Biotechnology is *strain/clone-dependent* — the industrial piracy and strain pilfrage is a major concern

The organisms is designed, moulded, modified and trained for producing the product of commercial interest. This makes the technology extremely dependent on the specialised organism. There are known incidences of unethical practices involving pilferage of strain/clones from one industry to the other, within India as well as abroad. A number of individuals from a specific country in the world, are known to approach the corporations with the pilfered strains from the western multinationals. However, generally the strains available with such vendors are outdated and discarded versions of the production strains of the multinationals.

(c) The process is extremely sensitive to microbial contaminations

The production of a desirable metabolite/product using microbial or cell culture clone, is carried out under highly specialised "aseptic envelope", due to the exacting needs of obviating the microbial contaminations. The cultivation media, more generally are extremely rich, to support growth of fast growing contaminating microbes. A contamination in a batch could lead not only to rejection and drainage of batch causing collosal monetory losses, but also leads to environmental problems of disposal. Typically a contaminated batch of penicillin could cause a direct material loss of at least Rs. 3 to 5 million, apart from the losses due to missing/rescheduling of a batch, power and energy input.

(e) The consumption of energy for fermentation, sterilisation and maintenance of sterile envelope is very high

The fermentation/biotransformation activity is relatively much low on consumption of energy, as compared to their chemical technology counterparts. However, a very large amount of energy (in form of steam) is used to maintain steam locks in the sterile envelope. We have witnessed that almost 2 to 2.5 times the amount of steam is used to maintain the sterile envelope around the penicillin fermenter, as against the one used for sterilisation. A batch of 150 M^3 fermenter typically consumes about 25 tonnes of steam for sterilisation. On the other hand almost 50 tonnes of steam is consumed throughout the batch, to maintain the sterile envelope.

(e) The product concentration is, in general very low

The product purification (DSP) costs, are any where in the range of 25 to 90% of the production costs. The recovery efficiencies, especially with the protein products, could be as low as 1% to 50%. This is due to the fact that the product concentration in the biotechnological processes could be anywhere between a few ppm to 1-4%. The exception is the case like citric acid and similar high volume fermentation processes, where the product concentration is as high 20-30%. Moreover, the low concentration necessitates multi-step recovery and purification process. This leads to loss of as much as 20-80% of the product, available for purification ex-fermenter. The recovery efficiencies, especially with the protein products, could be as low as 1% to 50%. This is a great commercial loss, especially for the very high value low volume products like interferons and interleukins.

(f) The "GMOs" require strict containment

The present day regulatory laws regarding use and dissemination of the products, manufactured using Genetically Engineered Micro-organisms, are fairly stringent. It is not possible to manufacture a r-DNA product using a GMO, without proper validation and certification by competent authorities. In certain cases containment levels of P-3 are desirable. The effluent as well as laboratory washings from such a plant/lab, which uses GMO, has to be thoroughly sterilised, and de-contaminated, before disposal into the main ETP. This restricts the dissemination and growth of the Biotech industry.

(g) The cost of the product is directly related to the process efficiency, and can nose-dive indefinitely with the advent of improved strains/processes

Let me offer you a hypothetical case. In a situation wherein the productivity of a certain strain for a certain product is let's say, 0.15 kg/M^3/hr. A better technology involves a strain which gives a productivity of 0.225 kg/M^3/hr, the manufacturer would be capable of dropping his prices by almost 25 to 33%. Further if his recovery efficiencies are close to 90% (as against 75% of the previous manufacturer), he gains further margin of almost 15%. The prices in the industry have been sliding down steadily, as a result of continuous upgradation efforts in the technology development. The cutthroat competition leads to sustained efforts in technology improvement. The original inventor/ discoverer, gains a lot of commercial advantage, being a pioneer, garnering very high value per unit, before competition sets in.

(h) The ensuing *i.p.r.* is a minefield

Any New Strain/Plasmid/Vector/ Process Develop May be Covered Under Some patent : There has been a flurry of activity in the field of filling patents during the "Black-box" period, by almost all the Western corporate and Academic researchers. I intend to give a little extra emphasis on this matter here, which I am sure, you

Table I. Publications in the field of Biotechnology and The Patent Craze

	Publication Frequency in Biotechnology				
Year	Total Publications	Patents	% Publications As patents.	r-DNA patents as % of total patents	
1982	8256	1831	22.0%	N.A.	–
1983	12180	2686	22.0%	N.A.	–
1984	13049	3377	25.9%	N.A.	–
1985	13366	3804	28.5%	N.A.	–
1986	13642	4072	29.8%	N.A.	–
1987	13601	4138	30.0%	N.A.	–
1988	13583	4276	31.5%	N.A.	–
1989	14382	4610	32.0%	1004	21.8
1990	15129	4940	32.7%	1217	24.6
1991	14688	3889	33.3%	1309	26.8
1992	14367	3785	33.3%	1356	28.3
1993	14090	4864	34.5%	1452	29.9
1994	16362	5352	32.7%	1637	30.6
1995	15717	5155	32.8%	1702	33.0
1996	14199	5484	38.6%	1999	36.5

are not going to find out of place. Let's look at the data in Table-1.

Look at the a few interesting trends in this data. On an average 13 to 14000 publications have been appearing in the field of biotechnology every year. The extent of patent publication has been steadily rising from 22% in 1982 to almost 38.6%. The scientists like us, have become extremely materialistic and possessive about the knowledge base developed! It is more obvious with the r-DNA researchers. A look at the data in fourth and fifth column of the table-1 proves the point. The publications in the field of recombinant DNA technology have risen from 21.8% of the total Biotechnology patents in 1989 to 36.5% by 1996!

Now let us look at another set of trend data.

Let us now note from Table II that except the HGH, all other products show above 40% of the publications in these products were published as patents.

Table II- The intensity of patent publication in the field of commercial Biotechnology.

Product	Publications	Patents	%
Interferon	2855	1251	43.8%
Insulin	2078	926	44.6%
Interleukin	3083	1331	43.2%
Human Growth Hormone	174	31	17.8%
Erythropoietin	385	175	45.5%

Let us now visualise the distribution of patents in various disciplines of Biotechnology.

Table III.

Discipline	Patents in the Discipline	% of Total
Fermentation	26963	41.6%
Recovery	2843	4.4%
r-DNA	14320	22.1%
Hybridoma/MAb	5566	8.6%
Enzymes	14150	21.8%
Others	946	1.5%
Total	64788	100.0%

It can be seen that almost 42% patents cover fermentation processes, while ~ 22% each were granted in the field of recombinant processes and enzyme based processes.

In conclusion there has been a rush to file patents for the findings, however trivial. These patents, the researchers feel, are their insurance policies for reaping the benefits, out of their lien on the knowledge. It would not be very easy for us Indians to develop our won clones for production of existing bio- active products, unless we have a substantial database to prove that the strain originates from the biodiversity of India. Every technique, process, equipment we attempt to use might have already been patented, making it susceptible to infringement suits. I have therefore titled this point as a "mine field", obviously meaning it literally, a dangerous path to chart for the Indian researchers, in the field of commercial Biotechnology.

4. OPPORTUNITIES

The biggest opportunity that Indian Biotechnclogy ventures can look forward to is in the field of its richest Biodiversity resource, second only to Brazil. *I will try to enumerate the possible opportunity zones, with a possible recommendation for every case.*

A. GENERATION OF MICROBIAL BIODERVISTIY BANK

There has been a commendable and laudable effort on part of a number of the culture collections in India to store and supply the microbial strains. Somehow their efforts were drifted towards acquiring parallelly deposited stains from other international culture collections. As an industrial scientist, I look forward to very active programs being undertaken by every University, Academic Institutions as well as Industries, to undertake screening of micro-organisms and deposition in a centralised culture bank. This deposition, should as far as possible, be carried out with either generalised restriction digests data or RFLP maps. The government could appoint a body, authorised to register the micro-organisms and their preliminary genetic information. This would help in generating a registered source of micro-organisms for any future activity without into the IPR mine-field. The interested industries, may then undertake screening programs for bioactive molecules using specialised techniques.

B. DEVELOPMENT OF COMBINATIONAL CHEMISTRY FOR SCREENING

The science of *combinational chemistry* should be developed fully by joint efforts of industry and Academic/ CSIR Institutions. This would help industry in setting-up highly specialised screening cells with an intention to isolate a target specific bio-active compound. The necessary legal protection of the intellectual property thus developed, can be legalised and formalised, keeping in view the investor's interest. The industry has to intelligently spread the contract research, in such a way that the information is shared, only on "need to know" basis.

C. GENERATION OF PLANT BIODERVISITY BANK

A very active program should be undertaken by the Agricultural Universities, Agriculture Research Institutions as well as Seed Industries, to undertake screening of Indian cultivars and medicinal plant varieities. These germplasm lines should be deposited in a centralised germplasm bank. This deposition, should, as far as possible, be carried out with data on protein banding, restriction digests or RFLP or AFLP MAPS. A centralised authority could register the plant germplasms/ seed material/DNA. The deposition of the elite lines may not be insisted upon, but their RFLP/AFLP maps may be registered. This would help in generating a registered source of plant varieties for any future activity without any risk of patent infringements. This would be useful not only for pharmaceutical industry, but also seed industry.

D. DEVELOPMENT OF CENTRALISED DATABASE FOR MEDICINAL PLANTS

There is a burning need for availability of on-line/CD Rom based information base on the ancestral knowledge base on our heritage like Ayurveda and Unani systems. The data base could describe almost all the bio-diversities of India, using documented, spoken, transmitted by generation, knowledge in this field. This would help Indian companies to develop products based on plants. Such information, product or processes could be patented world- wide to prevent pilferage/misuse by the other countries.

E. SCREENING PLANT AND MICROBE BASED PHARMACOLOGICALLY ACTIVE MOIETIES

The above data and knowledge base, could become handy for the interested researchers/industries, for undertaking screening programs for bio-active molecules using specialised techniques. A good combination of data base, combinatorial chemistry, biotransformation capabilities and specialised issue based high through-put screening techniques, would put India in general and Indian industry in particular on the World IPR map.

F. GENERATION OF VECTOR BANK

A large number of transformation/ integration vectors have been published and patented abroad. A dedicated effort by DBT, with contribution from industry, should be undertaken to isolate from natural Indian biodiversity/synthesise a large collection of vectors for r-DNA research. Such registered vectors, would be useful for patenting the resultant clones/ expression systems, without getting into major IPR problems. A mechanisms to register the genetic map/sequence of such vectors/clones needs to be designed.

5. THREATS

We have all been a party to the Research and Developmental activities in India for last twenty-five years. By design or by default, we had been actively engaged in re-inventing the wheel. It paid us, to reverse engineer a product or a process and achieve a great commercial success. This was possible due to our existing patent act. The mind set of scientist, engineers and industries needs to be re-oriented. They have to be made to realise the fact, that being "Copy Cats" would not take us any more to the path of commercial success in the 21st Century. The competition is extremely intense. Darwin's theory of "Survival of a the Fittest" has been never so apt in the past. One of the minor threat perception is existence of a possibility of engineering hazardous genes, especially by the evil elements within our fraternity. We must develop fool-proof regulatory safeguards to protect the society from such hazards.

The Remedial Action : The western world is at a formidable position of commercialising their research due to a very active collaboration between academia and industry . We in India have been repeatedly stressing the need for "Academia -industry Collaboration" since early eighties. What we have achieved, is far from desirable. Each party has been more critical of the other, for the mutual attitudes towards each other.

The major action which needs to be undertaken is hence, improvement in industry academia relations.

I suggest following measures being taken to alleviate the problem of institution industry interactions.

A. RESEARCH IN ACADEMIC/ INSTITUTIONS

It has been observed that a large number of laboratories and institutions carry out repetitive work/projects. This leads to waste of time as well as funds. The institutions and their laboratories should be identified for their core competence and funded for basic research in highly specialised and focused projects. This information of core competence and the on-going projects should be available on an on-line data base.

B. THE RESEARCH & DEVELOPMENT IN THE INDUSTRY

The industyr on the other hand tends to employ a Technocrat - Manager, who has enough knowledge of science and technology to understand the academic "High-Talk". At the same time, he neither has a manpower, nor resources to carry out basic/applied research. This technocrat could split his project into number of sub-projects and assign them to the specialised research group in academic/ government laboratories on commercial terms. He has to integrate the findings of sub-projects, undertake the development activities at his centre and commercialise the technology.

I am sure an immediate action plan on these lines could answer most of our woes and difficulties for the challengers of the 21st century.

EPILOGE

I have tried to cover the vast arena of S.W.O.T. Analysis of the Commercial Biotechnology. It would not be physically possible to cover all the aspects. I have tried to cover at least the major facets of the topic. I would appreciate criticisms and lively discussion on the aspects touched by me. I thank you all for your co-operation.

CHAPTER - 28

BIODIVERSITY, BIOTECHNOLOGY AND INNOVATIVE BIOCHEMISTRY IN THE DISCOVERY OF NOVEL SECONDARY METABOLITES

B.N. Ganguli

Emeritus Scientist of the C.S.I.R., Govt. of India at the U.D.C.T., Mumbai

1. INTRODUCTION - THE NEED

The immediate questions are : "*Why do we need novel secondary metabolites?*"

"*Where do we find them?" and finally "How do we detect, isolate and identify them?*"

The answer to the first question lies in a W.H.O. report published in 1996 and shown here in Tables 1 and 2. Obviously there are many chemotherapeutics already available. (Fig. 1). But these do have problems and in addition the pathogens have devised a variety of methods to become resistant. The answers to the other questions lie in the multidisciplinary approach outlined in the title of this article. It must be kept in mind that, besides Microbiology, there is very innovative and sophisticated Biochemistry, Chemistry and Pharmacology which is an integral part of the discovery of new chemical entities. It must also be remembered that, today, very elegant receptor based screens are available which can detect specific and low levels of pharmacologically active compounds. Thus, any secondary metabolite could be an antibacterial, an antifungal, an antiviral or a chemical moiety effective against metabolic diseases. For example diabetes, to name just one.

2. SOURCES - BIODIVERSITY

The source of new molecules lies in the exploitation of the biodiversity of micro-organisms and plants. Table 3 gives some figures which are indicative of the enormous potential that lies in the biodiverse kingdom. It also indicates that we have barely touched the surface of this biodiversity. There is an entire environment, namely the marin ecosystem, that has hardly been explored. Classical microbiology and new microbial techniques are needed to explore both the terrestrial and marine ecosystems.

Fig. 1. CHEMOTHERAPEUTICS AVAILABLE

1 Ceftadizime

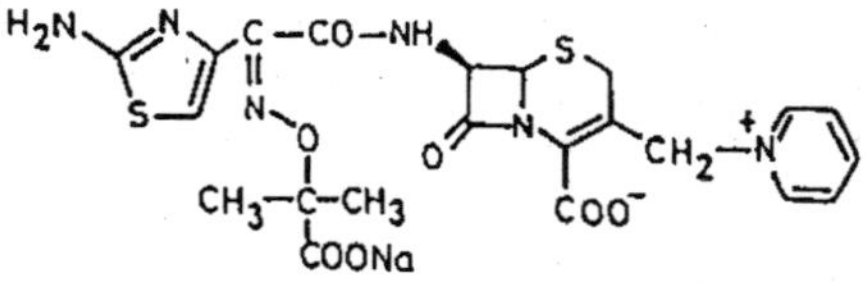

2 FCE-22101

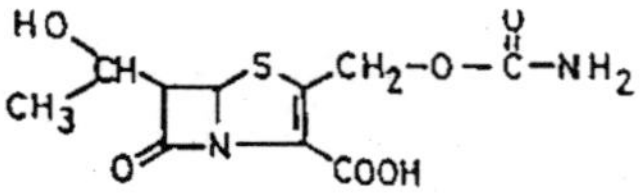

3 Moxalactam

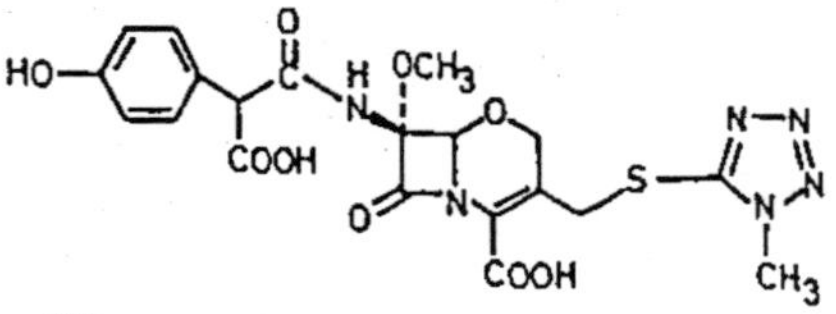

4 Imipenem

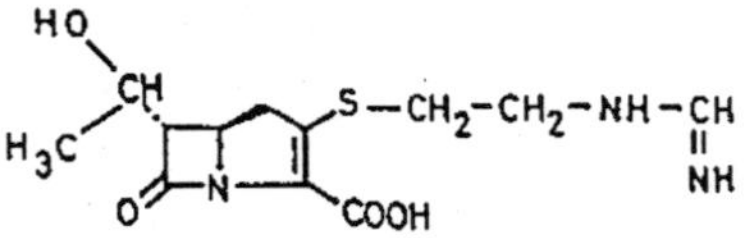

5 Ciprofloxacin

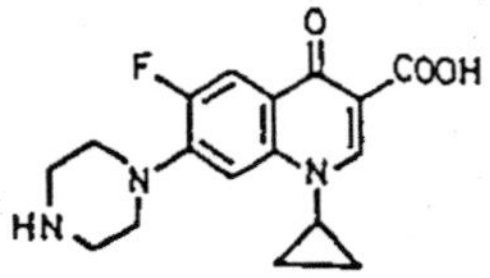

6 Teicoplanin

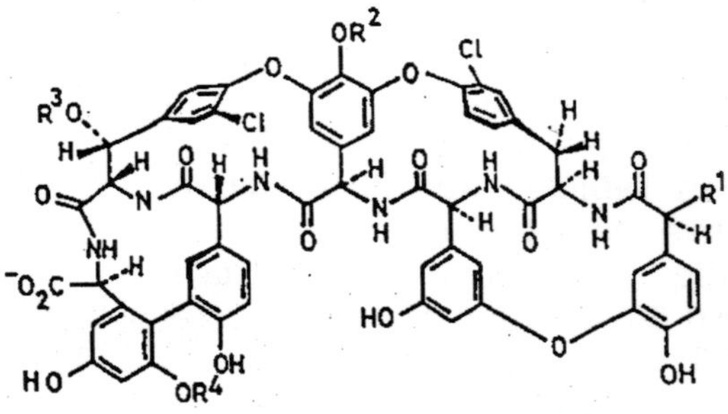

7 Roxithromycin

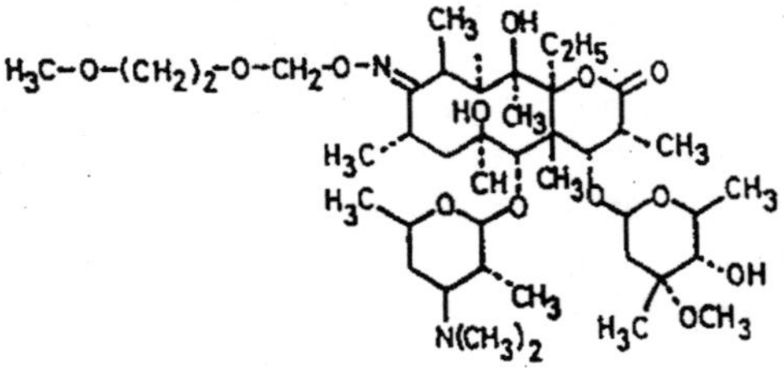

INFECTIOUS DISEASES : W.H.O. 1996 REPORT

Mortality : 17 million persons per year

30 new infections in the past 20 yrs.

No cure, no vaccines for many.

Table 1. Ten Biggest Killers in 1995 :

Pneumonia : 4.4 mio deaths

Diarrhoea, Cholera, Typhoid, Dysentery : 3.1 mio

Tuberculosis : 3.1 mio

Malaria : 2.1 mio

Hepatitis B : 1.1 mio

Measles : 1.0 mio

Neonatal tetanus : 0.46 mio

Whooping Cough : 0.355 mio

Intestinal Worms : 0.135 mio

Others : Hepatitis C, Meningitis, STD.

Problems :

(a) "*Old diseases-Old problems*" Immunisation, Sanitation, Clean water, Waste disposals, diet supplements.

(b) "*Old diseases - New Problems*" Resistance - Insect borne malaria Dengue, Food chain carrier

(c) "*New diseases - New problems*" Ebola virus, madcow disease (BSE) Creutzfeldt-Jacob brain disease

Table 2. New Diseases

1973 : Rotavirus : Infant diarrhoea

1976 : Cryptosporidium parvum : diarrhoea

1977 : Legionella pneuophilla : Legionnaires diseases

1977 : Ebola virus : Haemorrhagic fever

1977 : Hantaan virus : Haemorrhagic fever

1977 : Campylobacter Jejuni : diarrhoea

1980 : HTLV : Leukaemia

1982 : E. coli 0157-H7 Bloody diarrhoea

1982 : HTLV2 : Hairy Cell Leukaemia

1983 : Helicobacter pylori : Peptic ulcers

1983 : HIV

1988 : Hepatitis E : Jaundice epidemics in hot climates.

1988 : Human Herpes 6

1989 : Hepatitis C

1991 : Guanarito virus : Venezueian fever

1992 : Vibrio Cholerae 0139

1994 : Sabia virus : Brazilian fever

1995 : Human Herpes : In AIDs patients virus 8 : Kaposis syndromes

1996 : Resistant infective organisms

Example : MRSA, MDR mycobacteria, Fungal diseases.

Table 3. The Dimensions of Microbial Diversity

1. Global Biodiversity is around 3 to 500 x 10^6 species of prokaryotes and eukaryotes in 70 phyla.
2. Marine Biodiversity is 0.5-30 x 10^6 species of macrofauna alone.
3. Myxobacter have yielded 130 new compound of which 30 are new structures.
4. There are 24,000 low molecular weight compounds known which have potential pharmacological activity.
5. There are 250,000 - 750,000 species of plants known with 2 to 5 different endophytic fungi per plant. Therefore 0.5 to 3.5 million fungal strains exist.
6. 90-99 percent of micro-organisms still unexplored.

Major Policy Recommendations of the American Association of Microbiologists

1. Unless microbial exploration is undertaken the country will face scientific and industrial setbacks.
2. Identifying, developing, harvesting of microbial resources is of importance.
3. Primary funding of the above must come from federal agencies but also active participation by pharma and biotech companies is required.
4. Microbial explorative research must be supported for at least ten years to ensure a return on research investment.

3. EXPLOITATION - INNOVATIVE FERMENTATION

To exploit the microbial biodiversity, several steps need to be considered and are showed in Chart I. An unusual microorganism may produce a novel skeleton. On the other hand such microbes could be induced to produce secondary metabolites by biochemical manipulations. For instance, myxobacter do not need as high levels of oxygen as streptomyces. Glycopeptides are produced only under very high aeration conditions while macrolides may be produced under very little.

Chart II indicates how the emphasis is changing from, antibacterials in 1982-85 to pharmacologically active molecules in 1992-95.

4. CREATIVITY INDICES

A study of the creativity indices of actinomycetes and fungi are shown in Figures 2 and 3. In addition to Streptomyces, we see species of *Amycolatopsis*, *Micromonospora*, *Nocardia* and *Saccharopolyspora* are good producers. Similarly in fungi, other than the fungi imperfecti we see that species of *Trichoderma* and *Fusarium* are also good producers of new compounds.

5. DETECTION METHODS

In order to find or devise effective molecules against problem bacteria and problem metabolic diseases it is essential to know the biochemistry of the disease process. It is essential to know the biochemistry of the growth of the pathogen, the biochemistry behind its virulence. For example, methicillin resistant *Staphylococcus aureus* (MRSA) is a very intractable pathogen. Only Vancomycin has been used in such cases. But it is used parenterally, has severe side effects and today resistant strains are coming up. MRSA is resistant to Beta lactams because it synthesizes a new protein, PBP2' when exposed to Beta lactams. It uses this enzyme to continue to grow. Therefore, any screening method which can find an inhibitor of PBP2' will probably give a new molecule.

CHART – I
MICROBIAL DIVERSITY

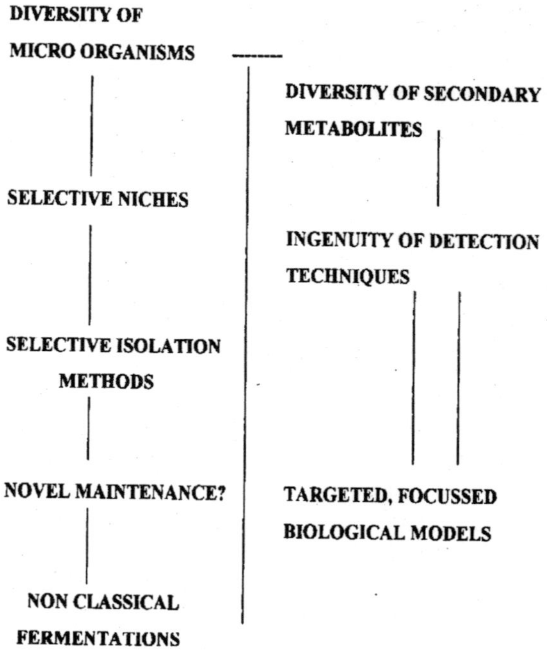

Fig. 2. Creativity Index of Selected Actinomycetes Taxa

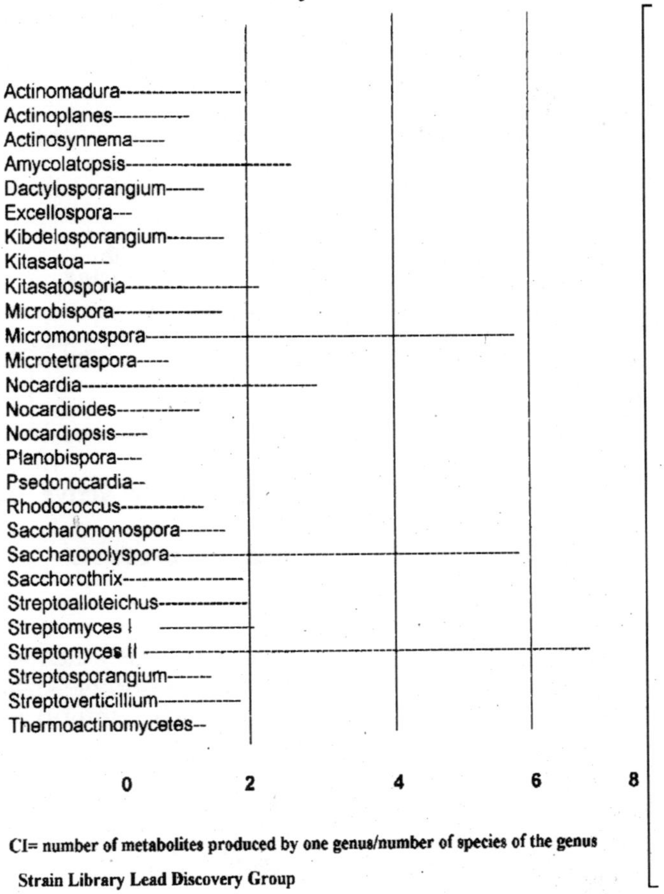

Fig. 3. Creativity Index of Selected Fungal Taxa and Groups

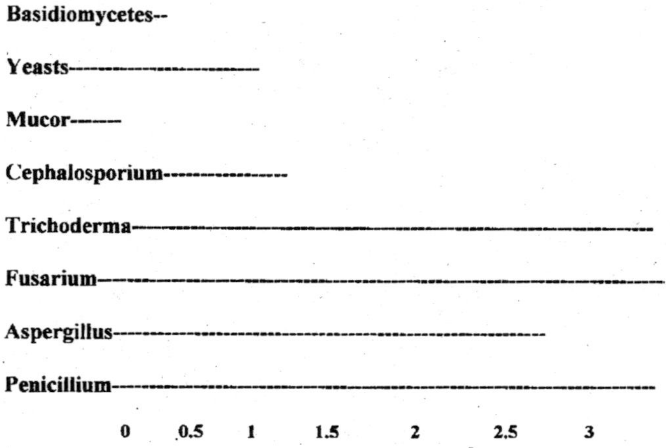

CHART – II
TRENDS OF NEW AGENTS IN J. ANTIBIOTICS - I

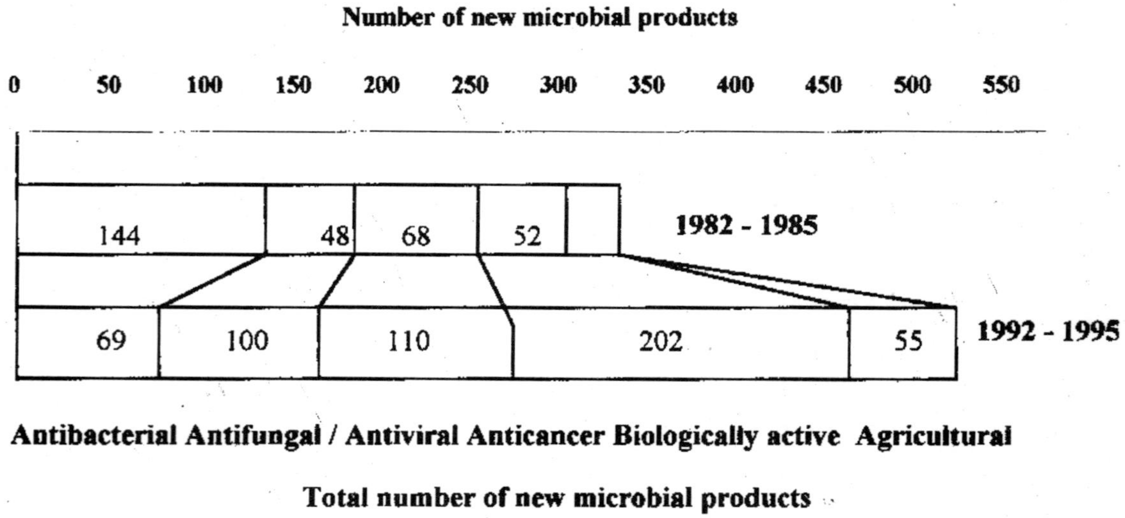

The screening method is simple. At 28°C, a MRSA strain synthesizes PBP2', at 42°C the same strain grows normally using normal PBPs. Therefore a differential assay at 20°C and 42°C should enable an investigator to detect PBP2' binding molecules.

Similarly, DNA gyrases are cold shock proteins, produced in larger amounts at 18°C - 20°C than at 28°C. Any compound binding to the gyrases can thus be detected again using a differential assay. Some detection methods are shown in Chart III. In metabolic diseases such as arthero-sclerosis, auto-immune disease (RA) and septic shock, the cytokine Interleukin-1 B is involved. There is a protease, called ICE i.e. Interleukin Converting Enzyme which cleaves inactive pre-interleuking 1B (33Kd) to the active IL-IB (17.5 Kd). If ICE could be inhibited, one could get a new therapeutic possibly.

Inhibitors of enzymes have given many promising molecules such as Penicillin G the transpeptidase inhibitor, Bestatin the cell surface amino-peptidase inhibitor, reverse transcriptase inhibitors against HIV such as Nevaparine and Protease inhibitors such as ABT-538 and L-73552 which are being tried as combination therapy.

CHART III

	Detection Method	Detects
1	Use of Beta-lactam super sensitive *E. coli* (Mutant)	Beta-lactams
2	Differential assay at 28°C and 42°C using *S. aureus* (MRSA)	Detects PBP2' binding compounds
3	Cold shock assay using Gram negatives at 18°C vs 28°C	For Gyrase binding molecules
4	Using N-acetyl-lysyl-D-ala-D-ala binding assay	Detects glycopeptides
5	Using induced resistance to MLS compounds in Gm + ves	Macrlides

Fig. 4. Selected new Microbial Metabolites

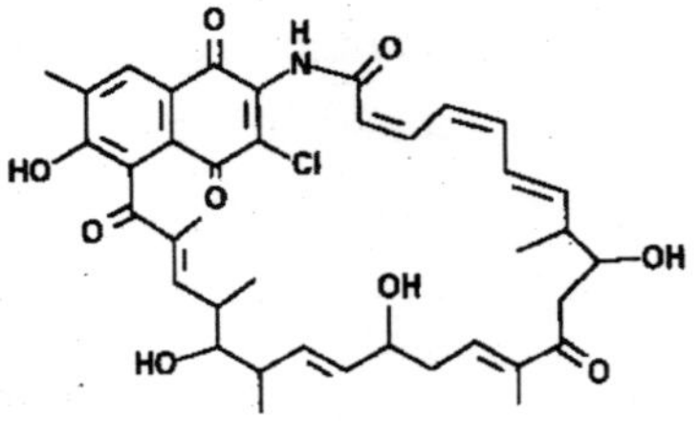

Naphthomycin H

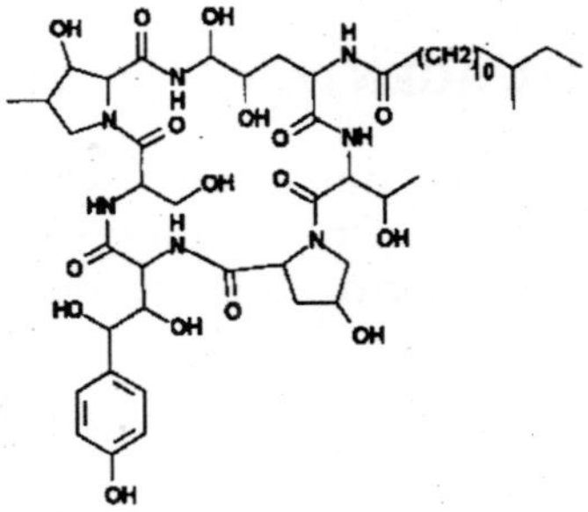

Mulundocandin

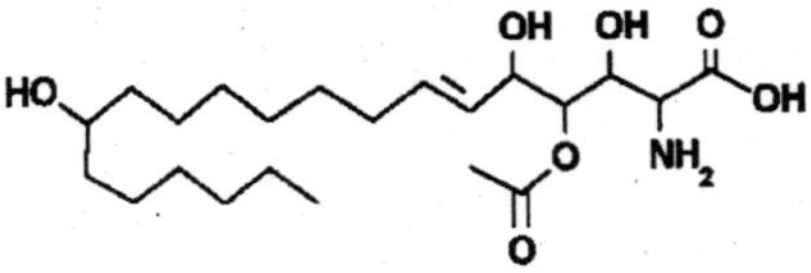

Fumifungin

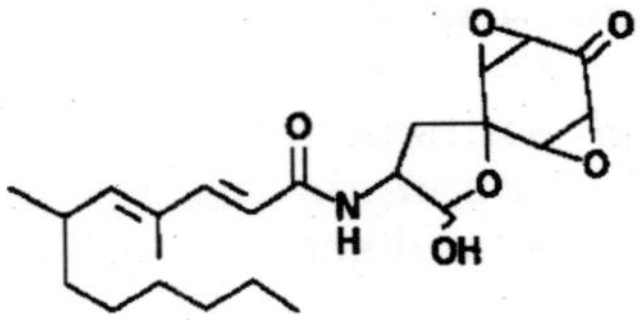

Aranorosin

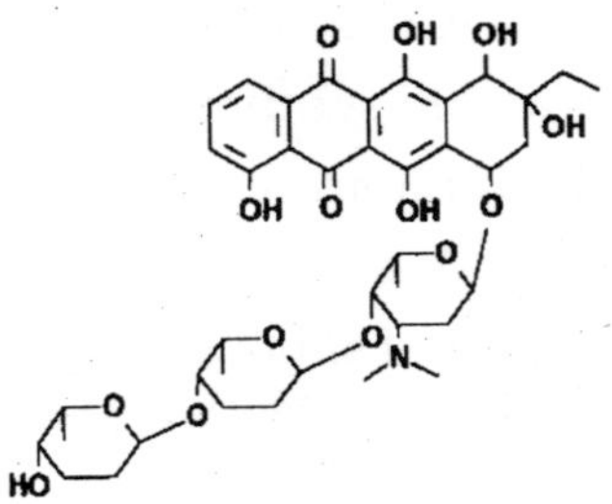

Kesarirhodin B

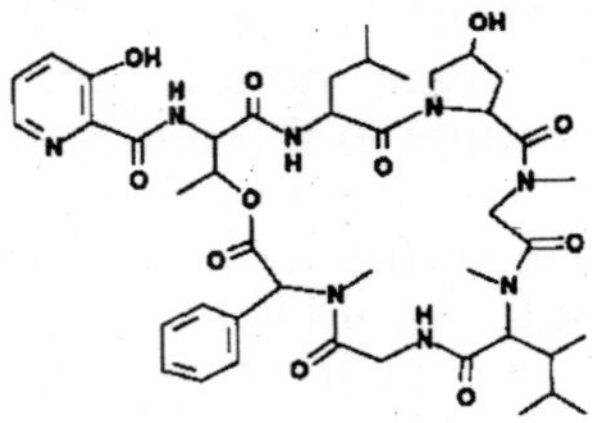

Grividomycin II

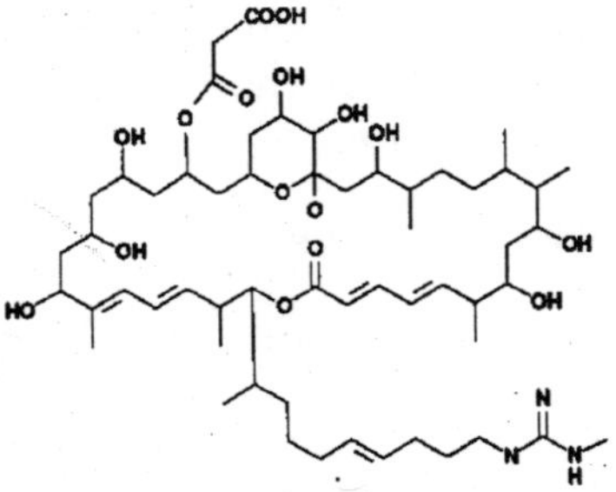

2-Demethylazalomycin F_{4a}

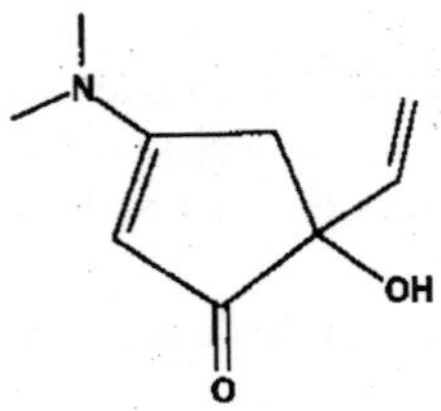

L 919867

Using such assays we isolated some interesting compounds whose structures are shown in Figure 4.

6. ANTIFUNGAL AGENTS

The need for non-toxic, systemic antifungal still exists. Only Amphotericin B is the naturally occurring polyene available today.

Today, the immuno-compromised patient is much more susceptible to fungal infections. Screening for new molecules uses new targets that are indicated in Table 4.

Metabolic Diseases : Screening of natural products from micro- organisms and plants of synthetic and semi-synthetic libraries of compounds against a whole range of metabolic diseases has become a major global effort of Pharmaceutical companies. For example, against non-insulin dependant diabetes, alzheimer's disease, lipid control, arthritis and inflammation, cardio-vascular diseases, hypertension and others. High thru'put screening, called HTS in microtitre plates, using robotics can handle 50,000 to 100,000 compounds in 6 months and even less. Some assays are shown in Table 5. Inhibitors of transcription are another area which, potentially, could lead to new drugs. Examples of gene expression modulating drugs are known such as Aspirin which inhibits IL 1 activation of prostaglandin synthesis, Cyclosporin A and FK-506 bind to proteins which regulate the activity of transcription factors. Nolvaden (Tamoxifen), Eulexin (Flutamide) regulate expression of steroid responsive genes. Ubimenix is an inhibitor of leucine amino peptidase and enhances the release of IL-1 and IL-2 from macrophage and spleen cells.

Table 4. New Anti-fungal Targets

1. Ser/Thr Protein kinases in cell wall biogenesis in *S. cerevisiae*. Their role in fungal morphogenisis, target may be the proteins they phosphorylate. Are they present in Candida?
2. Cell wall associated "*Adhesins*" These are Mannoproteins.
3. Secreted Aspartic Proteinase.
4. Cell wall Biosynthesis
5. Translation Elongation Factor EF-3.
6. Dimorphism : Yeast-Mycelial switch as target.

Table 5. Assays for Metabolic Diseases

1. RECEPTOR based screening assays.

Cytokine receptors, cell-cell adhesion molecules Solid phase ELISA

2. ENZYME inhibitor screens —

Glucose-6-Phosphate Translocase inhibitors

Urokinase inhibitors

Prolyl cis/ trans isomerase inhibitors

Prolein Kinase C inhibitors

Glutamine synthetase inhibitors

ICE inhibitors

3. Aptamer Approach

Selection of small pieces of DNA or RNA from random pools that bind to different types of molecules, Dyes, amino acids, proteins.

4. Transcription Assays

Screen for small molecular weight compounds that modulate the expression of a target gene.

7. CONCLUSIONS

Micro-organisms are the goldmine of secondary metabolites. Their enormous genetic biodiversity can be exploited to give an equally enormous metabolic bio-diversity. Provided we are able to isolate, culture and ferment the unusual micro-organisms, provided we are able to detect them by using very specific, target directed, innovative assays. It is not enough to be able to detect and isolate the metabolite of interest. Structure elucidation of a new molecule, available only in milligram quantities, needs the latest developments of structure elucidation chemistry. Such as tandem MS., 2D-NMR, FT-IR, Correlation Spectroscopy, X-ray Crystallography and supporting techniques.

This still leaves the entire profiling of the new compound through activity spectrum *"in vivo"*, pharmacokinetics, acute and chronic toxicity, and finally formulation, dosage and costs.

8. RECENT APPROACHES IN SCREENING CHANGES AS A RESULT OF—

1. **Advances in** *Gentic Engineering*

   Recombinant DNA Techniques and Molecular Biology Methods.

2. **Advances in** *Identification of Target Genes.* Expressed Sequence Tagging and Positional Cloning.

3. **Availability of** *Solubilised Receptors/ enzymes* and

   Matrix Metallo Proteinases.

4. **Combinatorial Chemistry**

   Structural Based Design.

   Directed Diversity Technology of 3D. Pharmaceuticals.

It may take anywhere from 8 to 12 years before a new molecule reaches the market and the costs range from 300 to 500 million dollars.

Therefore, pharmaceutical companies have the motto "Succeed or Fail Fast". If a new compound is launched six months ahead of a competition, the investment in Research and Development can be recovered.

Therefore, the entire approach to screening is changing. Contracting out research is getting popular. Mergers to pool resources are taking place. Use of combinational libraries of compounds, peptides, natural products, in high thru-put micro titre screens, specific target directed screens using r-DNA based products is the focus of International Drug Research.

REFERENCES

1. Strategies in the Discovery of Drugs from Natural Sources : Annual Reports in Medicinal Chemistry, Vol. 17, 301, 1982.
2. Mulundcandin, a new lipopeptide antibiotic *J. Antibiotics*, Vol. 40, No. 3, Pages 225-289, 1987.
3. A New Ansamycin Antibiotic, Naphthomycin H from Streptomyces sp. Y - 83, 40369 *J. Antibiotic*, Vol. 40, No. 7, 1987.
4. Fumifungin, a new antifungal antibiotic from *Aspergillus fumigatius* Fresnius 1863. *J. Antibiotics*, Vol. 40, No. 7, 1987.
5. Kesarirhodins, a new complex of anthracycline antibiotics. Abstracts of the "Tubingen" Resprache zur Chemie von Micro- organismen.
6. Aranorosin, a novel antibiotic from *Pseudoarachniotus roseus*. J. Antibiotics, Vol. 41, No. 12, 1780-86, 1988.
7. L-919867, an unstable antifungal metabolite from *Trichoderma koningii* J. Antibiotics, Vol. 49, Pg. 210-211, 1996.

8. A. Microbial Screening Programme : Scope and Limitations in progress in Industrial Microbiology. Vol. 27, 1989 (Elsevier)

9. Novel pharmaceutical compounds from marine bacteria. J. Appl. Bacteria. Vol. 67, Pgs. 461-470, 1989.

10. Strategies for the discovery of secondary metabolites from marine bacteria Ann. Rev. Micro. Vol. 49, Pgs. 559-84, 1994.

CHAPTER - 29

PHARMACEUTICAL POTENTIAL OF CYANOBACTERIA

T. Thirunalasundari and G. Subramanian

National Facility for Marine Cyanobacteria, Barathidasan University, Tiruchirapalli - 620 024 INDIA

1. INTRODUCTION

Actinomycetes and Hypomycetes have been the source of the majority of bioactive natural products currently known from microorganisms. The discovery of new compounds decreased steadily over the past few decades from these two groups. Hence it is important to discover new source of bioactive natural products to treat newly emerging infectious diseases and also those that have become resistant to currently available drugs. In the last decade, microalgae have drawn much attention as prospective and excellent source of biologically active constituents (25). The cyanobacteria or blue green algae, are a well recognized source of structurally unique and biologically active natural products (10, 29).

Although there are numerous publications on the therapeutic properties of marine algae, very little information is available in similar application of marine cyanobacteria.

2. CYANOBACTERIA

As a group, cyanobacteria or 'blue bacteria' or blue green algae or 'cyanoes' evolved nearly three billion years ago and they have been responsible for the initial oxygenation of the Earth's atmosphere. Cyanobacteria are a large and extraordinarily diverse group of prokaryotes. Based on their ultra structure they are classified as gram negative bacteria.

Cyanobacteria range from unicellular to multicellular, coccoid to branched filaments, nearly colorless to intensely pigmented, heterotrophic to autotrophic, psychrophilic to thermophilic, fresh water to marine and hypersaline. They are free-living and are also found as endosymbionts. They are distinct from other algae and archebacteria. They have unique characteristics. The products of cyanobacteria are also unique and that could make them indispensable as a source for new drugs, chemicals and bioactive compounds. Cyanobacteria perform various metabolic functions like other prokaryotic bacteria. However, they are exposed to diverse conditions in their natural habitat and hence they have evolved alternative metabolic pathways and elegant regulatory mechanisms to respond to adverse environmental conditions and nutrient availability. The importance of cyanobacteria has been realized since long. In recent years people throughout the world are focussing their attention towards cyanobacteria for their possible use for the photoproduction of biofuels, ammonia, various metabolites, vitamins, toxins, therapeutic substances and aqua or animal feeds [22].

A large number of companies and government organizations are interested in screening cyanobacteria for bioactive natural compounds. Areas being targeted are Cancer therapy (12, 32), antibacterials (18, 23, 30, 34), antifungal (6, 9 & 17), antiviral (15, 20), antineoplastic (3, 29) and others. The wide array of cyanobacterial strains in the world and the range of genetic diversity they possess predict success in finding drugs. The pharmacologically active substances can subsequently be chemically synthesized. However, the complexity of some of these molecules would favour their production by cyanobacterial cultures only. Some studies on the effect of culture conditions on the production of these substances have already been undertaken (5, 14, 21).

The generally low level of these bioactive substances in cells requires that they either will have to be of very high value or their production cost must be low.

3. PHARMACEUTICAL COMPOUNDS

High value fine chemicals from cyanobacteria are recently being targeted towards medical diagnosis. Phycobiliproteins, the water soluble, coloured, highly fluorescent compounds consisting of covalently attached prosthetic groups, are used in food, drug, cosmetic industries as natural colouring agents and in clinical diagnostic laboratories as fluorescent reagents. Phycoflours i.e. phycoerythrin and allophycocyanin are used in fluorescence activated cell sorting (FACScan), flow cytometric analysis and histochemistry. They can be coupled to proteins like monoclonal antibodies, avidin and streptavidin or the small molecules like biotin and digoxigenin. Hence, they can be used as immunological probes in medical laboratory diagnosis.

Microcystins and *Nodularins*, the two important hepatotoxins produced by *Microcystis* and *Nodularia*, caused death of livestock and have been implicated in reports of human toxicity. They cause necrosis of liver in the affected animals and death is due to hemorrhagic shock caused by pooling of blood in the liver [13].

Based on the reports of potent toxicity associated with cyanobacteria, screening programmes have started as early as 1970, to collect and examine materials from natural population of cyanobacteria for novel compounds with antimicrobial and antineoplastic effects [24, 31]. In 1981, an algal culture programme began at the University of Hawaii resulting in the screening of extracts from 400 species of cyanobacteria. The strains screened were from freshwater, terrestrial and marine environments.

Highly encouraging preliminary results on the antiviral, antibacterial and antifungal activities of different marine cyanobacterial strains have resulted in enlarging the scope of the work towards isolation, purification and identification of the active compound and also development of the novel drugs [36].

The effect of the water soluble fraction of the ethanol extract of 12 strains of marine cyanobacteria on the biochemical composition of the wistar strain male albino rats suggested the potential application of some of the marine cyanobacteria as pharmaceuticals/modulators of the regulatory system, though a few strains exerted minor toxic effects [39].

(a) Anticancer Compounds

The term cytotoxicity is used to indicate inhibition of the proliferation of tumor cells *in vitro* and is distinct from the term antitumor or antineoplastic, which refer to *in vivo* activity in animal model systems. Screening of cyanobacterial extracts for the presence of novel cytotoxins has been promising. Approximately 7% of the extracts tested inhibited proliferation of the KB cell line. The *Scytonemataceae* and *Stigonemataceae* were the major families of cytotoxins [29].

The extracts of the cyanobacterium *Spirulina* in which B carotene is a major component, inhibited growth and destroyed oral cancer cells *in vitro*. B carotene, a nontoxic carotenoid has a cytostatic and cytotoxic dose - dependent effect on hamster and human oral squamous cell carcinoma cell lines. *In vivo* studies using the hamster buccal pouch cancer model have shown tumor inhibition and regression following topical or local injection of B carotene. Regression was associated with an *in situ* increase in tumor necrosis factor in conjunction with a significant increase in peritoneal macrophage cytotoxicity [32].

Preliminary results from the Harward groups study suggests that a natural carotene protein complex prepared from algae may act as an immune system modifier that marshals the body's own intricate immune defence system to combat cancer. This promising substance trade marked 'PHYCOTENE' consists of B carotene, fifteen other natural carotenes and photodynamic pigments like chlorophyll and other porphyrins [32].

The commercial product of *Spirulina* Phycotene appears to accomplish cytotoxicity by some immunological means. The pigment complex was more effective than pure B carotene in some experiments on tumor cells in culture. This is an exciting development in cancer research.

B carotene and B carotene-rich algal extracts inhibit the development of cancer tumor and cause regression of existing tumors. Early results of the research at Harward show that B carotene causes the inhibition and regression of cancer in animals and in cultures of human and animal cancer cells. The research also shows that B carotene-rich extracts of *Spirulina* and *Dunaliella* supplied by Microalgae International were more effective than pure synthetic B carotene in killing cancer cells in culture in some experiments. The preliminary results suggest that B carotene and the algal product works at least in part by triggering the body's own immune system defense mechanism [32].

Marine microalgae in general and marine cyanobacteria in particular have been identified as the most signficant group of organisms from which new anticancer type natural products have been isolated largely through the pioneering efforts of Moore and Patterson (25,27,28, 29). Many of these bioactive compounds possess unprecedented structures. Structural novelty is combined with exceptionally high incidence of biological activity. To make a mention, an early survey of marine cyanobacterial crude extracts showed 6% to be cytotoxic to KB cancer cell lines, 9% to possess antifungal activity and more than 5% to possess antiviral activity to Herpes simplex type [25].

Barchi and his coworkers (2) cultured some cyanobacteria in the laboratory, the extracts of which displayed good anticancer activity. Ethanolic extract of cultured

Tolypothrix byssoideae was found to show significant inhibitory activity *in vitro* against KB and NIH/3T3 cells and *in vivo* against P-388 lymphocytic leukemia in mice (2). They have reported the bioassay directed isolation and identification of the active constituent tubericidin from *Tolypothrix byssoideae*.

Cyanobacteria beloging to the Oscillatoriaceae are potential sources of new antineoplastic agents. The crude extract of *O. acutissima* shows cytotoxicity against KB and NIH/3T3 cells and significant antineoplastic activity *in vivo* against murine lewis lung carcinoma. The lipophilc extract of *O. acutissima* contains two novel macrolides, acutiphycin and a 20,21-dihydroacutiphycin which exhibit cytotoxicity and antineoplastic activity (3).

Bioassay directed fractionation of extract of the cyanobacterium *Phormidium tenue* led to the isolation of three classes of glycolipids viz. monogalactosyl diacylglycerol [MGDG], diagalactosyl diacylglycerol (DGDG) and sulfoquinovasyl diacylglycerol (SQDG) as antitumor factors. In comparing the antitumor effect of the three classes of glycolipids with various acyl residues MGDG and DGDG inhibited tumor promotion more intensely than SQDG. With respect to MGDG and DGDG the inhibitory activity of each glycolipids possessing a single pair of acyl residues was also investigated after HPLC separation. Among the 17 tested MGDGs and DGDGs 3 of the DGDGs showed more potent inhibitory activity toward tumor promotion than the others [33].

In the anticancer drug discovery efforts focused in the Caribbean collection, the cyanobacterium *Lyngbya majuscula* showed strong cytotoxic effect against a vero cell line ATCC CCI 81 (12). Subsequent evaluation at National Cancer Institute, assay of 60 cell lines uncovered a potent antiproliferative and cytotoxicity activity showing some selectivity for colon, renal and breast cancer derived cell lines. The extract was also found to be toxic to brine shrimp (LC_{50} = 25 ng/ml). Using the brine shrimp assay to guide fractionation, a unique metabolite isolated [8-10% of the crude extract] was found to be responsible for the potent brine shrimp toxicity (LC_{50}= 3ng/ml) and mammalian cell antiproliferative activity (LC_{50} = 6.8 ng/ml) in the Chinese hamster Aux B1 cell line.

Bioassay guided fractionation of the organic extract of a Curaco collection of *Lyngbya majuscula* led to the isolation of a new lipid, curacin A, with exceptional brine shrimp toxic and antiproliferative activities. Its unique thiazoline - containing structure has been deduced from spectroscopic information. Pure curacin A is an antimitotic agent (LC_{50} values in three cell lines ranging from 7-200 nm) that inhibits microtubule assembly and the binding of Colchicine to tubulin (11).

In discovering new efficient drugs for the treatment of solid tumors in clinical condition, Corbett (8) developed a rapid inexpensive disk diffusion assay to find agents that exhibit greater cytotoxicity against solid tumor cells than leukaemia cells (solid tumor selectivity) and/or greater cytotoxicity against tumor cells than normal cells (tumor selectivity). This assay is equivalent to the assay commonly used in antifungal and antibacterial screens. Aulosirazole, the major cytotoxin in the cyanobacterium, *Aulosira fertilissima* (uH strain DO-8-1) shows solid tumor selective activity in the Corbett assay. Its structure has been determined to be 5-hydroxy -3 methoxynaphthol (2, 3-d) 1,2 - thiazole - 4, 9-dione by a combination of spectroscopic, notably NMR and X ray crystallographic methods (34).

Extracts of pure strains of cultured marine microalgae were evaluated in a series of mechanisms based anti-cancer screens, including protein kinase C (PKC) assay, protein tyrosine kinase (PTK) assay and inosine monophosphate dihydrogenase (IMPDH) assay. The extracts showed activity against all the three enzymes (12).

The recently discovered antiproliferative agents cryptophycin A, Curacin A (12), and notodion A (21) produced by *Nostoc* sp., *Lyngbya majuscula* and *Nostoc commune* respectively, all inhibit microtubule assembly, function and show promise as potentially useful anticancer agents.

(b) Antiviral Compounds

To discover new antiviral agents from natural sources, extracts from cyanobacteria were screened for anti HIV properties (7). Cellular extracts from cultured *Lyngbya langerhimii* and *Phormidium tenue* protected human lymphoblastoid T cells from the cytopathic effect of HIV-I infection. With the use of an *in vitro* anti-HIV-1 assay (39) to guide fractionation a new class of HIV-I inhibitory compound the sulfonic acid containing glycolipids were isolated. At non-cytotoxic concentrations the pure compounds were strikingly active against HIV-1 in cultured human lymphoblastoid T-cell lines as indicated by a tetrazolium - based micro culture assay as well as by correlative p 24 viral protein (38) and syncytium formation assays (26).

A recently developed tetrazolium based microculture assay was used to screen extracts of cultured cyanobacteria for inhibition of the cytopathic effects of HIV. A number of extracts were found to be remarkably active against HIV. This HIV-1 inhibitory compound was also a sulfonic acid containing glycolipid. The pure compounds were active against HIV-1 in cultured human lymphoblastoid CEM, MT-2, LDV-7 and 13-44 cell line as in p24 viral protein and syncytium formation assay (15).

Water extract of *Spirulina platensis* (*Arthrospira platnesis*) inhibits HIV-1 replication in human derived T-cell lines and in human peripheral blood mononuclear cells (1). The extract also blocked Raucher murine leukaemia virus (RLV) induced plague (1).

Phormidium valderianum BDU 20571, *Plectonema terebrans* BDU 30342 and *Pseudoanabena schmidlei* 30313 showed antihepatitis activity (35).

(c) Antibacterial Compounds

Weak antibacterial activity of cyanobacterial extract has been commonly observed (25, 30). *Nostoc muscorum* was reported to produce an antibacterial agent with broad spectrum activity against bacteria (5). The chloroform methanol extract of all the species of *Oscillatoria* (except *Oscillatoria formosa* BDU 40261) could effectively inhibit the gram positive bacterial pathogens *Staphylococcus aureus*. *Phormidium angustissimum* BDU 40061 and *Lyngbya* sp. BDU 30601 were possessing antibacterial compounds against gram negative and gram positive pathogens while *Spirulina subsalsa* BDU 30311 possesed antibacterial compounds against gram negative organisms (37).

Phormidium angustissimum BDU 40061 and *Lyngbya* sp. BDU 30601 are potential sources of new broad spectrum antibacterial antibiotics. (37) Extracts of *Synechococcus elongatous* BDU 30312, *Phormidium corium* BDU 30214 and *Dichothrix baueriana* BDU 40481 are active against the fungus *Candida* sp. (37).

(d) Food and health Products

The cyanobacterium *Spirulina* is a nutrient-dense material with over 60% good quality proteins, vitamins, B carotene and gamma linoleic acid which are mainly used as health, food and feed additives. Though initially considered as a protein source it has now been found to be a natural food with therapeutic value (41, 42). b-carotene which is identified as one of the top ten scientific achievements of 1991, has an important role in cancer treatment. *Spirulina* is rich in B carotene and in receiving a great deal of attention in India as in the international scene, *Spirulina* has been shown to produce hypocholesterolemic and hypoglycemic effects, increased lactation in nursing mothers. It also has chemopreventive and wound healing properties (19).

At CFTRI, Mysore, *Spirulina platensis* was used to biotransform codein to morphine, which is a valuable pharmaceutical compound with powerful analgesic, narcotic and antispasmodic action and is used to relieve pain.

Spirulina is currently used as a food supplement and as a source of nutrients like vitamins, essential amino acids, fatty acids. It is a source of dietary gamma linolenic acid (GLA), polyunsaturated fatty acid, like GLA which have been demonstrated to have therapeutic potential in activating or preventing various diseases and disorders (16).

In the last few years, there has been a great interest in commercialization and marketing of *Spirulina* as ethical drugs in health care. New Ambadi estate, Madras has set up a commercial unit in Tamil Nadu to produce about 20 tonnes of *Spirulina* per year which is being scaled up and has launched a product called Multinol in tablet form which is marketed in Tamil Nadu. Ballarpur Industries (Thapar Group, New Delhi) have set up a plant for 40 t/a at Nanjangud near Mysore. The other products marketed presently in India are Progen - a product from Spiro-tech-Lab, Madras for chemoprevention. Another pharmaceuticals company also at Madras (Natura, American Remedies) has introduced two products viz. Inspira (Powder and tablet form) as an ethical product sold through drug stores.

Studies on the effect of *Spirulina* on hyperlipidemia, adipohepatosis, tumor propagation, immune system response, radio protection, renal toxicity, intestinal flora, diabetics, obesity, hypertension and iron assimilation showed beneficial results. (4).

Based on the food intake, body weight change and the haematological characters in rats treated with the extracts of marine cyanobacterial species, showed that strains of species such as *Spirulina subsalsa, Oscillatoria salina* and *Phormidium valderianum* may proved to be potential organisms useful as feed/nutritional supplement, while all the species examined may in one form or other be useful pharmacologically (38).

A nutritive, non-toxic marine cyanobacterium *Phormidium valderianum* BDU 30501 in a dry pellet form proved superior to pondscum containing a mixture of fresh water algae, cyanobacterial pellet fed (as a sole energy source), tadpoles showed more growth than pondscum and control (40).

A marine cyanobacterial strain with reasonable high growth rate, nutritive value and acceptability by fishes and prawn has been identified and formulated as an aquaculture feed that has been successfully tested in the laboratory and being tested in the field (36).

The traditional use of cyanobacteria including food, energy source and biofertilizers flourish today and are begining to reap the benefits of modern molecular biology to enhance their performance. The first generation of high value added biproducts produced by cyanobacteria, including phycoflours are currently available in the market and is quite popular. Only a few cyanobacterial strains producing interesting compounds have been well characterised or exploited commercially. Screening programmes must identify useful products and the cyanobacterial strains that produce significant amounts of the product.

Projection for future research on pharmaceutical potential of cyanobacterial include discovery of novel bioactive metabolites, investigation on the pathways by which selected metabolites are biosynthesized and regulated and a precise elucidation of the mechanisms by which these novel compounds exert their biological effects.

REFERENCES

1. Ayehuine, S., Belay, A., Hu, Y., Baba, T., Ruprecht, T. 1996. 7th International Conference of the International Society of Applied Algology, Knysna, South Africa, 16th April 1996.
2. Barchi, J.J., Norton, T.R., Gurusawat, E., Patterson, G.M., Moore, R.E. 1983. Phytochemistry. 22 (12) : 2851-52.
3. Barchi, J.J., Moore, R.E. Patterson, G.M.L. 1984. J. Am. Chem. Soc., 106 : 8193-97.
4. Belay, A., Ota, Y., Miyakawa, K., Shimamatsu, H. 1993. J. Applied Phycology, 5 : 235-41.
5. Bloor, S., England, R.R. 1991. Enzyme microb. technol. 13 : 76-81.
6. Bonjouklian, R., Smitka, T.A., Doolkin, L.E., Molloy, R.M. Debono, M., Shaffer, S.A. Moore, R.E., Stewar, J.B., Patterson, G.M.L. 1991. Tetrahedron. 47 : 7739-50.
7. Boyd, M.R. 1988. In AIDS etiology, diagnosis, treatment and prevention, V.J. Jr. DeVita, S, Hellman, S.A. Rosenberg, PP. 305- 17 Philadelphia; Lippincott.
8. Corbett, T.H., Valeriote, F.A., Polin, L., Panchapor, C., Pugh, S., White, K. et al., 1992. In cytotoxic anticancer drugs : Models and concepts for drug discovery and development ed. F.A. Valeriote, T.H. Corbett, L.H. Baker, PP. 35-87. Kluwer Norwell. Academic Publishers.
9. Decaire, G.Z., Decano, M.S., Demule, M.C.Z., Dehalperin, D.R. 1990. Phyton. 51 : 1-4.
10. Gerwick, W.H., Jiang, Z.D., Agarwal, S.K., Farmer, B.T. 1992. Tetrahedron. 48 : 2313-23.
11. Gerwick, W.H., Protean, P.J., Nagle, D.G., Hamel, E., Blokhin, A., Slate D.L. 1994. J. Org. Chem. 59 : 1243-45.
12. Gerwick, W.H., Roberts, M.A., Protean, P.J. Chen, J.L. 1994. J. of Applied Phycology. 6 : 143-49.
13. Gorham, P.R., Carmichael, W.W. 1988. In : Algae and Human Affaris, C.A. Lembi, J.R. Waaland, Cambridge University Press. pp. 403-431.
14. Gromov, B.V., Vepritskiy, A.A., Titova, N.N., Mamkayeva, K.A., Alexandrova, O.V. 1991. J. Appl. Phycol. 3 : 55 - 59.
15. Gustafson, K.R., Cardellina, J.H., Fuller, R.W., Weislow, O.S., Kiser, R.F., Snader, K.M., Patterson, G.M.L., Boyd, M.R. 1989. J. Nat. Cancer Inst. 81 : 1254-58.

16. Harris, R.W.C., Key, T.J.A., Silcocks, P.B., Bull, D., Wald, N.J., 1991. Nutr. Cancer. 15 : 63-68.

17. Ishibashi, M., Moore, R.E., Patterson, G.M.L., Xu, C., Clardy, J. 1986. J. Org. Chem. 51 : 5300-06.

18. Kellam, S.J., Walker, J.M. 1986. Brit. Phycol. J. 24 : 191- 94.

19. Klasner, 1986. Biotechnology, 4 (11) : 947.

20. Knubel, G., Larsen, L.K., Moore, R.E., Levine, I.A. Patterson, G.M.L. 1990. J. Antibiot. 43 : 1236-39.

21. Kobayashi, J., Ishibashi, M., Murayama, J., Takamatsu, M., Iwamura, M., Ohizumi, Y., Sasaki, T. 1990. J. Org. Chem. 55 : 3421-23.

22. Mohanty, P. 1996. J. Sci. and Indus. Research. 55 : 553-54.

23. Miura, Y., Current topics in marine Biotechnology. Miyachi, S., Kanube I., Ishida Y. The Japanese Society for Marine Biotechnology, Tokyo, pp. 189-190.

24. Moore, R.E. 1981. Marine natural products : Chemical and Biological perspectives. Volume 4, ed. P.J. Schere. New York Academic Press. pp. 1-52.

25. Moore, R.E., Patterson, G.M.L. Carmichael, W.W. 1988. Biomedicinal importance of marine organisms. ed. D.G. fautin. Mem. Calif. Acad. Sci. 13 : 143-50.

26. Nara, P.L. Hatch, W.C., Dunlop, N.M., et al., 1987. AIDS Res. Hum. Reterovirus, 3 : 283-02.

27. Patterson, G.M.L., Baker, K.K., Baldwin, C.L., Bolis, C.M., Caplan, F.R., Larsen, L.K., Levine, I.A., Moore, R.E., Nelson, C.S., Tschappat, K.D., Tuang, G.D., Boyd, M.R., Cardellina II, J.H., Collins, R.P., Gustafson, K.R., Snader, K.M., Weislow, O.S., Lewin, R.A. 1993. J. Phycol. 29 : 125-30.

28. Patterson, G.M.L., Larsen, L.K. Moore, R.E. 1994. J. Applied Phycology. 6 : 151-57.

29. Patterson, G.M.L., Baldwin, L., Bolis, C.M., Caplan, F.R., Karuso, H., Larsen, S.K., Levine, I.A., Moore, R.E., Nelson, C.S., Tschappat, K.D., Tunag, G.D., Furusawa, E., Furusawa, S., Norton, T.R., Rayborne, R.B. 1991. J. Phycol. 24 : 530-36.

30. Reichett., J.L., Borowitzka, M.A. 1984. Hydrobiologia 116/117 : 158-68.

31. Rinehart, Jr. K.L., Shaw, P.D., Shield, L.G., Gloer, J.B., Harbour, G.C., Koker, M.E.S. et al. 1981. Pure. appl. chem. 53 : 795-17.

32. Schwartz, J., Shklar, G. 1986. Abstract 40th Annual meeting of the Americal academy of oral pathology, Toronto, Canada, May 13, 1996.

33. Shirahashi, H., Murakami, N., Watanabe, M., Nagatsu, A., Sakakibara, J., Tokuda, H., Nishino, H., Iwashima, A. 1993. Chem. Pharm. Bull. 41 (9) ; 1664-66.

34. Startman, K., Belli, J., Jensen, C.M., Moore, R.E., Patterson, G.M.L. 1994. J. Org. Chem. 59 : 6279-81.

35. Subramanian, 1993. In Proceedings of the National seminar on cyanobacterial research Indian scene pp. 86.

36. Subramanian, G., 1996. In : Proceedings of the International symposium

on cyanobacterial biotechnology. [in press].

37. Sundararaman, M. 1992. Ph.D. Thesis submitted to Bharathidasan University.
38. Sundararaman, M., Averal, H., I., Akbarsha, M.A., Subrramanian, G. 1994. Ann. appl. Biol. 125 : 195-06.
39. Sundararaman, M., Subramanian, G., Averal, H.I., Akbarsha, M.A. 1996. Phytotherapy Research, 10 : 9-12.
40. Uma, L., Palanisamy, M., Murugian, P., Akbarsha, M.A., Subramanian, G. Subramanian, G. 1996. In : Proceedings of the International symposium on cyanobacterial biotechnology [in press].
41. Venkataraman, L.V., Becker, E.W. 1986. In : Biotechnology and utilization of algae- The Indian Experience. 126-136. Published by Department of Science and Technology, New Delhi and CFTRI, Mysore, India.
42. Venkataraman, L.V., 1983. *Spirulina* - A monograph. Published by the Central Food Technological Research Institute, Mysore, India.

CHAPTER – 30

SYNTHESIS OF SURFACE ACTIVE COMPOUNDS BY MICROORGANISMS

Swaranjit Singh

Institute of Microbial Technology, Sector 39-A, Chandigarh 160 036

1. INTRODUCTION

Surface active compounds or biosurfactants are amphiphilic compounds produced by microorganisms that are capable of decreasing surface and interfacial tensions. Microbial interest in biosurfactants has been associated mainly with the problem of substrate transport during hydrocarbon fermentations. These biosurfactants may be extracellular or they may be associated with the cell-wall without being excreted into the culture broth. Emulsification results in a dispersion of small oil drops in water, which increases the surface area between the two phases. These interfacial surfaces are most efficient for microbial growth.

Different types of biosurfactants are reported in the literature which are produced by a variety of microorganisms. They are broadly grouped into the following chemical categories : carbohydrate - or amino-acid - containing or as a phospholipid, fatty acid or neutral lipid (13). The most common hydrophobic group is the hydrocarbon chain of a fatty acid. The polar or hydrophilic group includes a wide range of organic functional groups. Examples are the ester and alcohol functional groups of neutral lipids, the phosphate - containing portions of phospholipids, the sugars of glycolipids, or acetate groups. The microbial cell surface itself can be considered to be a biosurfactant (30). The cell surface is composed of a mosaic of hydrophobic and hydrophilic moieties which help in emulsification and reduction of surface and interfacial tensions.

2. HISTORY

Microbiological interest in biosurfactants as mentioned in the introduction is associated mainly with the problem of substrate transport during hydrocarbon fermentation. A large number of reports appear from authors all over the world on the production, degradation of biosurfactants and effect of different factors on their synthesis. Purification and characterization of surface active compounds composed of protein and lipid was reported by Cameotra & Singh (9). They observed that carbohydrate was not an essential component. Some workers have used thermophilic cultures for the purpose (11). Enhanced removal of Exon Valdez spilled oil from Alaskan gravel by a microbial surfactant was reported by Harvey *et al.*, (21). Production of lipids by bacterial strains showing emulsifying activity was isolated by different groups (3, 7, 24). Bryant reported an improved method for the isolation of an glycolipid surfactant from a Rhodococcus strain H13A (6). McElwee *et al.*, observed that the emulsifying activity of biosurfactant produced by *Pseudomonas aeruginosa* UG1 was not influenced by pH over the range 5 to 9. A linear relationship was observed between emulsifier concentration and emulsifying activity (33). A review listing the different chemical types of biosurfactants is available (62). The surface activity of Halobacteria and potential use in microbially enhanced oil recovery is discussed by Post and Al-Harjan (39). It was demonstrated that a yeast strain *Endomycopsis lipolytica* YM produced a stable biosurfactant if prior addition of a biosurfactant isolate of Pseudomonas PG-1 was added to the yeast culture (8). It is reported that uptake of n-alkanes above n-nonane by microorganisms is mediated by formation of emulsifiers (10).

3. APPLICATIONS OF SURFACE ACTIVE COMPOUNDS

There are a large number of industrial applications of biosurfactants. Petroleum production and enhanced oil recovery represents the largest future market for biosurfactants. The total market for surfactants totalled 1.6 billion kg. in 1983, and was predicted to reach 1.8 billion kg, by 1990 (32). The second largest market for surfactants is emulsion polymerization for paints, paper coatings and industrials coatings. Surfactants are also used in food and cosmetic industries. Hydrophobic pollutants found in petroleum hydrocarbons require emulsification before being degraded by microbial cells (23, 27, 35, 51). Surfactants increase the surface area of hydrophobic pollutants, thereby increasing their water solubility. Hence, the presence of surfactants may increase microbial degradation of pollutant in both soil and water. Biosurfactants are technically important because they exhibit surface active properties differing from synthetic surfactants. Biosurfactants are more effective for different specific pur- poses. In contrast to some chemical surfa- ctants, biosurfactants are biodegradable thereby reducing the potential for pollution (37, 50, 59). Biosurfactants have also been used in the formulation of poorly soluble organophosphorus pesticides. Chemical surfactants have been added to disperse oil spills, however their use has been criticized due to the toxicity and persistance of some of the surfactants (5). Therefore, bacterial surfactants have use in clean up of oil spills. Biosurfactants can also be used to remove oily waste from solid surfaces such as in oil tanker cleanup. For smooth flow of crude oil in pipelines in cold countries, biosurfactants are added to decrease the viscosity of oil.

Microbial surfactants compare favourably with synthetic surfactants because they are easy to produce, cheap, highly effective in low concentrations and are biodegradable. There is an ever increasing demand for biosurfactants in oil service industry. Their use in enhanced recovery of oil, tanker cleanup and in the oil slick dispersion is on the rise. Their use in enhanced recovery of oil, tanker cleanup and in the oil slick dispersion is on the rise. Apart from these applications biosurfactants have other uses in the field of herbicidal applications (2, 11) and removal of environmental pollution listed in Table 1 ,are production of some useful products by microbial growth on alkanes. The scope of microbial genetic and fermentation technology in the oil service industry will ultimately depend upon the willingness of the industry to explore the new research frontiers that could open up entirely new opportunities for developing microbial surfactant.

Table I. Production of some useful compounds from Alkanes

Compounds	Yield
Group I	
L-Glutamic acid	72-84.0 g/l
L-Lysine	10-34.0 g/l
L-Threonine	9-18.9 g/l
L-Valine	9.0 g/l
α-Ketoglutaric acid	65.8 g/l
Citric acid	62-183.0 g/l
Isocitric acid	40.0 g/l
Fumaric acid	50.0 g/l
Mallic acid	24.3 g/l
Erythritol	180.0 g/l
Mannitol	65.9 g/l
5'-Inosinic acid	1.9 g/l
Group II	
Vitamin B2 (Ribolavin)	281.0 mg/l
FMM + FAD	27.0 mg/l
Coenzyme A	80.0 mg/l
Cytochrome C	12.2 mg/l
Coenzyme Q	32.2 mg/l
Catalase	6200.0 U/ml
Uricase	140.0 U/l
Group III	
Tetradecane-1	
14-dicarboxylic acid	61.5 g/l
2-Methylisocitrate	35.0 g/l

4. ROLE OF BIOEMULSIFIERS IN HYDROCARBON UPTAKE

Emulsification may play an important role in the microbial uptake of hydrocarbons either by governing the interfacial area of the substrate to be contacted by cells or by controlling the transfer rate of the substrate into the aqueous phase. The effect of addition of various synthetic and natural surfactants on the growth of organisms was however found to be highly variable. Aiba *et al.*, 1969 (1) reported that addition of small amount of Tween 20 and Scourol 400 (polyoxyethylene sorbitan monolaurate) reduced the specific growth rate of *Candida guilliermondii* on n-alkanes. They observed that contact between yeast cells and oil droplets was prevented by the presence of these surfactants. Similarly, Mimura *et al.*, 1971 (34) reported that almost all the synthetic surfactants weakened the affinity of cells to oil but naturally occuring surfactants promoted the growth of yeast on n-hexadecane. Wang and Ochoa (56) however showed that the presence of artificial surface active materials greatly influence the interfacial area between oil and water and increased the specific growth rate. Whitworth *et al.* (58) also reported stimulating effect of several synthetic surfactants on the oxidation of n-dodecane by *Candida lipolytica.* Nakahara *et al.*, (36) observed that growth of *Pseudomonas aeruginosa* with a low affinity for hydrocarbon was stimulated by surfactants, the strongest activity being produced by polyethylene glycol alkyl aryl ether type of surfactants. Both inhibitory and stimulating effects of synthetic non-ionic surfactants on the growth of yeast on hydrocarbon mixture was reported by Tanaka & Fukui (54) and it was shown that the effect was dependent on the chemical structure of the surfactants. Similarly conclusion was also arrived at by Whitworth (57).

The production of bioemulsifiers by microorganisms growing on hydrocarbon substrates was widely reported. These bioemulsifier which are in various chemical forms - glycolipid, lipopeptide/lipoprotein, phospholipid, neutral lipid and protein were claimed to play a role in hydrocarbon uptake through emulsification of the substrate. As early as 1962, Tallock *et al.*, (53) reported the production of rhamnose lipid with hydrocarbon emulsifying properties in the culture of *Torulobsis magnolise* grown on hydrocarbon. Later ltoh *et al.*, (26) and Hisatsuka *et al.*, (22) observed the extracellular production of rhamnolipid by *Pseudomonas aeruginosa* grown on alkanes and showed that the growth of the organism on alkane was stimulated by the addition of rhamnolipid. Itoh & Suzuki (28) further showed that the growth of a mutant strain of *Pseudomonas aeruginosa* deficient in n-paraffin utilizing ability was stimulated by the addition of rhamnolipid indicating the important role of bioemulsifier in hydrocarbon uptake and growth.

From the emulsion layer of *Arthrobacter paraffineus* grown on n-alkanes, trehalose dimycolates (trehalose lipids) were isolated with mycolic acid (long chain α-branched β-hydroxy fatty acids) residues ranging from C_{28} to C_{38} by Suzuki *et al.*, (52) and it was suggested that these surfactant metabolites may play a role in the uptake of hydrocarbons by micro- organisms. Rapp *et al.*, (42) reported the production of 2.1 g per litre of surface- active 6,6-trehalose dimycolate by *Rhodo- coccus erythropolis* DSM 43215 grown on n-alkanes but the production was lower (0.4-0.8 g/L) when the organism was grown on glucose or glycerol. The trehalose lipid from *R. erythropolis* can drastically decrease the interfacial tension between n-alkane and water at a very low concentration. This was observed by Rapp *et al.*, (43). However, it was reported that with highly purified trehalose dimycolates, at least one additional compound was necessary for emulsification and emulsion stabilization of an n-alkane/water emulsion - Rapp *et al.*, (42). The finding may be significant in relation to the reports of the production of specific hydrocarbon solubilizing factors produced by hydrocarbon grown organism - Goma *et al.*, (19, 20), Roy *et al.*, (49), Reddy *et al.*, (44, 45). Other surface active glycolipids produced by microorganism grown on hydrocarbons are sophorose lipids produced by *Torulopsis gropengiesseri* - Jonnes *et al.*, (29) and mannosylerythritol lipid which was reported to be produced in large amounts (36 g/l in 4 days) by a mutant of Candida species from n-alkanes or triacylglycerols but not from carbohydrates Kowashina *et al.* (31).

There are a number of reports on the production of lipoprotein type of biosurfactants by hydrocarbon-grown microorganisms. Lguchi *et al.* (25) isolated an emulsifying factor produced by hydrocarbon assimilating *Candida petrophilum* and it was shown to be lipopeptide in nature. *Zajic et al.*, (61) described the extracellular production of a bioemulsifier by *Corynebacterium hydrocarboclastus* grown on hydrocarbons and it was found to be composed of protein (4.5%), lipid (25%) and carbohydrate (15.2%). The bioemulsifier was produced to a maximum amount of 5-6 g/l and because its emulsification properties were unaffected by high salt concentration and high level of water hardness, it was considered to have potential in enhanced oil recovery. Gerson & Zajic (18) reported biosurfactant production at a concentration of 50-80 times the critical micelle concentration by four species of *Corynebacterium-C, fasciens, C. hydrocarboclastus, C. lepus and C. xerosis* from kerosene. It was shown that *C. lepus* produces a mycolic acid containing lipopeptide which was postulated to enhance growth by emulsifying hydrocarbons - Cooper *et al.*, (14) Cooper, *et al.*, (15). It was shown that mycolic acids were the most important components for surfactant properties and these properties depend on the length of the alkyl chain and pH of the solution in a manner consistent with the hydrophilic/ lipophilic balance of these compounds Cooper *et al.*, (16).

Table II. Surface Active Compounds produced by Microorganisms

Compound	Microorganism
Rhamnolipid	*Torulopsis magnoliae*
–do–	*Pseudomonas aeruginosa*
–do–	–do–
Trahalose lipid	*Arthrobacter paraffineus*
–do–	*Brevibacterium sp.*
–do–	*Corynbacterium sp.*
–do–	*Nocardia rhodochrous*
–do–	*Rhodococcus erythropolis*
Sophorse lipid	*Torulopsis gropengisseri*
Mannosyleryt-hirtol lipid	*Candida sp.*
Lipopeptide	*Candida petrophilum*
–do–	*Corynebacterium lepus*
Peptide	*Endomycopsis lipolytica*
Protein, lipid, carbohydrate complex	*Candida lipolytica*
–do–	*Corynebacterium hydrocarboclastus*
–do–	*Pseudomonas sp.*
Lipopolysac-charide	*Arthrobacter sp. (Acinetobacter calcoaceticus)*
Neutral lipids	*Nocardia erythropolis*

Gerson *et al.*, (17) further showed that isonicotinic acid hydrazide inhibited the growth of *C. lepus* on hexadecane but not on fructose, though inhibition of the synthesis of mycolic acid containing lipopeptide bioemulsifiers. Cooper *et al.*, (12) reported the production of several types of surface active compounds by *Corynebacterium fasciens* CF 15 from alkanes and carbohydrates and showed that the lowering of surface and interfacial tensions were mainly due to a mixture of extracellular lipids but emulsion stabilization and de-emulsification properties were primarily associated with the bacterial cells. MacDonald *et al.*, (35) reported the production of surface-active lipids by *Nocardia erythropolis* (ATCC 4277) grown on hydrocarbons and showed that 90% of the activity was due to neutral lipids containing a fatty alcohol. *Candida lipolytica* was shown to produce an extracellular polymer with emulsifying properties when grown on n-C_{14} or mixture of linear hydrocarbons - Pareilleux (38). The polymer was further known to be complex molecule with a protein, a lipid and a carbohydrate moiety.

Roy *et al.*, (49) reported the isolation of an extracellular emulsifying factor from the culture of *Endomycopsis lipolytica* Y-13 grown on alkanes and showed its peptide character. It was reported that oil-degrading *Arthrobacter* sp. RAG-1 (later characterised as *Acinetobacter calcoaceticus* RAG-1) studied by Israeli investigators produced an extracellular non-dialyzable emulsifying agent when grown on hexadecane, ethanol or acetate medium - Reisfeld *et al.*, (46). The purified emulsifying factor was shown to emulsifying light petroleum oil, diesel oil and a variety of crude oils and gas oils but not aromatic or aliphatic fractions of crude oil. Binary mixtures containing an aliphatic and aromatic hydrocarbons were shown to be excellent substrates for emulsification by the emulsifier Rosenberg *et al.*, (47). The emulsifier was shown to be a lipopoly-saccharide with average molecular weight of 9.8×10^5. The major components of the emulsifier were found to be D-galactos-amine, an unidentified amino acid and fatty acids joined to the heteropolysaccharide backbone through O-ester linkages Zuckenberg *et al.*, (63). The fatty acid moiety which was reported to play an important role in emulsification contained unusual amounts of 3-hydro- xydedecanoic and 2-hydroxydedacanoic acids - Belsky *et al.*, (4). Rosenberg and Rosenberg (48) showed that a non-adherant mutant of *Acinobacter calcoaceticus* RAG-1 would not grow on hexadecane but after addition of the emulsifying agent produced by the adherent strain enabled the mutant to grow on hexadecane and concluded that adherence is a crucial factor in the growth of *A. calcoaceticus* RAG-1 on hexadecane in the absence of extracellular emulsification of the substrate.

Reddy *et al.*, (45) reported the production of a heat stable polymeric substance with strong hydrocarbon emulsifying and solubilizing properties by a Pseudomonas species PG-1 grown on pristane. The crude isolate of the polymer was found to contain 34% protein, 16% carbohydrate and 40% lipid. It was suggested that the isolate may contain at least two different compounds - one with hydrocarbon emulsifying property and the other with hydrocarbon solubilizing property. Recently, Pruthi & Cameotra (40) using three bacterial strains demonstrated that the maximum reduction of the surface tension of the medium can be used to detect maximum surface active compounds in the medium.

ACKNOWLEDGEMENT

The author acknowledges the help rendered by Mr. Sushil Kumar in excellent typing of this manuscript.

REFERENCES

1. Aiba, S., Moritz, V., Someya, J.I., Haung, K.L. 1969. J. Ferment. Technol. 47 : 203.
2. Banerjee, S., Dutta Gupta, S., Chakrabarty, A.M. 1983. Arch. Microbiol. 135 : 110.
3. Banerjee, S., Karns, J.S., Chakrabarty, A.M. US Plant US 50 : 13654.
4. Belsky, I., Gutnic, D.L., Rosenberg, E. 1979. FEBS Letters 101 : 175.
5. Brown, M.J. 1989, Biodegradation of oil in fresh water, ed. J. Green, M.W. Trett PP. 197. Elsevier Science Publishers Ltd.
6. Bryant, F.O. 1990. Appl. Environ. Microbiol. 56 (5) : 1494.
7. Cameotra, S.S., Singh, H.D., Hazarika, A.K., Baruah, J.N. 1983. Biotechnol. Bioeng. 25 : 2945.
8. Cameotra, S.S., Singh, H.D., Baruah, J.N. 1984. Biotechnol. Bioeng. 26 : 554.
9. Cameotra, S.S., Singh, H.D. 1990. J. Ferment. Bioeng. 69 : 341.
10. Cameotra, S.S., Singh, H.D. 1990. J. Biosci. 15 (4) : 313.
11. Chakrabarty, A.M. 1985. Trends in Biotechnol. 83 (2) : 32.
12. Cooper, D.G., Akit, J., Kosaric, J. 1982. J. Ferment. Technol. 60 : 19.
13. Cooper, D. Zajic, J. 1980. Adv. Appl. Microbiol. 26 : 229.
14. Cooper, D.G., Zajic, J.E., Gerson, D.F. 1979. Appl. Environ. Microbiol. 37 : 4.
15. Cooper, D.G., Zajic, J.E., Gracey, D.E.F. 1979. J. Bact. 137:795.
16. Cooper, D.G., Zajic, J.E., Denis, C. 1981. J. Am. Oil Chem. Soc. 58 : 77.
17. Gerson, D.F., Cooper, D.G., Ramsay, B., Zajic, J.E. 1980. Can. J. Microbiol. 6 : 1498.

18. Gerson, D.F., Zajic, J.E. 1979. Antonie Van Leeuwenhoek 45 : 81.
19. Goma, G., Pareilleux, A., Durand, G. 1973. J. Ferment. Technol. 51 : 616.
20. Goma, G., Pareilleux, A., Durand, G. 1974. Agr. Biol. Chem. 38 : 1273.
21. Harvey, S., Bashvili, I., Valdes, J.J. Karnvely, D., Chakrabarty, A.M. 1990. Bio/Technology. 8 (3) : 228.
22. Hisatsuka, K., nakahara, T. Sano, N., Yamada, K. 1971. Agr. Biol. Chem. 35 : 686.
23. Hisatsuka, K., Nakahara, T., Sano, N., Yamada, K. 1971. Agr. Biol. Chem. 35 : 686.
24. Horowitz, S., Gilbert, J.N., Griffin, W.M. 1990. J. Ind. Microbiol. 6 (4) : 243.
25. Lguchi, T., Tanaka, I., Oksawa, H. 1969. Agr. Biol. Chem. 33 : 1657.
26. Itoh, S., Honda, H., Tomita, F., Suzuki, T. 1971. J. Antibiot. 24 : 855.
27. Ito, S., Inoue, S. 1982. Appl. Environ. Microbiol. 43 : 1278.
28. Itoh, S., Suzuki, T. 1972. Agr. Biol. Chem. 36 : 2233.
29. Jones, D.F., Moore, R.H., Craley, G.C. 1970. J. Chem. Soc. London C. No. 12 : 1725.
30. Kosaric, N. Crairns, N.L., Gray N.C.C. 1987. Biosurfactants and Biotechnology, arcel Dikker Inc.
31. Kowashima, H., Nakahara, T., Oagaki, M., Tabuchi, T. 1983. J. Ferment. Technol. 61 : 143.
32. Layman, P.L. 1985. Chem. Eng. News 63 (3) : 23.
33. MacElwee, C.G., Lee, H, and Trevors, J.T. 1990. J. Ind. Microbiol. 5 (1) : 25.
34. Manimura, A., Watanbe, S., Takeda, I. 1971. J. Ferment. Technol. 49 : 255.
35. McDonald, C.R., Cooper, D.G., Zajic, J.F. 1981. Appl. Environ. Microbiol. 41 : 117.
36. Nakahara, T. Hisatswka, K., Minoda, Y. 1981. J. Ferment. Technol. 59 : 415.
37. Oberbremer, A., Muler-Hurtig, R., Wagner, F. 1990. Appl. Microbiol. Biotechnol. 32 : 485.
38. Pareilleux, A. 1979. Eur. J. Appl. Microbiol. 8 : 91.
39. Post, F.J., Al-Harjan, F.A. 1988. Appl. Microbiol. 11 : 97.
40. Pruthi, V., Cameotra, S.S. 1995. Biotechnol. Techniques 9 (4) : 271.
41. Premuzic, E.T., Liu, S.M. 1991. Resour. Conserv. Recycl. 5 (2-3) 277.
42. Rapp, P., Bock, H., Wray, V., Wagner, F. 1979. J. Gen. Microbiol. 115 : 491.
43. Rapp, P., Bock, H., Urban, E., Wagner, F. Gebetsberger, W., Shulze, W. 1977. Decheme-Monographien 81 : 177.
44. Reddy, P.G. Singh, H.D., Roy, P.K., Baruah, J.N. 1982. Biotechnol. Bioeng. 24 : 1241.
45. Reddy, P.G., Singh, H.D., Pathak, M.G., Bhagat, S.D., Baruah, J.N. 1983. Biotechnol. Bioeng. 25 : 387.
46. Reisfeld, A., Rosenberg, E., Gutnick, D. 1972. Appl. Microbiol. 24 : 363.

47. Rosenberg, E., Perry, A., Gibson, D.T., Gutnick, D.L. 1979. Appl. Environ. Microbiol. 37 : 409.

48. Rosenberg, M., Rosenberg, E. 1981. J. Bacteriol. 148 : 51.

49. Roy, P.K., Singh, H.D., Bhagat, S.D., Baruah, J.N. 1979. Biotechnol. Bioeng. 21 : 955.

50. Shoham, Y. Rosenberg, M., Rosenberg, E. 1983. Appl. Environ. Microbiol. 46 : 573.

51. Stucki, G.,.Alexander, G. 1987. Appl. Environ, Microbiol 53 : 1278.

52. Suzuki, T. Tanaka, K., Matsubara, I., Kinoshita, S. 1969. Agr. Biol. Chem. 33 : 1619.

53. Talloch, A.P., Spencer, J.F.T., Goria, P.A.J. 1962. Can. J. Chem. 40 : 1326.

54. Tanaka, A., Fukui, S. 1971. J. Ferment. Technol. 49 : 809.

55. Thomas, J.M., Trody, J.R., Amador, J.A., Alexander, M. 1986. Appl. Environ. Microbiol 52 : 290.

56. Wang, D.I.C., Ochoa, A. 1972. Biotechnol. Bioeng. 14 : 345.

57. Whitworth, D.A. 1974. Compt. Rend. Tray. Lab. Carlsberg. 40 : 33.

58. Whitworth, D.H., Moo-Young, M, Viswanatha, T. 973. Biotechnol. Bioeng. 15 : 649.

59. Zajic, J.E., Guignard, H., Gerson, D.F. 1977. Biotechnol. Bioeng. 19 : 1303.

60. Zajic, J.E., Guignard, H., Gerson. D.F. 1977. Biotechnol. Bioeng. 19 : 1285.

61. Zajic, J.E., Guignand, H., Gerson, D.F. 1977. Biotechnol. Bioeng. 19 : 1303.

62. Zajic, J.E., Seffens, W. 1984. CRC crit. Rev. Biotechnol. 1 : 87.

63. Zuckenberg. A., Diver, A., Peeri, Z., Gutnick, D.L., Rosenberg, E. 1979. Appl. Environ. Microbiol. 37 : 414.

CHAPTER - 31

CYCLODEXTRINASE ACTIVITY OF A NATURAL ISOLATE OF *Alkalophilic bacillus*

A.M. Sonawane, Nawani, and B.P. Kapadnis

Department of Microbiology, University of Pune, Pune-411 007, INDIA

ABSTRACT

A cyclodextrin glycosyltransferase (CGTase) producing alkalophilic bacterium was isolated on a HM-I medium of pH 10.4, was identified as a member of the genus *Bacillus*. Highest yield of the CGTase was 6.5 U/ml. The enzyme was active within pH range 5-8, with maximum activity at pH 5. The temperature optima for enzyme activity was 60°C. CGTase production was induced by starch. The enzyme activity was inhibited by glucose at 1.3 mg/ml. The activity of enzyme was accelerated in presence of Mg^{2+} and Mn^{2+} and inhibited in presence of Zn^{2+}, Hg^{2+}, Cu^{2+}, and Fe^{2+}.

1. INTRODUCTION

Cyclodextrins are non-reducing oligosaccharides $(C_6H_{10}O_5)_n$, with glucose units joined by α-1,4,-glucosidic linkage resulting in circular molecules. They have applications in food, pharmaceuticals and agrochemicals because of their unique properties. Cyclodextrins enhance bioavailability of some drugs. They stabilise substances against oxidation, decomposition and hydrolysis. Cyclodextrins can very well form inclusion complex with steroids and increase their water solubility. In many food products, β-cyclodextrin stabilises volatile compounds and improves stability of ingredients. The demand of β-cyclodextrin is rising sharply (2-5) due to their role in stabilising many synthetic pesticides, herbicides, insecticides and fungicides against light, heat, oxygen and ions. Cyclodextrins can be synthesised from various starchy materials by cyclodextrin glucanotransferase (CGTase), an inducible enzyme, whose production is strongly inhibited in the presence of simple sugars like glucose. *Bacillus cereus* RJ30 (NCIMB 13123) is the only known organism that can produce CGTase in the presence of glucose or xylose as a carbon source (6). CGTases have molecular weight ranging from 68,000 to 170,000 daltons. It may be single polypeptide or a dimer of two sub-units (7-9). Micro-organisms known to produce CGTase include almost all species of genus *Bacillus* (10), some species of Micrococcus (11), Brevibacterium, Coryne-bacterium (12), Thermoanaerobacterium (13), Clostridium (14), Klebsiella and Pseudomonas (15). Virtually all of these bacteria have been isolated from soil or plant sources. Also most of the organisms have been screened from extreme environments like hot springs (16), deep sea mud samples (17) and highly alkaline environments (18). A number of cyclodextrin producing alkalophiles grow well within pH 8-11 and temperature 28-67°C (18-21). The present work reports the general properties of Cyclodextrin glycosyltransferase produced by the bacterium isolated from an alkaline environment.

2. MATERIALS AND METHODS

(a) Screening and Isolation of CGTase producing Micro-organism

50 ml sterile HM-I broth in 100 ml conical flask inoculated with 5ml sample was incubated on shaker (200 rpm) at 37°C for 24 h. 0.1 ml growth from enriched broth was diluted to 10-1 to 10-8 using 0.9 ml saline blanks in eppendorf tubes. Every time after transfer of 0.1 ml aliquot to 0.9 ml saline blank, the tubes were vortexed to ensure homogeneous suspension. The HM-I plates inoculated with 0.1 ml aliquot from each dilution were incubated at 37°C for 24h. The colonies with yellow coloured halo around them, were selected and streaked on HM-I medium for single colony isolation. Well-isolated colonies were selected and subcultured on HM-I slants.

(b) Production and preparation of Crude Enzyme

Loopful culture from the HM-I slant was inoculated into 20 ml HM- II medium in 50 ml conical flask and incubated on shaker (200 rpm) at 30°C for 24 hours. This culture was transferred into 500ml flask containing 100 ml of HM-II medium. The flask was incubated at 30°C for 24 hours at 200 rpm. The culture was centrifuged (10,000 rpm for 20 min.) and clear supernatant was used as crude enzyme.

(c) Cyclodextrin glucanotransferase assay

Cyclodxtrin glucanotransferase activity was determined by Phenolphthalein assay method (8). Reagents used were 4mM Phenolphthalein stock in ethanol; 125mM Na_2CO_3 in distilled water; 0.05M Tris-HCl buffer, pH 5.0; β-CD (0-200 μg/ml) in buffer and working phenolphthalein solution made by adding 1ml of stock in 100 ml Na_2OH_3 just before starting the experiment.

For standard assay, 100 μl of appropriately diluted enzyme was incubated with 1ml starch solution (1% in Tris buffer, pH 5) at 60°C for 20min. The tubes were transferred to ice bucket. The reaction was stopped by adding 100μl of 1N NaoH. Four ml of working phenolphthalein solution was added to each tube To the control tube the enzyme was added after addition of 100(l of 1N NaOH. Tubes were vortexed and the absorbance of the test and control was immediately measured at 550 nm on spectrophotometer (Shimadzu model UV 1601) using water as a blank. Percent decrease in absorbance of sample with respect to control was calculated.

Percent decrease in absorbance =

$$\frac{\text{a control} - \text{a test}}{\text{a control}} \times 100$$

where ac and at are absorbance of control and test respectively.

The amount of Oβ-CD produced was extrapolated from the standard curve (0-200 μg/ ml). One CGTase unit was defined as the amount of enzyme that produced 1μm Oβ-CD/min. under standard assay conditions.

(d) Effect of temperature on enzyme activity

The reaction mixture containing 1% starch solution in 0.05M Tris- HCl buffer (pH 5) and 100 μl of appropriately diluted enzyme was incubated at temperatures ranging from 20 to 60°C for 20min. The reaction was stopped by adding 100μl of 1N NaOH at low temperature. Percent decrease in absorbance was calculated and enzyme activity at different temperatures was determined.

(e) Effect of pH on enzyme activity

The substrate (1%Starch) was prepared in 0.05M Tris-Hcl buffer of varied pH (4-10) separately. 100μl of crude CGTase and 1ml of starch solution (varied pH) were incubated at 60°C for 20 min. and enzyme activity at different pH was determined.

(f) Effect of glucose and metals on enzyme activity

Different concentrations of glucose from 0.1 to 2.0 gm were added to the enzyme reaction mixture and left at room temperature for 30 min. Subsequently the reaction mixture was incubated at 60°C for 20 min. and enzyme activity was determined.

Deferent metals were used at 0.02% concentration in enzyme mixture at pH 5.0 and tubes were left to stand at room temperature for 30 min. The reaction mixture was incubated at 60°C for 20 min. The enzyme activity was calculated with and without metals.

3. RESULTS AND DISCUSSION

(a) Characteristics of CGTase producing bacterium

The isolated bacteria was aerobic, gram positive, motile and rod shaped. The morphological and biochemical characteristics indicated that the isolate belongs to the genus *Bacillus*. The growth of organism was scanty in nutrient broth at neutral pH. The CGTase from alkalophilic *Bacillus* appears to have two distinct advantages over other bacterial CGTase Viz., first the

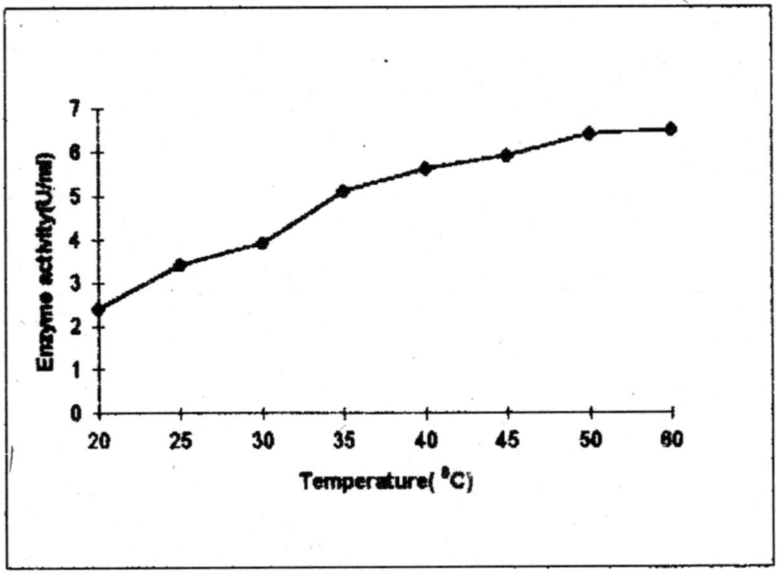

Fig. 1. Effect of temperature on enzyme activity

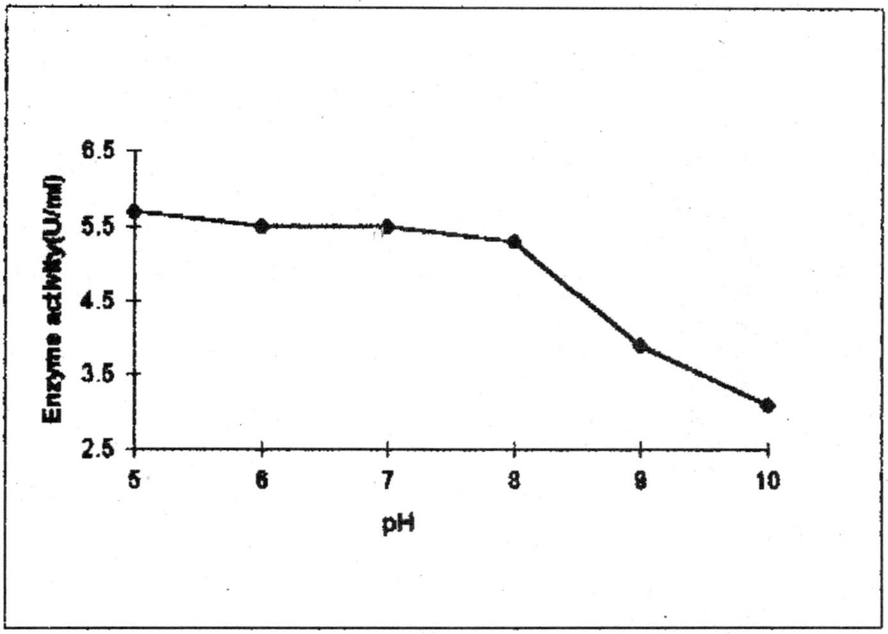

Fig. 2. Effect of pH on enzyme activity

CGTase from neutrophilic *Bacillus* sp. mainly produce α-cyclodextrin whereas those from alkalotolerant strains of *Bacillus* mainly produce β and γ-cyclodextrin (18,7). The β and γ-cyclodex- trins are of greater value because of their larger cavity size (3). Second, cultivation of bacteria at high pH offers some protection against microbial contamination. Moreover, at pH more than 8.0, retrogradation of starch is minimum (25), allowing use of high concentration of starch for the enzymic conversion.

The isolate produced CGTase within 24 hours of incubation as compared to 2-4 days and 40 hours reported in *Bacillus macerans* WKMW506 and *Bacillus circulans* respectively.

(b) Effect of temperature on enzyme activity

The effect of temperature on CGTase activity is shown in Fig 1. The CGTase activity was found to increase with an increase in temperature. The optimum for it's activity using soluble starch as a substrate was found to be 60°C. The activity was least at 20°C. Temperature optima of most CGTase are in the range of 45-70°C.

(c) Effect of pH on enzyme activity

The CGTase from *Bacillus* appears to have a wide pH optimum as shown in Fig. 2. The optimum activity was shown at pH 5.0 m the enzyme expresses more than 95% activity between pH 6 and 7. The activities at pH 9 and 10 were above 70% and 56% respectively.

(d) Effect of metals and different glucose concentrations on enzyme activity

The effect of different metals on the enzyme activity are shown in Fig. 3. The activity was strongly inhibited by Zn^{2+}, Hg^{2+} and Co^{2+} but was not strongly inhibited by Cu^{2+}, Ba^{2+} and FeS. The activity of enzyme was restored (70%) in presence of Mn^{2+}

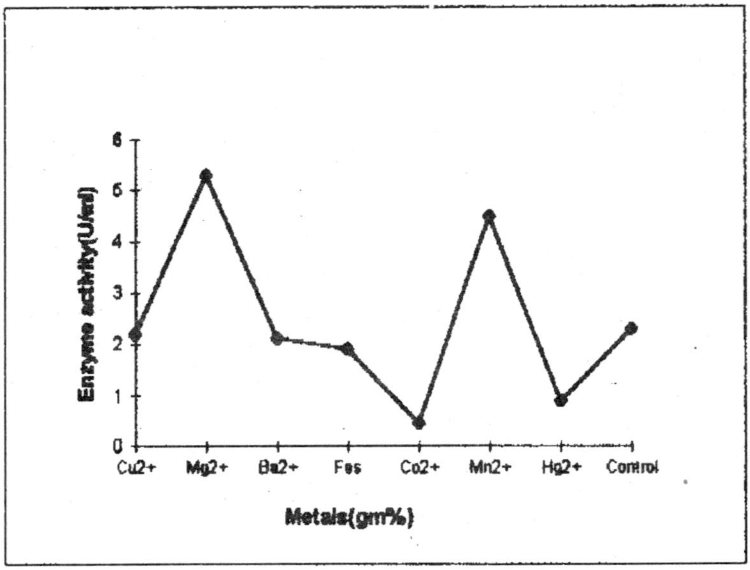

Fig. 3. Effect of metals on enzyme activity

while it showed maximum activity in presence of Mg^{2+}.

CGTase activity at different concentrations of glucose was studied. Negligible amount of activity was detected in presence of higher concentrations of glucose. About 82% and 50% enzyme activity was restored at 0.5 and 0.7 mg/ml glucose concentration respectively.

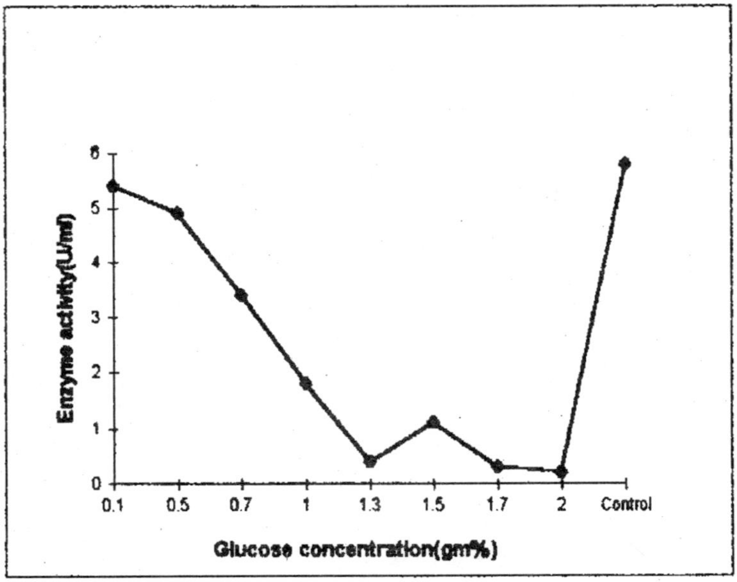

Fig. 4. Effect of glucose on enzyme activity

REFERENCES

1. Desai, L. and Pai, J. S. Ind. J. Microbiol. 36, (1996) 29-32.
2. Horikoshi, K. Process Biochem. 14, (1979) 26-30.
3. Szejtli, J. Cyclodextrins and their inclusion complexes. Akademiai Kiado, Budepest. (1982)
4. Bender, H. Adv. Biotechnol. Processes, 6, (1986) 31-71.
5. Hedges, A. R. Cyclodextrin: Production, Properties and applications. In: Starch hydrolysis products. Schenk, F. and Hebeda, R. E. (Eds.) VCH publishers, New York, (1992).
6. Jamuna, R.; Saswathi, N.; Sheela, R. and Ramkrishna, S. V. Appl. Biochem. Biotechnol., 43, (1993) 163-176.
7. Kobayashi, S.; Kainuma, K. and Suzuki, S. Carbohydr. Res., 61, (1978) 229-238.
8. Aoki, H.; Yao, D. and Misawa, M. Prog. Biotechnol. 3, (1987) 81-93.
9. Yan, T. and Lin, T. Biotechnol. Tech., 9, (1995) 215-220.
10. Makela, M. J. and Korpela, T.K. J. Biochem. Biophys. Methods, 15, (1988) 307-318.
11. Yogi, Y.; Kouno, K. and Inui, T. EP 17, (1979) 247.
12. Amano-Pharma JP 97, (1979) 077.
13. Novo WO 8, (1987) 903, 421.
14. Starnes, R. L. WO 9, (1991) 109, 962.
15. Allenza, P.; Cliff, C.G. and Morrella, M.J. US, 5008, (1991) 195.
16. Kaneko, T.; Yoshida, M.; Yamamoto, M.; Nakamura, N. and Horikoshi, K. Starch, 42, (1990) 277-281.
17. Georganta, G.; Kaneko, T.; Nakamura, N.; Kudo, T. and Horikoshi, K. Starch, 45, (1993) 95-99.
18. Nakamura, N. and Horikoshi, K. Agric. Biol. Chem., 40, (1976a) 753-757.
19. Nomoto, M.; Sheu, D. C.; Chen, S. J.; Yen, T. M.; Liao, C. W. and Yang, C. P. Agric. Biol. Chem., 48, (1984) 1337- 1338.
20. Pogsawadi, P. and Yagisawa, M. J. Ferment. Bioeng., 65, (1987) 463-467.
21. Horikoshi, K.; Nakamura, N.; Matsuzawa, N. and Yamamoto, M. Industrial production of Cyclodextrins. In: Proceedings of first international symposium on Cyclodextrins, (1981) pp 77-87. Szejtli, J. Ed. Reidel, D. publishing company, Dordrecht.
22. Goel, A. and Nene, S. N. (1995b) Modification in the method for spectrophotomectric estimation of B-CD. Starch (In press).
23. Yu, E. K. C.; Aoki, H. and Misawa, M. Appl. Microbiol. Biotechnol. 28, (1988) 337-339.
24. Steighardt, J. and Kleine, R. Appl. Microbiol. Biotechnol., 39, (1993) 63-68.
25. Bovetto, L. J.; Backer, D. P.; Villette, J. R.; Sicard, P. J. and Bouquelet, S. J. L. Biotechnol. Appl. Biochem., 15, (1992a) 48-58.

26. Bender, H. Enzymology of the cyclodextrin. In: Proceedings of the first International symposium on cyclodextrins, (1981b) 77- 87. Szejtli, J. (Ed.) Reidel, D. publishing company, Dordrecht.

27. Nakamura, N. and Horikoshi, K. Agic. Biol. Chem., 40, (1976b) 935-941.

28. Nakamura, N. and Horikoshi, K. Agr. Biol. Chem., 40, (1976) 1785-1791.

CHAPTER - 32

STUDIES ON MICROBIAL DEGRADATION OF POLYSTYRENE : LACTOSE, POLYSTYRENE, STARCH COMPOSITES

*S. Mishra, N.R. Talele and N.V. Phirke**

University Department of Chemical Technology,
* School of Life Sciences, North Maharashtra University, Jalgaon - 425 001 (INDIA)

ABSTRACT

Several million tones of plastics waste is added every year to nature, which has created severe environmental hazards. Flexibility, lightness, resistance to water microbial attack and also excellent mechanical properties comparable to metals and cheapest cost are few of the advantageous properties of synthetic polymers, which drove the whole human race to use it as per their requirement in day-to-day life especially for packaging. "Use and Throw" nature of modern society and limitations in re-cycling of plastics has created troubles in waste management. From the point of view of plastics applicability and its waste management, biodegradable polymers have become a new interest. Our studies emphasise on designing such a polymer which can minimize disposal and degradation problems. Blending of cheap carbohydrate material with synthetic polymer shows a promising solution to protection of environment. The starch and lactose materials from biological origin are used as a filler to enhance the biodegradability of polystyrene. These blends were subjected to microbial degradation in soil for different period of time.

1. INTRODUCTION

Flexibility of plastics is merely unaccountable and so its applications have proved to be a curse. Plastics have gained a unique position in packaging technology with 25% rise per annum thus causing its accumulation. An easy solution to tackle the problem of plastics degradation is to shift to suitable conventional material (paper) as far as packaging is concerned. Such change if followed abruptly may again create problems considering the energy and sudden deterioration of the resources ultimately affecting environment. So the cohesion of polymer (non-degradable) with biological material (degradable) is needed. Microorganisms are well known for their capability to degrade variety of materials. The barrier in between the plastics and microorganisms should be minimized. Starch is a natural polymer. The case of simplicity of processing, availability and economically viable qualities make the starch to be used as the filler in preparation of polymers. This confers the property of biodegradability to polymers to a large extent. The starch varies in grain size and can be obtained from potatoes (1). It is obvious that the percentage of filler affects the mechanical properties of polymer. The shift of mechanical properties should serve the purposes of use.

When the organic materials like carbohydrates are attacked by microorganisms, the matrix (polymeric material) may be degraded, resulting in reduction of important mechanical properties such as flexibility, impact strength and toughness (2).

Many microorganisms can derive soluble nutrients from insoluble organic materials by action of extracellular enzymes such as esterase and amidase (hydrolytic) and others catalyzed oxidative or reductive reactions. The combined activity of the enzymes result in the breakdown of some part of polymer. Fillers such as silica, clay, carbon black, glass fiber and asbestos are inert and do not support microbial growth. But organic fillers may make up as much as 50% by weight rendering the whole composite, susceptible to microbial attack. Considering biodegradability, polymers can be classified in three groups based on microbial attack viz. susceptible polymers, variable and resistant polymer (3).

Susceptible polymers again can be classified in two types with two different approaches. The first is polymeric macromolecules supporting growth of organisms (via degradation by enzymes e.g. polyesters, polyurathane) and second composite materials consisting of polymeric matrix and a dispersed phase (e.g. starch) (3). Polymer composites containing corn starch and poly (ethylene-co-acrylic acid) have been reported previously (4). Some microorganisms have been reported to degrade polyhydroxy alkanoates (PHAs) and on its exploitation, rapid evaluation plate test for the biodegradability of plastics have been developed (5).

In the present work composite of polystyrene with potato starch as well as lactose prepared with varying ratios were employed for biodegradation.

2. MATERIALS AND METHODS

(a) Preparation of polystyrene-starch composite

Polystron 678 SFI having a specific gravity of 1.05 and an MFI of 15 gm/10 min was used for preparation of polystyrene-starch composite (PSC). Polystyrene was added in benzene in the ratio of 10:90 and kept overnight for proper dissolution. Starch (readymade from Hi- media) was added in the polymer solution. With mechanical stirrer, it was allowed to homogenize for four hours. The homogenous mixture was poured in a silicon oil coated dish with 5 cm diameter. It was protected well from dust and other particulate matter. It was allowed to stand for 4 days for solvent evaporation. Then, the sheets were removed from dish and dried in an oven at 50°C for 30 minutes. In this fashion composites with 10%, 15%, 20%, 25% and 30% were prepared. The same procedure was followed for preparation of blends of Polystyrene-Lactose composite (PLC). The lactose was obtained from Amul.

(b) Soil burial degradation

The weighed amount of PSC\PSL sheets were burried in the soil kept in the wooden boxes (45 cm x 45 cm x 45 cm) with the soil depth of 30 cm. The soil was added with some quantity of water to maintain moisture, and were allowed to stand for 2 to 10 weeks.

After the fix interval of time PSC\PLC were removed from soil and washed to make it free from microbial culture and soil particles and dried in an oven at 50°C for 15 min. The weight loss was noted.

3. RESULTS AND DISCUSSION

Fig. 1 shows that the percent of starch based composite degradation increases with increase in incubation time from 2 to 10 weeks. As the percentage of starch increases in PSC, the mechanical properties change (6). The increase in degradation occurred with increasing amount of starch in composites. A remarkable difference was observed between 15% and 20% starch containing composites during 2-10 weeks. However more degradation was observed with increasing starch but the rate of degradation with increase in starch was maximum between 15% and 20% of starch containing composites.

Fig. 2 shows degradation of composite bearing polystyrene and lactose in varying ratios of polystyrene : lactose. The degradation of composites increases with increase in amount of lactose and with increase in duration 10% and 20% lactose bearing composites showed sharp increase up to 10 weeks period of time, while others exhibited sharp increment up to 8 weeks (15%, 20%) and 6 weeks for 30% of lactose in composites. Further the rate of degradation decreases with increase of in time, unlike the starch composites.

On comparing the results of starch with lactose it was observed that the lactose

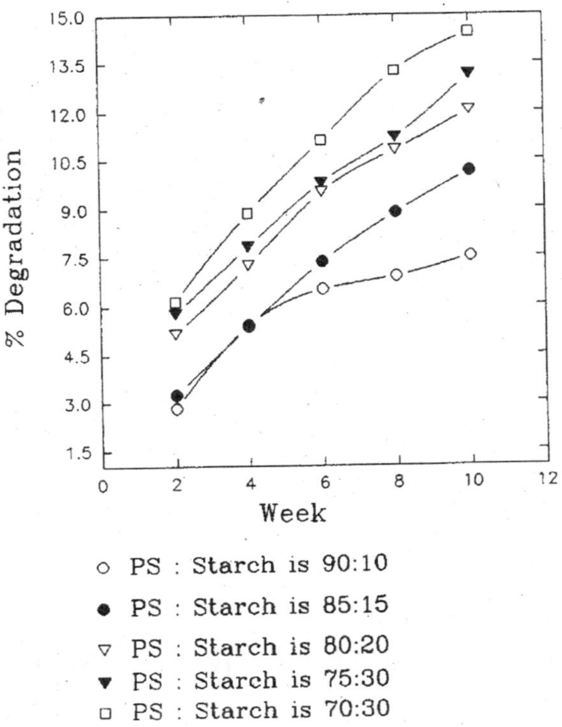

Fig. 1 % Degradation of polystyrene : starch composite at different period of time.

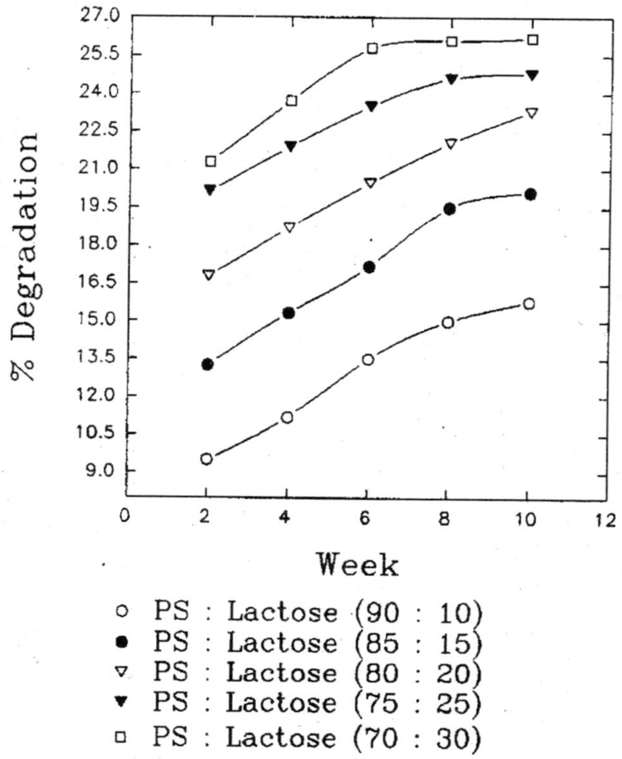

Fig. 2 % Degradation of polystyrene : lactose composites at different period of time.

bearing composites degrade faster than the starch bearing composites. The reason may be that the lactose is highly prone to microbial degradation due to its simple structure as compared to starch.

ACKNOWLEDGMENT

Authors are thankful to UGC for providing financial support to carry out this work.

REFERENCES

1. Griffin G J L & Hashami S A, *Iranian J. of Science and Technology*, 1 (1992) (1), 45.
2. Upsher F J, In : Proceedings of the conference held by *The Royal Australian Chemical Institute*, (1976) 173.
3. Andreopoulos A G and Theophanides T, *J. of Elastomers and Plastics*, 26 (1994) 308.
4. Perrone C, *Rinf*, 37 (1989) 38.
5. Augusta J, Muller R J & Widdecke H, *Appl Microbial Biotechnol*, 39 (1993) 673.
6. Wagner P A, Little B J, Hart K R & Ray R I, *International Biodeterioration & Biodegradation*, (1996) 125.

CHAPTER – 33

CITRIC ACID PRODUCTION BY *Aspergillus niger* IN SOLID STATE FERMENTATION

Luciana P.S. Vandenberghe[1,2], *Carlos R. Soccol*[*1], *Ashok Pandey*[1] *and J.M. Lebeault*[2]

[1]Laboratorio de Processes Biotechnologicos, Universida de Federal do Parana, CEP 81531-970, Curitiba-PR-Brazil;
[2] Laboratoire de Proceeds Biotechnologiques - Genie Chimique, Universite de Technologie de Compiegne/Centre de Recherches de Royallieu BP 529/60205 Compiegne cedex - France

ABSTRACT

Solid state fermentation was carried out to evaluate three different agro-industrial wastes, viz., sugarcane bagasse, coffee husk and cassava bagasse for their efficiency to produce citric acid by a *Aspergillus niger*. Cassava bagasse best supported mould's growth, giving highest yields of citric acid among the tested substrates. Results showed that the fungal strain has good adaptation for the substrate (cassava bagasse) which resulted in high protein content (23.07%) in the fermented matter. Citric acid production reached maximum (88-g/kg dry matter) when fermentation was carried out with a substrate having initial moisture of 62% at 26°C for 120 h.

1. INTRODUCTION

Citric acid is an important commercial product with its global production reaching 736,000 tons per year. Almost entire quantity of citric acid is produced by fermentation, mainly through submerged fermentation of starch or sucrose based media, using the filamentous fungus *Aspergillus niger*. Food industry is the largest consumer of citric acid, consuming almost 70% of total production, followed by about 12% by pharmaceutical industry and 18% for other applications (2). There is an annual growth of 3.5-4.0% in demand/consumption rate of citric acid.

Solid state fermentation (SSF) offers numerous advantages for the production of bulk chemicals and enzymes (3-7). In past one decade or so, there has been an increasing number of reports on the use of solid state fermentation processes for the production of a number of microbial products (8-11). This is partly because solid state processes have lower energy requirements and produce much less wastewater and environmental concerns regarding disposal of solid wastes. Different agro-industrial residues, like apple pomace, coffee husk, wheat straw, pineapple waste, cassava bagasse, banana, sugar beet cosset, kiwi fruit peel, etc. have been investigated with SSF technique for their potential to be used as substrate.

Cassava bagasse is a solid residue coming from the starch extraction process, which is generated during the separation stage. It contains fibrous material and about 40% starch (6). Thousands of tons of cassava bagasse are discarded daily by the cassava processing industries, leading to a big environmental problem. Thus, there is an urgent need to find suitable application and disposal of this waste. One alternative for its economic utilization is to use it as substrate in fermentation processes for the production of value added products like enzymes and biomass. The aim of this study was to compare citric acid production from cassava bagasse and two other substrates (sugarcane bagasse and coffee husk) in solid state fermentation using a fungal culture of *Aspergillus niger*.

2. MATERIAL AND METHODS

(a) Fungus

Seven strains of *Aspergillus niger* were screened for citric acid production in liquid culture and one strain NRRL 2001 was chosen. The strain was maintained at 4°C on Czapeck Dox Broth slants and renewed once in each month.

(b) Inoculum

A. niger spores were produced in Czapeck Dox Broth with agar (50 ml) held in 250 ml Erlenmeyer flask by incubating it at 28°C for eight days. Spore suspention was prepared by adding 25 ml distilled water with Tween-80 (0.1%) and was stored at 4°C for a maximum of two weeks.

(c) Substrate

Three substrates, viz., sugarcane bagasse, coffee husk and cassava bagasse were tested for their efficacy to produce citric acid. Sugarcane bagasse was obtained from the local market of Curitiba, PR, Brazil; coffee husk was obtained from an industry of Curitiba, PR, Brazil, and cassava bagasse was supplied by Companhia Lorenz (Quarto-Pontes, SC, Brazil. All the substrates were ground in a mill to a particle size between 0.8- 2.0 mm and dried at 55-60°C for 12 h.

(d) Fermentation

Liquid nutrient medium containing (g/L) $ZnSO_4 \cdot 7H_2O$ (0.2) and $FeCl_3 \cdot H_2O$ (0.014) was sterilized at 121°C for 15 min. After cooling, methanol (4%) was added aseptically in it (12). Media prepared with sugarcane bagasse and coffee husk were supplemented with 40% starch (corresponding to the starch content of cassava bagasse). The spore suspension, containing 10^8 spores/ml, was mixed with the nutrient medium and then inoculated with the substrates (10^7 spores/g of dry substrate inoculum size). Initial moisture content of the substrates were 90, 65 and 62% for sugarcane bagasse, coffee husk and cassava bagasse, respectively.

Fermentation was carried out in column fermenters of 4 cm diameter x 20 cm length (13). Each column was charged with the inoculated moist solid medium and incubated at 26°C in a water bath. The columns were supplied with humidified air at a flow rate of 65 ml/min. Fermentation was carried out for stipulated period and the results reported are the average of two sets of the experiments. Samples were withdrawn as whole column.

(e) Analytical assays

Samples (5 g) were mixed well with 50 ml of distilled water to extract citric acid and sugars. Filtrate obtained was subjected to high performance liquid chromatograph analysis using a Shimadzu LC-10AD HPLC. A temperature of 60°C and 5 mm H_2SO_4 as the mobile phase at flow rate of 0.6 ml/min. were used. Citric acid and sugars were detected in the column elluent by differential refratometer (Shimadzu RID-10 A).

Protein content was determined by Stutzer method (14). Other parameters like pH, moisture and starch content were determined as described by Soccol (15).

3. RESULTS AND DISCUSSION

(a) Citric acid production with different supports

Table 1 presents the results of solid state fermentation with different supports for citric acid production. As is apparent, there was high sugar consumption by *A. niger* in all the three substrates (90, 97 and 87% corresponding to sugar cane bagasse, coffee husk and cassava bagasse, respectively) which indicated that the strain had a good affinity for these substrates. Citric acid production was highest for cassava bagasse (88 g/kg dry matter- DM against 48.76 for sugarcane bagasse and 12.72 g/kg for coffee husk). Based on these results, cassava bagasse was then chosen for further studies.

It has been reported that cassava bagasse contained many components like calcium, phosphorus, vitamin B2, thiamine and niacin (16). Higher citric acid production by cassava bagasse in comparison to other two substrates (coffee husk and sugarcane bagasse) might be due to the presence of the components in the substrate.

(b) Kinetics of citric acid production with cassava bagasse

Fig. 1 shows the pattern of fungal growth as monitored by protein content in the fermenting substrate and change in moisture content (humidity) during the 120 h of fermentation. Protein content increased from 13.07 to 23.06%, showing an increase of more than 90%. There was a gradual increase in the moisture content of the substrate but it has no inhibitory impact on the fermentation.

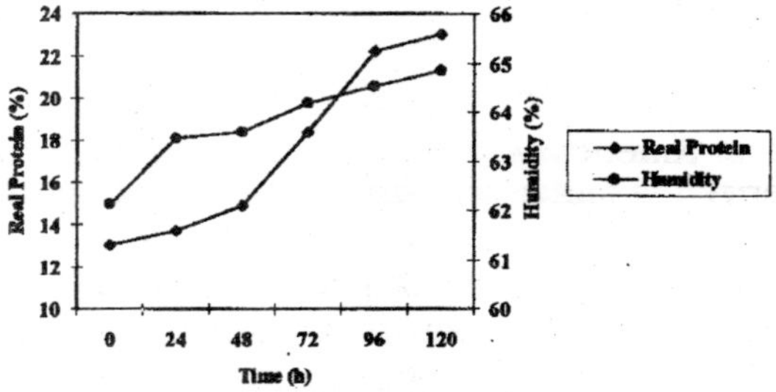

Fig. 1. Kinetics of citric acid production : Changes in protein and moisture contents

It is evident from Fig. 2, citric acid production started in the initial period of fermentation (at around 12 h) and reached maximum at 120 h. Final concentration of citric acid, i.e. 88 g/kg DM corresponded to a yield of 20% conversion factor (based on consumed sugars). During the fermentation, there has been drop in the pH of the fermenting substrate which was obviously due to the production of citric acid.

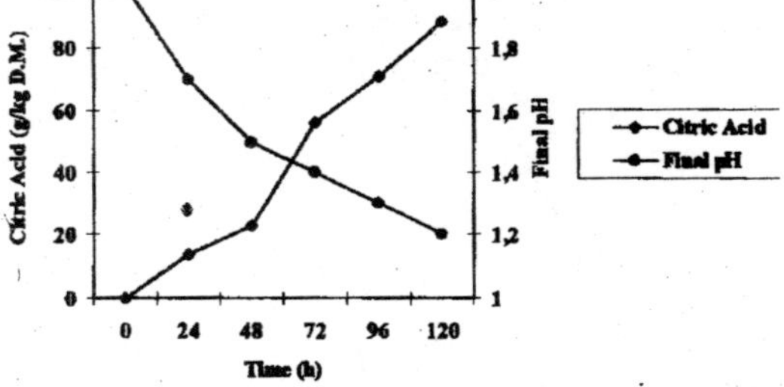

Fig. 2. Kinetics of citric acid production form Cassava bagasse in SSF

Data represented in Fig. 3 shows consumption of sugars and starch, available in the substrate by *A. niger*. There is good proportionate utilization pattern of both the compounds, which indicated good efficiency of the fungal culture.

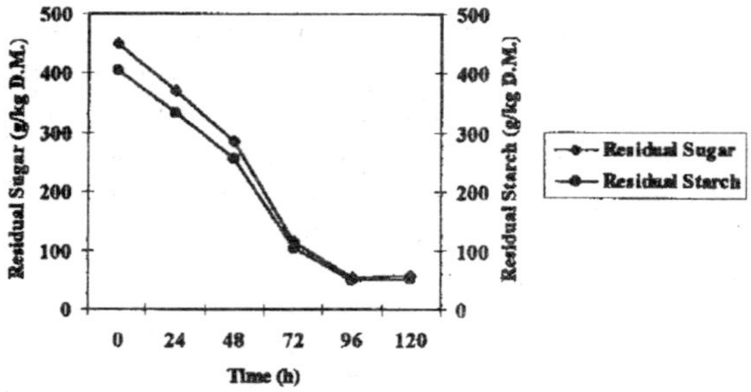

Fig. 3. Kinetics of citric acid production : Pattern of sugar and starch consumption in SSF

Generally, it is not possible the protein content of the substrate to a high level in citric acid fermentation. Many authors have reported that high concentration of citric acid is reached only under limited growth (17). In contrast to this, we have achived good growth of fungal culture, as indicated by high protein content of the substrate with citric acid production. High protein content makes the possibility of utilizing the fermented substrate (after citric acid extraction) as animal feed.

pH and other process parameters are important conditions for solid state fermentation and these must be rigorously defined for each different substrate and microorganisms in order to have good productivity. Citric acid accumulated in a high concentration (94.54 g/L) with an initial pH of 4.5.

A final citric acid concentration of 100 g/kg of Kiwifruit peel was obtained with solid state fermentation by *A. niger* NRRL 567 (18). When apple pomace was used as the substrae, citric acid production reached 83 g/kg sugar consumed with an initial sugar content of 123.4 g/kg apple pomace and a pH of 3.6.

Table 1. Citric acid production with different substrates (120 h)

Support	Final humidity %	Consumed sugars g/kg D.M.	Citric acid g/kg D.M.	Conversion Factor %
Sugarcane bagasse	84.70	334.32	48.76	14.58
Coffee husk	56.03	415.58	12.72	3.06
Cassava bagasse	64.89	392.00	88.00	20.55

Initial moisture %
Sugarcane baggase : 90.0
Coffee husk : 65.23
Cassava bagasse : 62.00
Initial sugar, g/kg DM
Sugarcane baggase : 370.0
Coffee husk : 428.30
Cassava bagasse : 450.50

4. CONCLUSION

This study showed the possibility of producing citric acid from three different substrates (Coffee husk, Sugarcane bagasse and Cassava bagasse) available in Parana State. Best citric acid production was obtained with Cassava bagasse. Although conversion factor based on sugars consumed was low, a reasonably good titres of citric acid were achieved. Attempts are being carried out to improve the yields further.

ACKNOWLEDGEMENTS

One of us (LPSV) thanks CAPES for the award of a fellowship.

REFERENCES

1. Quimica e Derivados, Estudo preve escassez mudial. Setembro, (1997).
2. Shah N D, Chattoo B B, Baroda Kothari R M, Patiala & Hegde V M, Starch,

45 (1993) 104.

3. Hammaci H & Hang Y D, Biotechnol Tech, 3 (1989) 51.
4. Pandey Ashok, *Process Biochem*, 27 (1992) 109.
5. Pandey Ashok. *In - Solid State Fermentation.* (Pandey A) Wiley Eastern Publishers, New Delhi (1994) pp 3.
6. Soccol C R, *Contribuicao ao estudo da fermentacao no estado solido em relacao com a producao de acido fumarico, biotransformacao de residuo solido de mandioca por Rhizopus e Basidiomacromicetos do genero Pleurotus.* Curitiba, 1994. Tese (Professor Titular) - Universidade federal do Parana, 228 p.
7. Raimbault M, Roussos S & Lonsane B K, *Advances in solid fermentation*, Chapter 2, (1997).
8. Roussos S, Hannibal L, Aquiahuatl M A, Trejo M & Marakatis S. *J Food Sci Technol*, 31 (1994) 316.
9. Saucedo-Castaneda G, Trejo-Hernandez M R, lonsane B K, Navarro J M, Roussos S, Dufour D & Raimbault M, *Process Biochem*, 29 (1993) 13.
10. Oriol E, Schettino B, Viniegra-Gonzalez G & Raimbault M, *J Ferment Technol*, 66 (1988) 57.
11. Nampoothiri K M & Pandey Ashok, *Biotechnol Letts*, 18 (2) (1996) 199.
12. Kolicheski M B, Producao de acido citrico por fermentacao no estado solido utilizando como subtrato bagaco de mandioca, Curitiba, 1995. Tese (Mestre) Universidade Federal do Parana. 115 p.
13. Raimbault M, & Alazard D, *Eur J Appl Microbiol Biotechnol*, 9 (1980) 99.
14. Vervack W, Analyse des aliments. Methodes courantes d'analyses. Laboratoire de Biochimie de la Nutrition. Luvain-la Neuve, U.C.I. (1973).
15. Soccol C R, Physiologie et metabolisme de Rhizopus en culture solide et submergee en relation avec la degradation d'amidon cru et la production d'acide L (+) lactique. These de doctorat, Universite de technologie de comiegne, France (1992) 218 p.
16. Hohnholz J H, Appl Geog Develop, 16 (1980) 117.
17. Grewal H S & Kalra K L, *Biotechnol Adv*, 30 (1995) 209.
18. Hang Y D & Woodmans E E, *Biotechnol Lett*, 6 (11) (1994) 763.

CHAPTER – 34

SELECTION OF SUITABLE SUBSTRATES FOR MASS MULTIPLICATION OF *Glomus mosseae* AND MIXED VAM FUNGI ISOLATED FROM COAL MINE OVER- BURDEN

*K.K. Chandra and Jamaluddin**

State Forest Research Institute, P.O. - Polipothar, Jabalpur - 482 008 (M.P.)

ABSTRACT

Different sterilized substrates including soilrite, mixed soilrite, vermiculite, overburden soil, flyash and their combinations were used in various proportion to compare with the traditional substrate i.e. Sand : Soil (1:1). Sand : Soil : Mixed soilrite (1:1:1) was found to be the best substrate for mass multiplication of both the inocula *G. mosseae* and Mixed VAM fungi in *Panicum maximum* trap plants. This exhibited maximum percentage of root colonization and spore population. Mixed soilrite : sand: soil combination having low nutrients and organic matter supported inoculum production, phosphorus status in substrates have played a major role in multiplication of VAM fungi in *Panicum maximum.*

1. INTRODUCTION

Mycorrhizal fungi are likely to be the most beneficial in revegetation sites enhancing the establishment and growth of plants under low fertility environment by increasing nutrient uptake. The efficient mycorrhizal fungi stimulated plant growth have been reported by several workers (1-3). The inoculation of VAM fungi reduces the need of fertilizers (4). The most difficult problem with VAM multiplication is that they can not be multiplied on artificial media. Selection of suitable substrate for inoculum production of VAM is also important (5) as well as selection of suitable host plant (6). The present study was conducted to select the suitable substrates for mass multiplication of *G. mosseae* and mixed VAM (*G. intraradices, G. mosseae, Acaulospora* sp.) isolated from coal mine ovderburden of Kusmunda, Korba (MP).

2. MATERIALS AND METHODS

Different substrates viz., soil, sand,

soilrite, mixed soilrite, vermiculite, overburden soil and fly-ash were used in this study. The substrates were autoclaved for 2 hour at 15 psi for 2 times, then filled into plastic buckets of 3 kg capacity and kept on iron bed. The buckets were arranged in Randomized Block design with 3 replicates.

Seeds of *Panicum maximum* were surface sterilized with 0.1% mercuric chloride worked in sterile distilled rotor and allowed to germinate on sterilized filter paper placed in a moist petridish. After 10 to 20 days, seedlings were transferred in sterilized substrate as per Mosse and Phillips (6) and inoculed by placing them in young actively growing lateral roots of trap plant. After 60 days of plant growth, root infection by VAM was examined by staining roots in trying blue (7) to assess the development of inoculated VAM species in roots and spore population/100g soil was also counted by removing a portion of soil and separating the spores by wet sieving and decanting technique (8). The selected substrate was used for maintaining the cultures of VAM and purity and progress were checked after 120 days and 180 days of inoculation.

3. RESULTS AND DISCUSSION

The results presented in Table 1 reported that Sand + Soil + mixed soilrite (1:1:1 by volume) was found to be the best substrate for both inocula which gave highest percentage of infection (38.38% for *G. mosseae* and 56.06% for mixed VAM fungi) after 60 days of inoculation. The sporulation in this substrate was also higher as it yield 236 spores/50 g substrate in case of *G. mosseae* and 567.70 spores/50 g. substrates in case of mixed VAM fungi on the basis of wet weight of substrate. Least sporulation and colonization was recorded in sand + flyash + soilrite substrates (68 spores/50 g substrates with 10.34% root colonization for mixed VAM). Other substrates like sand + soil + vermiculite and sand + soil + soilrite also expressed their suitability for production of both inocula.

Sand + soil with mixed soilrite imparted highest root colonization (65.51% and

Table 1 : Effect of different substrates on sporulation and root colonization of *Glomus mosseae* and mixed VAM fungi in *Panicum maximum* after 120 and 180 days of inoculation.

S. N.	Substrates	Ratio volume	G. mosseae after 120 days			G. mosseae after 180 days			Mixed VAM 120 days		Mixed VAM 180 days	
			No. of spores/ 100 g soil	Rela. (%) of *G. mosseae*	Root infection (%)	No. of spores/ 100 g soil	Rela. (%) of *G. mosseae*	Root infection (%)	No. of spores/ 100 g soil	Rela. (%) of *G. mosseae*	No. of spores/ 100 g soil	Root infection (%)
1	Sand+soil	1:1	539.00	62.90	54.08	965.00	47.83	61.72	1076.0	56.33	1538.0	56.17
2	Sand+Soil + Soilrite	1:1:1	893.70	70.30	65.51	2073.0	68.21	71.88	1464.0	65.75	4328.0	74.96
3	Sand + Soil + M. Soilrite	1:1:1	1407.3	77.30	64.74	2654.0	72.94	66.20	3146.0	74.33	4924.0	83.66
4	Sand + Soil + Vermiculite	1:1:1	666.30	69.91	59.76	838.00	64.47	64.30	1229.0	57.53	1599.0	58.76
	CD at 5%		461.22	NS	NS	804.78	NS	NS	1359.1	NS	1454.1	NS
	Standard Error ±		200.01			349.00			589.35		630.57	

Table 2 : Nutrient status of different substrates used in multiplication of *G. mosseae* and Mixed VAM in *Panicum maximum* Plants

S.N	Substrates	N (%)	P (%)	K (%)	Zn (PPM)	Cu (PPM)	Mn (PPM)
1	Soilrite	0.10	0.08	0.43	15.31	2.75	43.20
2	Mixed Soilrite	0.06	0.09	0.21	8.02	1.41	24.00
3	Vermiculite	0.06	0.02	0.33	25.56	2.45	66.48
4	Soil	0.19	0.20	0.28	0.81	1.20	1.82
5	Fly-ash	0.03	0.02	0.93	1.13	1.51	23.71
6	Overburden materials	0.16	0.07	0.16	24.37	1.93	57.27

71.88% for *G. mosseae* after 120 and 180 days of growth, respectively) and sporulation (1407/100 g substrate and 2654/100 g substrate after 120 and 180 days of inoculation). Mixed VAM inoculum with same substrate exhibited 74.33% and 83.66% root colonization with sporulation (3146/100 g and 4924/100 g substrate) after 120 and 180 days of inoculation. The relative spore percentage of *G. mosseae* decreased by increasing the period of inoculation. All the treatments yielded sporulation which was significant at 5% probability level while root colonization percentage was found non significant.

The suitability of Sand + soil + mixed soilrite (1:1:1) substrates might be due to low organic matter and nutrients (P 0.009% and K 0.21%) in mixed soilrite (Table 2) and in soil nutrients (N 0.19, P 0.20% and K 0.28%). According to Sylvia (9, 10) the ideal substrate for VAM production should be low in organic matter and available nutrients. The mixed soilrite, soil and sand (1:1:1) found to be the best substrate for the large scale production of *G. mosseae* and mixed VAM fungi.

REFERENCES

1. Powell C L, Groters M & Metcahf D M, New Zealand Jour of Agric Res, 23 (1980) 107.
2. Jasper D A, Robson A D & Abbott L K, New Phytologist, 118 (1991) 471.
3. Srinivas K, M.Sc. Thesis Tamilnadu Agric Univer, Coimbatore (1987) P. 76.
4. Mange J A, Davis R M, Johnson E L V & Zentmyer G A, Calif Agric, 32 (1978) 32.
5. Bagyaraj D J, In : VAMycorrhizas Powell CL and Bagyaraj DJ eds., CRC press, Florida, (1981) P. 131.
6. Mosse B & Philips J M, Jour Microbial, 69 (1971) 157.
7. Phillips J H & Hayman D S, Trans Br Mycol Soc, 55 (1970) 158.
8. Gerdemann J H & Nicolson T H, Trans Br Mycol Soc, 46 (1963) 235.
9. Sylvia D M, New Phytol, 95 (1984) 728.
10. Sylvia D M & Jarstfer A G, In : Management of mycorrhiza in Agric Hort and forestry Robson A.D, Abbott L K & Malajczuk ed.) Kluwar Acad, Publication, London, (1994) 231.

CHAPTER - 35

CITRIC ACID PRODUCTION FROM AGRICULTURAL WASTES BY *Aspergillus niger* UNDER SOLID STATE FERMENTATION

Lingappa K. and Chandrashekhar Naik

Dept. of Studies in Microbiology Gulbarga University, Gulbarga 585 106 INDIA

ABSTRACT

Investigations were carried out to determine citric acid production from fortified and unfortified substrates of sunflower seed cake and watermelon fruit peels through solid state fermentation, by two strains of *Aspergillus niger*. Under optimal conditions strain KLCN 2 produced larger amount of citric acid from sunflower seed cake waste fortified with 2% methanol, with a yield of 112.0. g/kg based on the amount of sugar consumed.

1. INTRODUCTION

Citric acid is the most important organic acid produced by modern biotechnological means (1). It has a multitude of usage (2-5), mainly in food and pharmaceuticals industries (6). Though it occurs naturally in good concentration in citrus fruits, the existing demand is almost entirely met by fermentative production. About 4 lakh tons are produced per year through processes involving *A. niger* (7)

Citric acid primarily produced by submerged fungal fermentation of sucrose or molasses medium (2). Recently some attempts have been made to produce citric acid by solid state fermentation by using agricultural wastes (8-12).

In India, water melon and sunflower have been used for the preparation of various products viz., pickles, sweets, re-freshing beverages, Jams, Jellies and oil seed or fodder crop and silage respectively (13-14), resulting huge quantity of wastes, containing abundant amounts of fermentative sugars (Table-1). There is every likely-hood of serious environmental hazards if these wastes are disposed without being utilized for some other purposes. As such, efforts have been made to utilize the watermelon fruit peels as well as sunflower seed cake in the production of citric acid through solid state fermentation employing the local strains of *A. niger*.

2. MATERIALS AND METHODS

Watermelon peels and sunflower seed cake used in this study were procured from local market. They were dried and ground as described by Hang (15).

Table 1. Biochemical composition of sunflower seed cake and water melon fruit peel

Sl No.	Chemical composition	Sunflower seed cake (%)	Water melon fruit peel (%)
1.	Moisture	3.7	95.5
2.	Reducing Sugars	10.8	4.0
3.	Fat	7.4	0.2
4.	pH	6.2	5.9

Two strains of *A. niger* isolated from agricultural soils and rotten fruits as well as *A.niger* MTCC1781 procured from IMTECH Chandigarh were maintained on PDA slants. A spore inoculum of each strains was prepared as described by Hang (15).

In the fermentation studies, 40 gm. of each dried wastes were separately rehydrated with distilled water in 500ml Ehrlenmyer flask in triplicates to have 65% moisture content. The media was inoculated with 0.2 ml of spore inoculum of each strains separately and flasks were incubated at 30°C for a period of 7d. Further in another set of experiments 2% of methanol was added to the dried wastes before rehydrating and herein the moisture level was adjusted to 65% and inoculated the media were with 0.2 ml of spore inoculum of each strains separately. The inoculated flasks in triplicate were incubated at 30°C for 7d.

The fermented medium after 7d. incubation period was extracted for citric acid with distilled water as described by Prapulla (16). The citric acid estimation was done as per Marrier and Bouler (17).

3. RESULTS AND DISCUSSIONS

The sunflower seed cake and the fresh water melon peels had moisture content of 3.7% and 95.5%; pH 6.2 and 5.9; reducing sugars 10.8% and 4.0% and fat content of 7.4% and 0.2% respectively.

Three strains of *A. niger* employed in the present study viz., KLCN 1, KLCN 2 and MTCC 1781 were capable of producing good yield of citric acid from substrates (Table 2). However sunflower seed cake proved to be a better substrate than the water melon fruit peel waste for citric acid production. The citric acid produced from sunflower seed cake ranged from 72.0 to 93.0 gm/kg of substrates by the three strains, while it ranged from 67.0 to 72.0 gm/kg of substrate of water melon peel wastes. The higher concentration of reducing sugars (10.8%) and Fats (7.4%) in the sunflower seed cake appears to enhance fungal growth and ultimately citric acid production in uniformity with earlier reports (18-19). Fats are broken down by *A. niger* to glycerol and fatty acids entering the cycle directly by formation of acetyl CoA (19). Further, fats and oils are reported to be good carbon sources (20).

The addition of 2% methanol to the substrates enhanced citric acid production by the three strains under study. Though methanol is not assimilated by *A. niger* strain and its role in stimulating citric acid production is not exactly known, supplementation of substrates with methanol has become a generally accepted practice in citric acid production (19).

Table 2. Yield of Citric Acid (gm/kg) by different organisms during Solid State Fermentation (7d. Incubation Period)

Organi-sms	Sunflower Seed Cake		Water Melon Peels	
	Without Methanol	With Methanol	Without Methanol	With Methanol
KLCN-1	7.20	9.00	6.70	7.10
KLCN-2	9.30	11.20	7.20	8.10
MTCC-1781	9.10	11.10	7.10	8.00

Further it has been observed that amongst the three strains of *A. niger* employed in the present study, strain KLCN2 appears to be better than KLCN 1 and slightly better than MTCC 1781. Citric acid production by *A. niger* KLCN 2 and *A. niger* MTCC 1781 are almost similar under both conditions i.e., with and without the addition of methanol to substrate. At the same time, the extent of beneficial effects of methanol was more in case of citric acid production from sunflower seed cake than of water melon peel wastes.

The present study indicates that the sunflower seed cake and the water melon fruit peels can be beneficially subjected to SSF for citric acid production by *A. niger* strains and also that supplementation of the substrates with methanol enhance the citric acid production. Sunflower seed cake is better substrate than the watermelon fruit peels.

REFERENCES

1. Rohar M., Kubicek C.P. and Kominek J. In: Biotechnology (HJ Rehm and Reed G eds.) weinheim; verlag chemie. Vol.3a (1983) 419-454.
2. Kapoor K.K., ChaudharyK. and Tauro P. In: (Prescott and Dunn's) Industrial microbiology Reed G. AVI Publishing Co. Inc. Westport. CT. (1982) 709-747.
3. Irwin W.E., Lockwood L.B. and Zienty M.F. In: Kirk- Othmer Encyclopedia of chemical Technology. (Standen HF, Mark JJ, Mcketta JR and Othmer DF eds.) Wiley New York (1967) 417-481
4. Schmidt T.R. "The use of citric acid in the canned fruit and vegetable Industry" Miles laboratories, Inc. Elkhart Indiana. (1983)
5. Kubieck C.P. and Rohr M. Crit Rev Biotechnol (1986) 331-373
6. Manonmani H.K. and Sreekantiah K.R. J Fd Sci Technol. 25(3) (1987) 159-161
7. Matic Legisa and Majda Gradsnik-Grapulin Appl Envi Microbiol (1995) 2732-2737
8. Lakshminarayana K., Chaudhary K., Ethiraj S. and Tauro P. *Biotechnol. Bioeng.*17, (1975) 291-193
9. Hamissa F.A. and Radwan A. J Gen Appl Microbiol 23, (1977) 23-27
10. Hildegard Kiel, Rumia Guvrin and Yigal Henis Appl Envi Microbiol 42(1) (1981) 1-4
11. Hang Y.D. and Woodams E.E. Biotech Lett 6 (11) (1984) 763-764
12. Hang Y.D. and Woodams E.E. Biotech Lett 17(4) (1985) 253- 254.
13. Wealth of India Raw Materials. CSIR Volume II (1950) 185-188
14. Wealth of India Raw Materials. CSIR Volume V (1959) 18-25

15. Hang Y.D., Luh B.S. and Woodams E.E. Adv Apl Microbial 38 (1987) 99-147

16. Prapulla S.G. In: (Short- term course on solid state fermentation. February 11th-28th (1991) 7.1 - 7.2

17. Marier J.R. and Boulet M. J Dairy Sci 4 (1958) 1683-1692

18. Nancy F., Mills B., Helen Trumpy and Brenda M. Palmer J Gen Microbiol. 30 (1963) 365-379

19. Bhatt C.S., Agarwal P.K. and Vishwanathan L. Commercial manufacture of citric acid. VII: Effects of some fats and oils on citric acid fermentation by *Aspergillus niger*. VIII: Effects of alcohol's and crude extracts on the production of citric acid by *Aspergillus niger*. 44th convention of sugars Technologists Association of India. IG and 75G (1980)

20. Manonmani H.K. and Sreekantiah K.R. Process Biochem (1987) 92-94.

CHAPTER - 36

DETECTION AND DEVELOPMENT OF LACTOBACILLI AND THEIR BIOPROCESSING FEATURES

M.B. Sulochana[1], B. Sharada[3] and A. Dayanand[2]

1. Department of Biotechnology and Microbiology
2. Department of Microbiology,
Gulbarga University, Gulbarga - 585 106, INDIA.
*Smt. V.G. Women's College, Gulbarga.

ABSTRACT

Lactobacilli were detected from various natural and house made fermented materials. *Lactobacillus maltaromicus*, *L. plantarum* and *L. amylophilus* were the three prominent mesophilic and homofermentative isolates obtained from cereals and millets. *L. maltaromicus* has shown greater physiological potentiality giving maximum amount of lactic acid and high yield. Positive mutants of Lactobacilli were obtained employing UV radiations (254 nm) to achieve the higher physiological potentiality. pH 6, temperature - 37°C and inoculum size of 10% were the three important parameters regulated in the batchwise bioprocess to examine the physiological potentiality of the Lactobacilli.

1. INTRODUCTION

Lactic acid is an industrially important product that has a great market value due to its attractive properties. It finds wider applications in food, pharmaceutical textile, leather and plastic industries. The detection and development of newer microbial types for the desired attributes from the natural substrates is warranted for the industrial production of lactic acid (1).

Detection technology of microorganisms is much concerned with the selection of appropriate organisms of desirable attributes which constitutes a critical feature of exploratory research. Detection including isolation, characterization and screening of microorganisms from naturally occuring processes and natural materials of various habitats, has always been the most powerful means for obtaining useful isolates for scientific and commercial purposes (2). Although many investigators (3-7), have studied the isolation, enumeration and characterization of lactic acid bacteria from different sources, a lot of potential sources of different habitats are yet to be explored.

This study describes the isolation, characterization and screening of lactic acid bacteria from various natural and house made fermented materials. The detected isolates were further developed by employing UV radiations for the high yield of the end product. The prominent bioprocessing features of the detected isolates of Lactobacilli have also been observed.

2. MATERIALS AND METHODS

About 20 different natural and house made fermented materials were used as sources for the detection of Lactobacilli. All the samples used for the isolation were broadly grouped as (1) Vegetables (2) Fruits (3) Dairy products (4) Cereals and (5) Millets. MRS agar was the specific nutrient medium used for the isolation of Lactobacilli (8). The ultra violet source of 254nm wavelength was employed for the development of the isolates of Lactobacilli.

The natural and house made fermented materials were collected in suitable sterile containers from the place of their availability. All the collected samples were processed separately according to their specific requirements and homogenised mixture was used as inoculum. Colony characters, microscopic features, biochemical and sugar fermentation tests were conducted according to the Bergey's Manual of Systematic Bacteriology.

The lactic acid produced by the detected isolates of Lactobacilli was estimated by spectrophotometric method (9). The selected isolates of Lactobacilli were exposed to the ultra violet radiations for different range of periods to obtain the positive mutants (10) for high yield of lactic acid. The important bioprocessing features of the Lactobacilli for the production of lactic acid in a batchwise bioprocess were also observed.

3. RESULTS AND DISCUSSION

The standard nutritional attributes (11) of various sources used for the isolation of Lactobacilli are recorded in Table 1. Among the different groups of sources, cereals and millets have indicated high amount of carbohydrate, protein and mineral contents. Fruit samples are the next, to indicate high amount of carbohydrate in comparison with the vegetables and dairy products. However, dairy products and vegetables have shown more amount of protein content than the fruits. No much variation appears in the mineral contents amongst the vegetables, fruits and dairy products, except ber fruit (0.3g) and buttermilk (0.1g). Arginine, Histidine and Lysine are the most common amino acids reported in vegetables, cereals and millets. Curds and milk have also shown the presence of similar amino acids. However, only guava fruit has shown the presence of Lysine, Tryptophan and Methionine amino acids.

Microscopic and biochemical tests of the detected isolates of Lactobacilli are presented in Table 2. All the detected isolates of Lactobacilli were Gram +ve, rod shaped and negative for catalase, nitrate reduction and arginine tests. Acid production from twelve different sugars by all the detected isolates of Lactobacilli are as recorded in Table 2. The selected representative isolates of Lactobacilli, the sources from which they have been obtained and their physiological potentiality indicating lactic acid produced and lactic acid yield, before and after exposure to UV radiations are presented in Table - 3.

Table - 1. Nutritional attributes of different sources used for the detection of Lactobacilli.*

Sl. No.	Samples	Nutritional attributes**			
		Carbohydrates (g)	Proteins (g)	Minerals (g)	Amino acids (mg/gN)
VEGETABLES					
1.	Cabbage	4.6	1.8	0.6	Arg, His, Lys
2.	Cucumber	2.5	0.4	0.3	Arg, His, Lys
3.	Potato	22.6	1.6	0.6	Arg, His, Lys
4.	Tomato	3.6	1.9	0.6	Arg, His, Lys
FRUITS					
1.	Ber	17.0	0.8	0.3	NR
2.	Gauva	11.2	0.9	0.7	Lys, Tyrp, Met
3.	Pomegranate	14.5	1.6	0.7	NR
4.	Sapota	21.4	0.7	0.5	NR
DAIRY PRODUCTS					
1.	Buttermilk	0.5	0.8	0.1	NR
2.	Curds	3.0	3.1	0.8	Arg, His, Lys
3.	Milk	5.0	4.3	0.8	Arg, His, Lys
4.	Whey	NR	NR	NR	NR
CEREAL GRAINS					
1.	Maize	66.2	11.1	1.5	Arg, His, Lys
2.	Rice	78.2	6.8	0.6	Arg, His, Lys
3.	Sorghum	72.6	10.4	1.6	Arg, His, Lys
4.	Wheat	71.2	11.8	1.5	Arg, His, Lys
MILLETS					
1.	Ragi	72.0	7.3	2.7	Arg, His, Lys
2.	Bajra	67.5	11.6	2.3	Arg, His, Lys
3.	Varagu	65.9	8.3	2.6	Arg, His, Lys
4.	Italian millet	60.9	12.3	3.3	Arg, His, Lys

NR - Not Recorded.
* Gopalan *et. al.*, (11)
** All the values are per 100 gms of edible portion.

Microbial biotechnology is the utmost important to obtain the large scale industrially useful products through interventions of cultures, substrates and bioprocess parameters. The detection of Lactobacilli from various sources has been well documented. Stirling and Whittenburry (5), Dick *et al.*, (2), Yimin *et al.*, (6) and Suresh *et al.*, (10), have detected the Lactobacilli from plants, grasses and cereal plants, traditional food fermentation processes, forage crops and grasses and naturally fermenting vegetables respectively. In addition to this, our study significantly reveals the detection of Lactobacilli from the grains of cereals and millets. We could detect 345 isolates of Lactobacilli from 20 different sources.

The prominent mesophilic and homofermentative isolates of Lactobacilli were chiefly detected from the grains of maize and sorghum. Most probably the

Table 2. Characteristic features of the detected isolates of Lactobacilli.

Sl. No.	Features	Isolates of Lactobacilli											
		GM Ca 22	GM Cu 26	GM Po 34	GM Be 5	GM Gu 9	GM Sa 15	GM Ma 60	GM So 69	GM Wh 76	GM Ra 102	GM Ba 105	GM Ba 108
I	Microscopic												
	Gram staining	+	+	+	+	+	+	+	+	+	+	+	+
II	Biochemical												
	Catalase	–	–	–	–	–	–	–	–	–	–	–	–
	Nitrate	–	–	–	–	–	–	–	–	–	–	–	–
	Arginine	–	–	–	–	–	–	–	–	–	–	–	–
III	Fermentation												
1.	Arabinose	–	W	–	–	–	–	–	–	–	W	W	–
2.	Cellobiose	W	+	–	+	+	+	+	+	+	+	+	+
3.	Galactose	–	+	+	+	+	+	+	+	+	+	+	+
4.	Lactose	–	+	–	+	W	W	+	+	+	+	+	+
5.	Maltose	W	+	+	+	+	W	+	+	+	+	+	+
6.	Mannitol	–	+	–	+	–	–	+	+	+	+	–	–
7.	Melibiose	–	+	–	+	–	W	+	+	+	+	W	W
8.	Raffinose	–	+	–	+	–	W	+	+	+	+	W	W
9.	Ribose	–	+	–	+		–	+	+	+	+	–	–
10.	Salicin	–	+	–	+		+	+	+	+	+	+	+
11.	Sucrose	+	+	–	+	–	+	+	+	+	+	+	+
12.	Xylose	–	W	–	–	–	–	–	–	–	W	W	–

W - Weak Reactions.

carbohydrate and protein contents are the vital components in providing the suitable C:N ratio for the favourable growth of Lactobacilli. Histidine, Arginine and Lysine which are the three important amino acids present in maize and sorghum, must have also had their influence on the favourable growth of Lactobacilli. However, the influence of these individual amino acids on the growth of Lactobacilli is yet to be studied.

The isolate *Lactobacillus maltaromicus* detected from maize and sorghum has produced high amount of lactic acid (27 g) and also indicated high yield of lactic acid (67.5%), followed by the isolate *L. plantarum* detected from wheat and ragi. The least amount of the end product (24 g and 25 g) and yield (60.0% and 62.5%) were observed in *L. amylophilus* and *L. plantarum* respectively obtained from bajra. All the isolates from all the sources have shown significant increase in the amount of lactic acid produced and the yield Table - 3, after the exposure to the UV radiations for 120 seconds. However, the consistency of producing lactic acid and the yield by the isolates after exposure to the UV radiations is yet to be confirmed.

Table 3. Yield potentiality of the isolates of Lactobacilli before and after exposure to Ultra Violet radiations.

Sl. No.	Sources	Isolates	Before exposure to UV radiations		After exposure to UV radiations	
			LA Produced (g/1)	LA Yield (%)	LA Produced (g/1)	LA Yield (%)
1.	Maize	*L. maltamoricus*	27	67.5	30	75.0
2.	Sorghum	*L. maltamoricus*	27	67.5	30	75.0
3.	Wheat	*L. plantarum*	26	65.0	29	72.5
4.	Ragi	*L. plantarum*	25	62.5	28	70.0
5.	Bajra	*L. plantarum*	25	62.5	28	70.0
6.	Bajra	*L. amylophilus*	24	60.0	27	67.5

LA - Lactic acid.

Bioprocess is the central part of modern biotechnology, where in the status of the process is changing rapidly with various prominent process parameters such as pH, temperature, inoculum size (1). pH - 6, temperature - 37°C and inoculum size of 10% are the given controlled parameters in batch-wise bioprocess to evaluate the physiological potentiality of the detected isolates of Lactobacilli in the form of lactic acid produced and the yield.

REFERENCES

1. Bull, AT, In Comprehensive biotechnology, (Murray Moo - Young edn.), pp. 3-5. Pergamon Press, New York (1985).
2. Dick, JC, Van den Breg, Annelies Smits, Bruno Pot, Aat M Ledeboer, Karel Kersters, John, MA, Verbakel C, Theo Verrips, Food Biotech 7(3) (1993). 189.
3. Allen LA & Harrison, *Ann Appl Biol* 23 (1936) 546.
4. Kroulik JT, Burkey, LA & Wiseman, HG, *J Dairy Sci* 38 (1955) 256.
5. Stirling, AI & Whittenbury, R, *J Appl Bacteriol* 26 (1963) 86.
6. Yimin Cai, Sadahira, Ohmomo & Sumio Kumai, J Japanese society of Grassland science, 39(4) (1994) 420.
7. Suresh, ER, Rajashekhara, E & Ethiraj, S, *Ind J Microbiol* 36 (1996) 185.
8. de Man JC, Rogosa, M & Sharpe, ME *J Appl Bacteriol* 23 (1960) 130.
9. Kimberley, ACCT, *Appl Biochem Biotechnol* 56 (1996) 49.
10. Thomas, BW, *Ind J Microbiol* 36 (1996) 185.
11. Gopalan, C, Rama Sastri, BV & Balasubramanian, SC, Nutritive value of Indian Foods : National Instiutte of Nutrition. Hyderabad, India (1996).

CHAPTER - 37

CHARACTERIZATION AND DETERGENT COMPATABILITY OF THERMOSTABLE ALKALINE PROTEASE FROM *Bacillus* sp. IS-3

J.K. Gupta, L.K. Purva Gupta and S.K. Soni

Department of Microbiology, Panjab University, Chandigarh - 160 014, INDIA

ABSTRACT

Bacillus sp. IS-3 produced high levels of alkaline thermostable protease. The kinetic studies of partially purified protease indicated the maximum activity at 60°C, pH 10.5, promoted further by the presence of Mn^{2+}, Fe^{3+}, and Ca^{2+}. The enzyme was found to be thermostable and pH stable, retaining almost 100% activity in the pH range 7.5-11.5 after 10 min. and at 50- 60°C after 1h. Further, the enzyme showed compatability with various commercial detergents and was found to improve the cleansing performance towards blood stains, when supplemented in detergent powder, Surf™.

1. INTRODUCTION

Proteases are one of the most importatnt group of commercial enzymes with major applications in food, pharmaceutical and detergent industries. Today, the use of microbial proteases as active ingredients for laundary detergents is the single largest application of enzymes with sales turnover of 89% followed by microbial rennets (8%) used for cheese making (1).

Thermostable alkaline proteases have high commercial potential where their ability to remain active in the presence of denaturing agents, bleach activators, organic solvent and harsh pH conditions is exploited. A number of thermostable, industrially important microbial proteases have been extensively studied and reported from mesophilic and moderately thermophilic organisms (2) However, there is still a need to discover new proteases with novel properties to cater for the rapid growth in industries based on enzyme technology.

In this paper we report the characterization and compatability of a thermostable alkaline protease from *Bacillus* sp. IS-3 with commercial detergents.

2. MATERIALS AND METHODS

Microorganisms : The bacterial strain, *Bacillus* sp. IS-3, used in the present study, was capable of producing high yield of alkaline protease as reported earlier (3) This was isolated from soil at 50°C and pH 10.5.

Culture media : The organism was maintained on nutrient agar slants and stored at 4°C. The production medium used had the following composition (g l^{-1}) : Glucose, 5.0; Defatted soybean meal, 5.0; $CaCl_2$. $2H_2O$, 0.20; $MgSO_4.7H_2O$, 0.20; KH_2PO_4, 1.0; Sodium acetate, 0.05. pH was adjusted to 10.5 with separately sterilized 20% Na_2CO_3.

Protease Production : Twenty ml of production medium, taken in different sets of 100 ml Erlenmeyer flasks, was inoculated with 1.0 ml of the inoculum (A_{600} 0.8) obtained from 12h old shake culture in nutrient broth. The flasks were then incubated for 72h at 37°C in New Brunswick Gyrotary water bath shaker (Frequency 3.3 Hz). These were then centrifuged (10000 g, 20 min, 4°C) and the cell free supernatant was used for determing the protease activity.

Enzyme assay : Protease activity was measured by modified Anson's method (4) using Hammerstein casein.

Enzyme unit : One unit (U) of protease actvity was defined as amount of enzyme required to release 1 umole of tyrosine per min under the assay conditions (60°C; pH 10.5).

Partial Purification of protease : The cell free supernatant was ultrafiltered using millipore assembly with a membrane of cut-off of 10 kDa (Amicon). The filtrate and retentate were then analyzed further for its protein content and enzyme activity.

Protein content of both crude and partially purified enzyme was estimated by method of Lowry *et al.*, (5).

Characterization of protease : The partially purified protease enzyme characterized with respect to the following parameters :

1. Temperature and pH activity profiles were studied in the temperature range of 37-40°C (pH 10.5) and pH range of 7.5 - 12.5 (60°C) using buffers of different pH (7.5-8.5, Tris buffer, 50 mM and 9.5-12.5, glycine; NaOH buffer, 50 mM).
2. The effect of various metal ions and inhibitors on enzyme activity was studied by incubating the enzyme with these (listed in Table 1) for 20 min and then assay was performed at 60°C, pH 10.5 by standard procedure.
3. pH stability profile was studied by incubating the enzyme at 60°C for 10 minutes with buffers having pH range of 7.5-12.5. After incubation, residual activity was measured at pH 10.5 by standard assay procedure.
4. Thermostability profile of the enzyme was studied by incubating at various temperatures (50-80°C) for different intervals of time (0-60 min) in presence of 5 mM $CaCl_2$. The residual activity (%) was determined under standard assay conditions (60°C, pH 10.5).
5. In order to assess the suitability of the enzyme for use as laundary detergent additive, its compatibility was determined with various commercial detergents including Ariel™, Nirma™, Wheel™, JVG™, Surf™, 555™ and Swan™. The detergent powders were diluted in distilled water to give final concentration of 7 mg/ml. Enzyme was incubated with each detergent at 60°C, pH 10.5 for different intervals of time (0-30 min) and the residual activity

determined under the standard assay conditions.

6. Application of the protease as a detergent additive was studied on white cotton cloth pieces (3 x 3 cm) stained with human blood (20 µl), taken in separate flasks. To these flasks was added distilled water, detergent, surf™ (7 mg ml⁻) and detergent + enzyme (3.2 mg protein/ml). These flasks were kept at room temperature at 37°C for 5-10 min. After incubation cloth pieces were taken out, dried and examined visually.

3. RESULTS AND DISCUSSION

The temperature activity profile exhibits that enzymes activity increased progressively upto 60°C where maximum activity was obtained (Fig. 1a). This declined on further increasing the assay temperature and 60% of the maximum activity was left at 70°C. Optimum temperature of 60°C has also been reported for protease activity in several *Bacillus* spp. (6-14).

pH activity profile shows that only 8% of maximum activity was obtained at pH 6.5. This increased with the assay pH upto 10.5 which was found to be the pH optima. Further increase in pH caused the slight decline in the activity (Fig. 1b). These findings are in accordance to several earlier reports showing pH optima of 10.0-10.5 of protease enzyme in *Bacillus* sp. GX 6638, *Bacillus* sp. AH-101, *Bacillus subtilis* (natto), *Bacillus* sp. B18', *Bacillus sphaericus* BSE18, *Bacillus subtilis* ATCC 14416 (4,6,16,17). Two important alkaline proteases of detergent industry viz. subtilisin Carlsberg produced by *Bacillus licheniformis* and subtilisin Novo or Bacterial Protease Nagase (BPN) produced by *Bacillus amyloliquefaciens*, also showed maximum activity at 60°C and pH 10.5 (15)

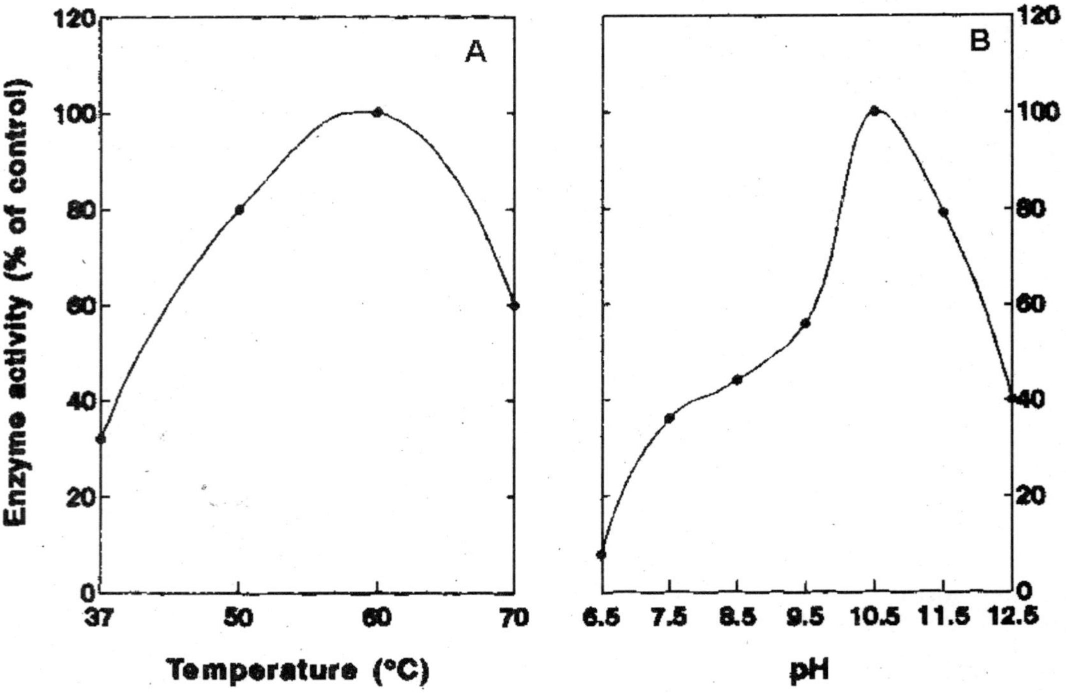

Figure 1. Effect of (A) temperature and (B) pH on protease activity.

The enzyme activity was promoted by Mn^{2+}, Mg^{2+}, Ca^{2+} and Fe^{3+} while there was no effect of K^{+}, Ni^{2+} and urea on enzyme activity. Zn^{2+} and Cd^{2+} resulted is slight fall in protease activity (Table 1). Ca^{2+}, Fe^{2+}, Fe^{2+} and Mn^{2+} have already been shown to enhance the protease activity in various Bacillus and Thermus spp. (14,18,19). Role of Ca^{2+} to stabilize the protease enzyme was reported in case of subtilisins (20). EDTA, PMSF, SDS, $(NH_2)_2SO_4$, $HgCl_2$ and CH_2CO-OPb resulted in great reduction of protease activity. Inhibition by PMSF and EDTA showed that our enzyme might be a metal ion-dependent serine protease. Debette (21) reported an extracellular alkaline serine protease from *Xanthomonas maltophila* which like metalloprotease was rapidly inactivated by EDTA but the activity was partially restored by Ca^{2+}. Stefantisi *et al.*, (22) also reported PMSF sensitive, Ca^{2+} dependent serine protease from *Lactobacillus bulgaricus*.

Table 1. Effect of various metal ions and effector molecules on protease activity.

Metal ions/ effector molecules	conc. (mM)	Protease activity (%) of control
EDTA	5	8.33
$CaCl_2$	5	129.60
SDS	3.5	29.16
$(NH_4)_2$ SO_4	40	33.33
PMSF	2.5	8.33
Urea	1000	100.00
$MgSO_4$	2	104.16
KH_2PO4	2	100.00
$(Fe_3)_2$ SO_4	1	154.16
$MnSO_4$	2	166.66
$CdCl_2$	1	80.00
$AgNO_3$	1	72.00
$NiCl_2$	2	100.00
$CoCl_2$	1	56.00
$ZnSO_4$	2	80.00
CH_3COOPb	2	44.00
$HgCl_2$	2	24.00
None (control)		100.00

The enzyme showed high pH stability, being 100% stable upto 10.5 and 94.44% at pH 11.5. Further increasing the pH to 12.5, resulted in sharp decline in protease activity (Fig. 2). The ability of this enzyme to withstand alkaline conditions is highly desirable characteristic for its use in laundary detergents. These observations confirm the earlier findings on protease enzyme of *Bacillus alcalophilus* sub sp. *halodurans* exhibiting 100% stability over pH range 5.0-10.0 Enzyme activity was reduced to 50% at pH 11.0 and was completely inactivated at pH 13.0 after 10 minutes of incubation at 55°C (8). Takami *et al.*, (4) reported alkaline serine protease from *Bacillus* sp. AH-101 showing stability over wide range of pH 5-13 after 10 minutes of incubation at 60°C. Enzyme showed 80% activity in the pH range 5-9 which then increased to 90% at pH 10.0-11.5. Comparative to these findings, protease produced by our strain *Bacillus* sp. IS-3 was much more stable under high pH conditions retaining 100% activity upto pH 10.5 and 94.44% activity at pH 11.5.

As indicated in Figure 2b, enzyme retained 100% activity at 50°C and 60°C after 1 h. Protease activity was reduced to half at 70°C after 20 minutes of incubation and was inactivated at 80°C. Thermostability studies of proteases in the published literature showed enzyme with half life of 25 min at 60°C (6), 10 min at 60°C (10,15) and 23 min at 60°C (12). Yamagata *et al.*, (23) also reported thermsotable protease having 100% stability upto 55°C for 10 min. Kobayashi *et al.*, (20) showed that alkaline serine protease of *Bacillus* sp. KSM-K16 remained stable for 10 min at 50°C and 60°C. Ferrero *et al.*, (24) also reported thermostable protease from *Bacillus licheniformis* MIR29 showing 100% stability upto 70°C for 10 min. On comparing these findings our protease has shown much higher thermostability retaining 100% activity upto 60% after 1 h of incubation.

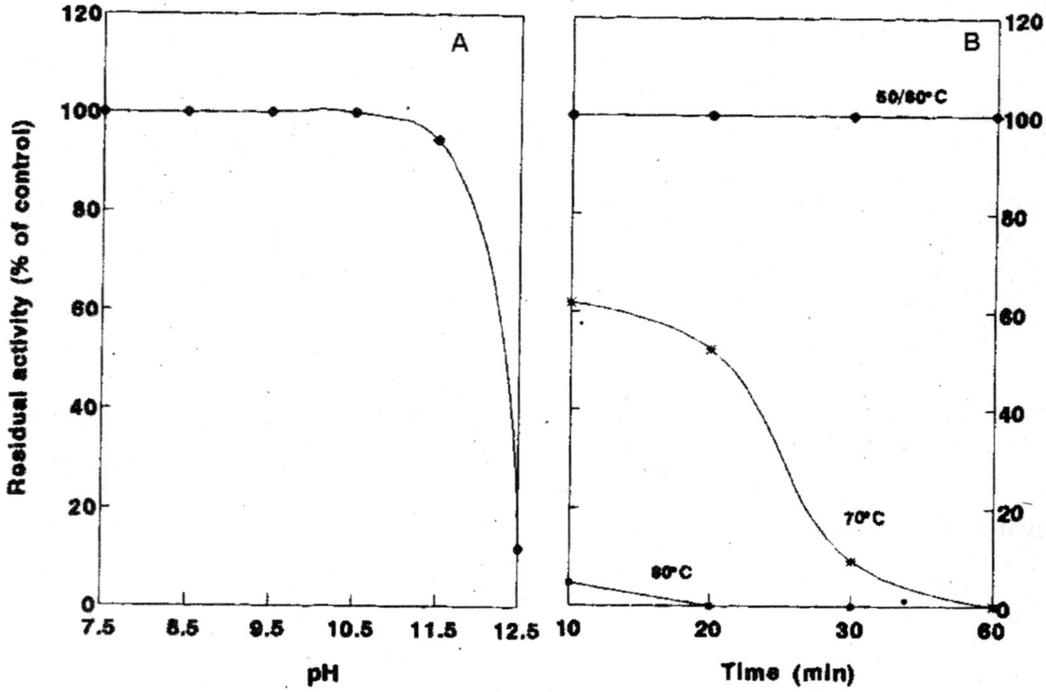

Figure 2. (A) pH Stability and (B) temperature stability profile of protease from *Bacillus* sp. IS-2.

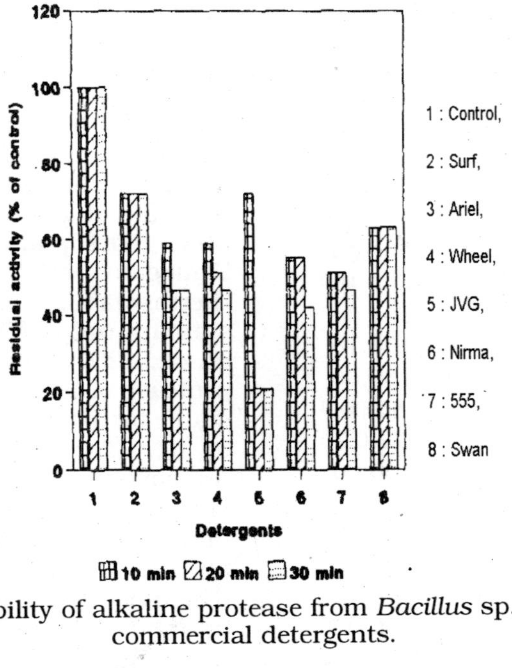

Figure 3. Compatability of alkaline protease from *Bacillus* sp. IS-3 with various commercial detergents.

The protease enzyme after 10 min of contact retained more than 70% activity in the presence of detergents, Surf^{TM} and JVG^{TM} and more than 63% its activity in the presence Swan^{TM}. Enzyme retained 59% activity in the presence of Nirma^{TM} and 51% in 555^{TM} (Fig. 3). The maximum stability was observed in Surf, as the enzyme retained 72% of its activity after incubation with detergent at 60°C for 30 min. Bhosale *et al.*, (25) reported high activity alkaline protease from *Conidiobolus coronatus* (NCL 86.8.20) showing compatability at 50°C, in the presence of 25 mM $CaCl_2$, with variety of commercial detergents. Enzyme retained 16% activity in Revel^{TM}, 11.44% activity in Ariel^{TM} and 6.6% activity in Wheel^{TM}. Comparing these results, our enzyme was much more stable in commercial detergents, retaining 72% activity at 60°C in Surf and 63.2% of its activity in a local made crude type of detergent Swan^{TM}.

As the protease enzyme produced by our isolate, *Bacillus* sp. IS-3, was thermostable and pH stable, it was used as additive in detergent to check the contribution of protease towards the improvement in washing performance of detergent. The supplementation of the enzyme in detergent surf significantly improved the cleansing performing towards blood stain.

REFERENCES

1. Godfrey T & West S, In : Industrial Enzymology (Godfrey T & West S, eds), Stockholm Press, New York (1996) 1.
2. Coolbear T, Eames CW, Casey Y, Daniel RM & Morgan HW, J Appl Bacteriol, 71 (1991) 252.
3. Purva, Soni SK, Gupta LK & Gupta JK, Indian J Microbiol, 38 (1998) in press.
4. Takami H, Akiba T & Horikoshi K, Appl Microbiol Biotechnol, 30 (1989) 120.
5. Lowry OH, Rosebrough NJ, Farr AL & Randall RJ, J Biol Chem, 193 (1951) 265.
6. Durham DR, Stewart DB & Stellway EJ, J Bacteriol, 169 (1987) 2762.
7. Kalisz HM, Adv Biochem Engg Biotechnol, 36 (1988) 3.
8. Takii Y, Kuriyama N & Suzuki Y, Appl Microbiol Biotechnol, 34 (1990) 57.
9. Fujio Y & Kume S, World Microbiol Biotechnol, 7 (1991) 12.
10. Chu MI, Lee C & Li TS, Enzyme Microb Technol, 14 (1992) 755.
11. Peek K, Daniel RM, Monk C, Parker L & Coolbear T, Eur J Biochem, 207 (1992) 1035.
12. Kwon YT, Kim JO, Moon SY, Lee HH & Rho HM, Biotechnol Lett, 16 (1994) 413.
13. Razak NA, Samad MYA, Basri M, Yunus WMZW, Ampon K & Salleh AB, World J Microbiol Biotechnol, 10 (1994) 260.
14. Kobayashi T, Hakamada Y, Hitomi J, Koike K & Ito S, Appl Microbiol Biotechnol, 45 (1996) 63.
15. Horikoshi K, In : Microbial enzyme and Biotechnology (Horikoshi, K, ed) Academic Press. London (1990) 275.
16. Kato T, Yamagata Y, Arai T & Ichishima E, Biosci Biotech Biochem, 56 (1992) 1165.

17. Fujiwara N, Masui A & Imanaka T, J Biotechnol, 30 (1993) 245.
18. Coolbear T, Daniel RM & Morgan HW, Adv Biochem Engg Biotechnol, 45 (1992) 58.
19. Dhandapani R & Vijayaragvan R, World J Microbiol Biotechnol, 10 (1994) 33.
20. Takii Y, Taguchi H, Shimoto H & Suzuki Y, Appl Microbiol Biotechnol, 27 (1987) 186.
21. Debette J, Current Microbiol, 22 (1991) 85.
22. Stefantisi D, Sakellaris G & Garel JR, FEMS Microbiol Lett, 128 (1955) 53.
23. Yamagata Y & Ichishima E, Current Microbiol, 30 (1955) 357.
24. Ferrero MA, Castro GR, Abate CM, Baigori MD & Sineriz F, Appl Microbiol Biotechnol, 45 (1996) 327.
25. Bhosale SH, Rao MB, Deshpande VV & Srinivasan MC, Enzyme Microb Technol, 17 (1955) 136.

CHAPTER - 38

In vitro PRODUCTION OF ALKALOIDS OF *Rauwolfia serpentina* BENTH

N.B. Singh and B.N. Pandey

Plant Physiology Lab., Department of Botany, University of Allahabad, Allahabad - 211 002.

ABSTRACT

Production of alkaloids of *Rauwolfia serpentina* Benth. from callus culture has been attempted. Profuse callusing was observed both from leaf and stem segments on MS and SH media supplemented with phytohormones IAA, BAP and 2,4-D. SH medium gave better result and 2, 4-D gave maximum callusing. Growth indices were calculated to show that after a certain concentration, phytohormones affect negatively the growth of callus. Alkaloids were extracted from dried and powdered callus tissue with boiling methanol and chromatographed on thin layer silica plates to show the presence of two major and at least five minor alkaloids.

1. INTRODUCTION

Two aspects of plant tissue culture are of immediate commercial application (1). One is micropropagation of elite varieties of plants, which is a flourishing industry today (2). The other aspect, which is gaining momentum and has a great potential, is *in vitro* production of vitamins, pigments, alkaloids, food flavours and other useful secondary metabolites (3,4). Economically, tissue culture for natural products may not be as expensive as the synthetic production of various commercial substances, or pharmaceuticals (5). Further, as the natural habitats for wild plants become endangered and make it difficult to acquire plant-derived chemicals, it has become pertinent to develop alternative sources of important natural products. And here, plant tissue culture has the potential as an alternative to traditional harvesting for production of secondary plant metabolites (6,7).

Here is a report of callus culture of *R. serpentina* for production of its alkaloids. *R. serpentina* is a tropical plant and is well-known for its hypotensive and tranquillising effects, but now it is threatend due to overexploitation and degradation of its natural habitats. Therefore, plant tissue culture can relieve pressure mounted on this endangered species by providing an alternative for production of it alkaloids.

2. MATERIALS AND METHODS

The plant material was procured from the garden of Botanical Survey of India, Allahabad. Pieces of stem and leaf were washed under running tap water to remove dirt and then washed with a mild detergent and rinsed. Pieces were surface sterilised with 0.2% (w/v) mercuric chloride and rinsed several times with sterile distilled water. About 10 mm of internodal stem pieces and 50mm^2 leaf pieces were aseptically cut and cultured on MS basal medium supplemented with 2% sucrose. pH of the medium was adjusted to 5.8 before autoclaving. Agar (0.8%) was used for solidifying the medium. The cultures were maintained at 28± 1°C under 16 hours photoperiod.

Three hormones, viz., IAA (indole-3-acetic acid), BAP (6- benzylaminopurine) and 2,4-D (2,4-dichlorophenoxy acetic acid) were tested for their effectiveness for callus growth on two basal media, (MS) and (SH). Pieces of callus from established callus cultures were transferred to the two media supplemented with hormones. After four weeks, fresh weight and dry weight of the callus were recorded for calculating growth indices. Growth index was calculated for each callus by the following formula :

$$\text{Growth index} = \frac{\text{Dry weight of callus} - \text{Dry weight of explant}}{\text{Dry weight of explant}}$$

For determining alkaloids present in the callus, tissue was dried, powdered and extracted in boiling methanol. The filtered extract was chromatographed on thin layer silica gel with chloroform : methanol (1:1) as solvent. Presence of alkaloids was confirmed by observation under UV light, followed by spraying with ferric chloride perchloric acid reagent by the coloured spots on the chromatogram (8).

3. RESULT AND DISCUSSION

A good callusing has been observed (Table 1). The colour of stem callus was pale yellow while that of leaf callus was greenish yellow. The texture of the callus was slightly compact with nodular surface which was more prominent in stem callus.

Result of the effect of different phytohormones on the growth of callus from leaf and stem is shown in the table 1.

Data clearly shows that SH medium is better for growth of callus, both from stem and leaf. It is important to note that up to 2 or 4 mg/1 concentration, phytohormones promote growth beyond which the effect declines.

The chromatogram showed that at least two major and five minor alkaloids are present in the extract of stem and leaf

Table 1. Growth of Callus from Leaf and Stem

Conc. (mg/l)	GROWTH INDEX											
	Leaf Callus						Stem Callus					
	MS Basal			SH Basal			MS Basal			SH Basal		
	IAA	BPA	2,4-D	IAA	BAP	2,4-D	IAA	BAP	2,4-D	IAA	BAP	2,4-D
0.5	0.6	0.8	1.2	0.8	1.2	2.4	0.4	0.8	1.1	1.3	1.4	1.8
1.0	1.2	1.4	2.8	1.3	1.9	3.9	0.9	1.1	1.9	1.8	1.9	3.2
2.0	1.8	2.8	4.9	2.1	4.6	5.2	2.4	3.1	3.9	1.8	3.8	3.8
4.0	1.3	4.2	6.2	1.6	2.9	7.0	1.9	2.6	4.1	1.2	3.2	4.7
8.0	0.4	2.9	3.9	1.1	1.8	3.8	1.1	0.7	2.1	0.7	1.1	1.2

callus. Detailed characterisation, identification and quantitative estimation is still in progress.

The first report on tissue culture of *R. serpentina* appeared in 1964 (9). Since then there have been several reports on different aspects, ranging from micropropagation to alkaloids production. Alkaloids production has been successfully attempted from callus culture and suspension culture. But commercial production of *R. serpentina* has to wait till economically feasible procedure are developed.

REFERENCES

1. Bajaj YPS (ed) Biotechnology in agriculture and forestry : Medicinal and aromatic plants I. Springer-Verlag; Berlin Hedelberg. (1988).
2. Bhojwani, SS and Rajdan, MK, Plant tissue culture : theory and practice 2/e. Elsevier, Amsterdam. (1996).
3. Collins, HA and Watts, M, In : Handbook of plant cell cultue, vol. I : Techniques for propagation and breeding. MacMillan, New York. DA Evans *et al.*, (eds) (1983).
4. Constabel F and Vasil IK (eds), Cell culture and somatic cells genetics of plants, Vol. 4 : Cell culture in phytochemistry. Academic Press, Inc., San Diego (1987).
5. Fujita, Y, In : Plant tissue culture : Application and limitations. Ss Bhojwani (ed), Elsevier, Amsterdam. (1990).
6. Fowler MW, *Pharma J*, 224 (1980) 39-40.
7. Morris P, Scragg AH, Stafford A & Fowler MW (eds.), Secondary metabolism in plant cell culture. Cambridge Univ. Press, New York. (1986).
8. Harborne JB, Phytochemical methods. Chapman and Hall, London. (1973).
9. Mitra, GC and Kaul KN, *Indian J Expt Biol*, 2 49-51. (1964).

SECTION – 3

ENVIRONMENTAL BIOTECHNOLOGY

CHAPTER - 39

STUDIES ON LEACHING OF COPPER FROM MOSABONI ORE USING THERMOPHILIC BACTERIA

Asha Gupta and Kumar Birendra

Central Mining Research Institute, Dhanbad-826 001 INDIA

ABSTRACT

Thermophilic leaching bacteria were isolated from different sources i.e., mine water from Amjhore pyrite mine, Mosaboni mine (two samples collected from two different locations), one sample also collected from hot spring water at Rajgir (Bihar). One sample of sludge from Mosaboni mine was also taken as source of thermophilic bacteria. These organisms were grown in Sulfolobus media at temperature 70°C and pH 1.8. Mosaboni ore (containing 1% copper) was leached with all the isolates at 70°C and keeping pH below 1.8. Maximum recovery of copper that is 33% (0.33 g/l) was obtained with thermophiles isolated from Mosaboni sludge in 75 day. Leaching experiments were also conducted with first recultured strains of all the above isolates. During this experimentation same isolate gave the highest recovery of copper 40.46% (0.4046 g/l) in 48 days. This shows that leaching rate is higher with first recultured strains than the strains of first isolate.

1. INTRODUCTION

The activity of *Thiobacillus ferrooxidans* declines markedly above 45°C, most if not all leach dumps can reach temperatures as high as 80°C, and it was suggested that such temperature increases were associated with bacterial oxidation of ferrous iron and metal sulphides (1). During past two decades, numerous thermophilic microorganisms were isolated from hot springs and acidic hot springs. Spherical, thermophilic microbes have been identified as Sulfolobus (*Sulfolobus acidocaldarius*) or Sulfolobus-like microbes capable of growing at temperatures of 45 to 80°C, at pH values ranging from roughly 1 to 6. Similar thermophilic, acidophilic micro-organisms were also isolated from a volcanic hot springs. Rod-shaped thermophilic microbes (including thermophilic thiobacilli) have also been isolated and rod-shaped thermophiles as well as spherical-shaped Sulfolobus have been found to co-exist in certain hot springs.

The *Sulfolobus*-like microbes have been demonstrated to be capable of oxidising ferrous iron and metal sulphides (2), particularly chalcopyrite in low-grade waste containing pyrite. This finding, in addition to the fact that the reaction rate roughly doubles for each 10°C rise in temperature, has provided considerable incentive for the development of high-temperature bacterial leaching regimes, as well as efforts to stimulate increased leaching by inoculation with thermophilic micro-organisms.

In view of these features, attempts were recently made to isolate thermophilic micro-organisms from a large, copper-bearing waste leaching experimental columns that was designed to provide test data for leaching model based in part upon Kennecott's Midas dump (3).

For both chalcopyrite and concentrate, tests were carried out using a mixed, moderately thermophilic culture. Copper dissolution in excess of 95% was achieved in all cases, and 98% in most, careful control of the pH value is needed to prevent co-precipitation of iron with the copper (4).

It was demonstrated in shake-flask experiments that the sulphur-oxidising microbiota of municipal sludge can be used at 53°C (5) for heavy-metal leaching. Five sludges were tested and the average final pH, oxidation/reduction potential and SO_4 concentration after 30 days were 2.8, 237 mV and 5668mg/l respectively. Ferric

Table 1. Copper leached out per 100g of ore with thermophiles isolated from Amjhore pyrite mine + hot spring water (Rajgir, Bihar)

Days	Leaching with strains of first isolate			Leaching with strains of first reculture		
	pH	pH adjusted to	Copper [g]	pH	pH adjusted to	Copper [g]
0	2.41	–	0.0000	2.06	–	0.0000
1	n.d.	n.d.	n.d.	3.54	1.88	n.d.
2	n.d.	n.d.	n.d.	2.88	1.77	n.d.
8	5.36	2.40	0.0000	2.81	1.07	0.0653
13	n.d.	n.d.	n.d.	1.45	1.45	0.0972
20	n.d.	n.d.	n.d.	1.62	1.62	0.1450
24	n.d.	n.d.	n.d.	1.83	1.83	0.1702
31	3.97	1.80	0.0451	1.87	1.87	0.2225
37	n.d.	n.d.	n.d.	1.93	1.93	n.d.
44	2.88	1.98	0.0849	n.d.	n.d.	n.d.
45	n.d.	n.d.	n.d.	1.94	1.94	0.3000
48	n.d.	n.d.	n.d.	1.94	1.94	0.3277
50	2.67	1.17	0.1052	n.d.	n.d.	n.d.
62	1.81	1.81	0.1503	n.d.	n.d.	n.d.
72	n.d.	n.d.	n.d.	1.98	1.79	0.3301
73	1.80	1.80	0.1826			
79	2.14	1.84	0.2224			
90	1.97	1.97	0.2643			
114	2.03	1.68	0.3073			

n.d. = not done

chloride was added to enhance the redox potential and to lower pH, which resulted in average value of 409mV and 1.86 respectively. The average solubilisation of Cu after ferric chloride addition was 65.90 ± 15.10%.

2. MATERIALS AND METHODS

(a) Ore

The ore was collected from 11th level 110N Room and Pillar stope of Badia Section of Mosaboni Mine, Hindustan Copper Limited, India. Ore contained quartz - 65%, Flaky minerals (Biotite, Chlorite, Mica, etc.) - 25%, Apatite - 2%, Tourmalite - 2%, other minerals (Zircon, Epidote) - 3% and opaque ores (Pyrite, Pyrrhotite, Chalcopyrite, Iron oxide, etc.)- 3%. The copper content was around 1%.

(b) Media

Sulfolobus media of composition $MgSO_4.2H_2O$ - 0.5 g/l, KH_2PO_4-0.5 g/l, $(NH_4)_2SO_4$-0.5 g/l, Yeast Extract-0.2 g/l, Cystene-0.004 g/l and pH Between 2-2.5 was used

Table 2. Copper leached out per 100g of ore with thermophiles isolated from Mosaboni water [a]

Days	Leaching with strains of first isolate			Leaching with strains of first reculture		
	pH	pH adjusted to	Copper [g]	pH	pH adjusted to	Copper [g]
0	2.41	–	0.0000	2.06	–	0.0000
1	n.d.	n.d.	n.d.	3.94	1.98	n.d.
2	n.d.	n.d.	n.d.	3.00	1.72	n.d.
8	n.d.	n.d.	n.d.	2.82	1.34	0.0413
13	n.d.	n.d.	n.d.	1.74	1.74	0.0662
16	5.14	1.72	0.0007	n.d.	n.d.	n.d.
20	n.d.	n.d.	n.d.	2.00	2.00	0.1067
24	n.d.	n.d.	n.d.	2.16	2.16	0.1291
28	2.54	1.88	0.0500	n.d.	n.d.	n.d.
29	2.22	1.69	n.d.	n.d.	n.d.	n.d.
31	n.d.	n.d.	n.d.	1.98	1.98	0.1700
35	2.26	1.71	0.0800	n.d.	n.d.	n.d.
37	n.d.	n.d.	n.d.	2.26	1.75	0.2015
40	2.00	2.00	0.1025	n.d.	n.d.	n.d.
47	2.15	2.15	0.1325	n.d.	n.d.	n.d.
48	n.d.	n.d.	n.d.	2.14	1.40	0.27
51	2.16	2.16	n.d.	n.d.	n.d.	n.d.
58	2.00	2.00	0.1350	n.d.	n.d.	n.d.
64	1.73	1.73	0.2035	n.d.	n.d.	n.d.
72	1.93	1.93	0.2468	1.70	1.70	0.3527
75	n.d.	n.d.	0.2597			
99	2.09	1.86	0.3138			

n.d. = not done

Table 3. Copper leached out per 100g of ore with thermophiles isolated from Mosaboni water [b]

Days	Leaching with strains of first isolate			Leaching with strains of first reculture		
	pH	pH adjusted to	Copper [g]	pH	pH adjusted to	Copper [g]
0	2.41	–	0.0000	2.06	–	0.0000
1	n.d.	n.d.	n.d.	3.92	2.02	n.d.
2	n.d.	n.d.	n.d.	2.97	1.84	n.d.
8	n.d.	n.d.	n.d.	2.66	1.39	0.0312
13	n.d.	n.d.	n.d.	2.00	2.00	0.0620
16	5.14	1.60	0.0006	n.d.	n.d.	n.d.
20	n.d.	n.d.	n.d.	1.99	1.99	0.0978
24	n.d.	n.d.	n.d.	2.18	2.18	n.d.
28	2.87	2.87	0.0500	n.d.	n.d.	n.d.
29	2.95	1.67	n.d.	n.d.	n.d.	n.d.
31	n.d.	n.d.	n.d.	1.87	1.87	0.1552
35	2.31	1.55	0.0800	2.33	1.64	0.1851
40	2.00	2.00	0.1028	n.d.	n.d.	n.d.
47	2.18	2.18	0.1330	n.d.	n.d.	n.d.
48	n.d.	n.d.	n.d.	2.16	1.61	0.2505
51	2.18	2.18	n.d.	n.d.	n.d.	n.d.
58	2.06	2.06	0.1350	n.d.	n.d.	n.d.
64	2.36	1.67	0.2043	n.d.	n.d.	n.d.
72	2.14	1.46	0.2500	1.85	1.60	0.3703
75	n.d.	n.d.	0.2693			
99	1.85	1.60	0.3145			

n.d. = not done

(c) Source of Organisms

Hot spring water (Rajgir, Bihar), mine effluent (Amjhore pyrite mine), Mosaboni mine water [a] passing through working stope has been collected at lower level of the same stope, Mosaboni water [b] percolated through old worked out stopes and lying stagnant on level, and Mosaboni mine sludge (11th level, 110 Room and Pillar stope) served as inoculum for the isolation of thermophilic leaching bacteria.

(d) Isolation of thermophilic leaching bacteria from different sources

For the isolation of thermophilic leaching bacteria, five 250-ml Erlenmeyer flasks, each containing 100 ml of Sulfolobus media of pH 2.41, 1g ferrous sulphate and 0.5 g of sulphur powder were taken. Flask 1 was inoculated with 5 ml of hot spring water, similarly flasks 2, 3, and 4 were inoculated with 5 ml of Amjhore mine water, Mosaboni mine water [a] and [b] respectively. Flask 5 was inoculated with 5 ml filtered solution of 1g sludge taken in 10 ml distilled water. All the flasks were incubated at 65-70°C in an oven. After 15 days, ferrous sulphate got oxidised to ferric sulphate and pH dropped to 2.00 in all the sets. The average cell count after 15 days was in the range of 10^{10} to 10^{11} cells/ml.

(e) First reculturing

The method opted was similar as mentioned for the isolation, this time inoculation was done with 5 ml culture of

Table 4. Copper leached out per 100g of ore with thermophiles isolated from Mosaboni sludge

Days	Leaching with strains of first isolate			Leaching with strains of first reculture		
	pH	pH adjusted to	Copper [g]	pH	pH adjusted to	Copper [g]
0	2.41	–	0.0000	2.06	–	0.0000
1	n.d.	n.d.	n.d.	4.10	1.50	n.d.
2	n.d.	n.d.	n.d.	2.70	1.71	n.d.
8	n.d.	n.d.	n.d.	2.60	1.17	0.0850
13	n.d.	n.d.	n.d.	1.96	1.96	0.1348
16	5.41	1.70	0.0007	n.d.	n.d.	n.d.
20	n.d.	n.d.	n.d.	1.92	1.92	0.2300
29	3.28	1.63	0.0712	n.d.	n.d.	n.d.
31	n.d.	n.d.	n.d.	1.79	1.79	0.3200
35	2.34	1.64	0.1098	n.d.	n.d.	n.d.
37	n.d.	n.d.	n.d.	2.19	1.89	0.3573
40	2.03	2.03	0.1380	n.d.	n.d.	n.d.
47	2.03	2.03	0.1800	n.d.	n.d.	n.d.
48	n.d.	n.d.	n.d.	2.08	1.54	0.4046
51	2.01	2.01	n.d.	n.d.	n.d.	n.d.
58	2.00	2.00	0.2400	n.d.	n.d.	n.d.
64	2.36	1.71	0.2742	n.d.	n.d.	n.d.
72	2.23	1.67	0.3100	1.96	1.85	0.3457
75	n.d.	n.d.	0.3301			
99	2.30	1.80	0.2492			

n.d. = not done

respective isolated culture. Oxidation of ferrous completed within 7 days, pH dropped from initial 2.2 to 1.92 and cell count was in the range of 10^{10} to 10^{11} cells/ml.

(f) Leaching experiments

Batch leaching was performed in 250-ml Erlenmeyer flasks containing 10g of ore, 100 ml Sulfolobus medium, and 5 ml of respective culture. Initially four sets were made with isolated cultures : set-1 contained mixed culture isolated from hot spring water and Amjhore pyrite mine water, set-2 and set-3 were inoculated with 5 ml culture isolated from Mosaboni water [a] and [b] respectively and set-4 was inoculated with culture isolated from Mosaboni sludge. All the flasks were incubated at 65-70°C. Leachants were analysed periodically for copper content by AAS (GBC-902, Australian). pH was also measured periodically and adjusted to required value to keep the copper in solution.

Similarly four more experiments were set with cultures of first reculturing.

3. RESULTS AND DISCUSSIONS

Results of copper leached out per 100g of ore has been depicted in Table 1, 2, 3, and 4. From the results it is inferred that rate of leaching is better with cultures of first reculturing of all the isolates compared to that of original isolated cultures. Maximum copper leached out from 100g of ore, with ore : leachant ratio 1:10, is 0.4056 g in 48 days. There is depletion of the copper content in the solution (Table 4). This may be due to the co-precipitation of iron with the copper. Therefore, it is advisable at this concentration, copper content may be cemented out from the solution using iron filings or recover it by solvent extraction/ electrowinning process.

ACKNOWLEDGEMENT

The authors are grateful to the authorities of Indian Copper Complex, Mosaboni (HCL), for providing ore, water and sludge samples.

REFERENCES

1. Cathles, L.M., and APPS. J.A., Metall. Trans., Vol. 6B. (1975) pp. 617-624.
2. Lyalikova, N.N.; Microbiologia, Vol. 29, (1960) pp. 29-37.
3. Miller P.C., Winby R., Preps. Soc. Min. Engrs. AIME, no. 94-97, pp. 9, (1997).
4. Murr, L.E., "Theory and practice of copper sulphide leaching in dumps and *in-situ*", Mineral Sci. Engng, Vol. 12, no.3, July, (1980).
5. Shooner, F., Tyagi, R.D., Appl. Microbiol. Biotechnol., Vol. 45, No.3, (1996) pp. 440-446, April.

CHAPTER - 40

INFLUENCE OF INOCULUM TYPES ON BIOEXTRACTION PROFILE OF COPPER AND ZINC FROM GMDC CONCENTRATE IN STR

D.R. Tipre[1] , *S.B. Vora*[2] *and S.R. Dave*[1]

1. Department of Microbiology, School of Sciences, Gujarat University, Ahmedabad - 380 009, INDIA.
2. Gujarat Mineral Development Corporation, Ahmedabad - 380 009, INDIA.

ABSTRACT

This investigation reports the influence of various types of inoculum viz., 1% Fe (II) grown culture, 10 and 20% supernatant and/or leachate, 10 and 15% only slurry used for copper and zinc extraction from GMDC polymetallic concentrate containing sphalerite, chalcopyrite, chalcocite, galena and pyrite as main constituents. All the experiments were carried out in 5 dm^3 Stirred Tank Reactor (STR) with working volume of 3 dm^3 having 20% (w/v) pulp density. Copper and zinc extraction rate varied from as low as 0.073 and 1.677 $g.l^{-1}.d^{-1}$ to as high as 0.47 and 4.142 $g.l^{-1}.d^{-1}$ respectively with normal flora of the concentrate and 20% (v/v) leachate as inoculum. Highest redox potential 435 mV, Fe (II) 2.585 $g.l^{-1}$ and pH 2.11 were recorded on 10th day of the leaching when 300 cm^3 supernatant, 3000 cm^3 1% Fe (II) grown culture and 600 cm^3 leachate were used as inoculum respectively. More than 20% (>600 cm^3) leachate as inoculum showed deleterious effect on metal extraction. The reaction time of 8 to 27 days were required to achieve the best metal extraction with respect to the type of inoculum used. Under the experimental conditions, 20% leachate as inoculum was found to be the best for the metal extraction. Maximum 4.138 and 44.37 $g.l^{-1}$ copper and zinc extraction was achieved after 15 and 27 days of contact time. The highest extraction corresponds to 73.95 and 82.76% for zinc and copper respectively. Detailed extraction profile would be discussed.

1. INTRODUCTION

The large amount of biomass needed to intensify the oxidation process can be obtained by different methods. But obtaining a concentrated biomass of metal leaching bacteria using the classical method for industrial purposes is very complicated and almost impossible. In order to obtain 1 g (25-30 % efficiency) of biomass about 100g of Fe^{2+} iron is to be oxidised (1). On the other hand, when inoculum of low cell concentration was used for reactor having high solids concentration, it resulted in a low cell-to-solid ratio and longer lag time during leaching process (2,3). Yunker and Radovich (4) have also proposed that initial bacterial concentration of 10^8 to 10^9 cell. ml^{-1} should be used for inoculation. Dave et al. (5) have also reported that 11 mg cellmass as inoculum/20 g sphalerite showed increase in zinc extraction, above this the leaching rate is decreased.

When Fe^{2+} grown culture was used as inoculum for metal extraction, it required some lag period to start the metal solubilization activity. While culture grown on similar substrate required no adaptation time. The reproducibility of the results of microbiological leaching of metal sulphides requires three to four transfers of the microorganisms to the specific substrate prior to experimentation (6). But when leaching bacteria were grown on solid substrate like chalcopyrite or pyrite, nearly ninety percent of the cells were reported to remained attached to the substrate (7,8). This fraction of the biomass may be responsible for the direct attack on sulphide mineral (9). When inoculum is prepared from suspension containing solid substrate, it offers three possibilities, like only supernatant, only slurry and leachate (mixture of above two) as inoculum. Inspite of this fact, there has been no clarity regarding the type of inoculum to be used at STR level leaching experiments.

Thus attempts were made to study the influence of various inocula viz., supernatant, slurry and leachate prepared from organisms grown on polymetallic concentrate used as substrate for the extraction of metal ions from GMDC concentrate.

2. MATERIALS AND METHODS

(a) Bacteria

Four strains of *Thiobacillus ferrooxidans* isolated from hot-water spring, lignite sample, chalcopyrite mine and polymetallic ore sample were maintained in liquid 9K medium (10). The cultures were harvested when more than 95 % Fe^{2+} was oxidised and preserved by the method described by Dave (11).

(b) Consortium

All the four strains were mixed in equal proportion to obtain 10^9 cells. ml^{-1}. The consortium was exposed to increasing doses of copper, lead and zinc for the respective metal tolerance development. The consortium adapted and mutated during many years of experiment on complex polymetallic sulphide concentrate has been used in the experiments.

(c) Polymetallic concentrate

The polymetallic concentrate was supplied by M/s. GMDC (Gujarat Mineral Development Corporation, Multimetal Project, Ambaji, Gujarat, India). The major constituents of the concentrate were sphalerite, chalcopyrite, galena and pyrite containing zinc (30 %), copper (2.5 %), lead (13 %), iron (8 %) and sulphur (28 %). The concentrate having mix particle size ranging between -150 and + 400 mess (B.S.S.) was used in all the experiments.

(d) Stirred Tank Reactor (STR)

STR designed in our laboratory was used for the laboratory scale-up study of bioextraction of metals from the polymetallic concentrate. Bioextraction experiment was scaled up from 100 cm^3 systems in shake flask study to 3 dm^3 STR having controlled agitation and aeration facilities.

(e) Preparation of inoculum

The consortium was grown on 20 % (w/v) polymetallic concentrate. After completion of leaching, system was allowed to remain at steady state condition for two hours, which resulted in separation of slurry and supernatant. Leachate represents unsettled system i.e. mixture of slurry and supernatant. The process was varied with respect to the type of inoculum used as mentioned in Table 1. Final system was adjusted to 3 dm^3 in each set with the medium.

Table 1. Inoculum and experiment design variation

No.	Inoculum		Aeration	Baffles
	Type	Amount		
1.	Natural flora	–	–	–
2.	1%Fe^{2+} grown	3.0 dm^3	–	–
3.	Wet slurry	60 g	–	–
4.	Wet slurry	90 g	–	–
5.	Supernatant	10%(v/v)	–	–
6.	Leachate	20%(v/v)	–	–
7.	Leachate	20%(v/v)	+	–
8.	Leachate	10%(v/v)	+	+
9.	Leachate	20%(v/v)	+	+

(f) Bioleaching study

Stirred tanks of 5 dm^3 capacity with a working volume of 3 dm^3 with and without baffles was operated using 20% (w/v) pulp density in all the experiments. Medium used throughout the process was composed of 0.5 and 0.125 $g.l^{-1}$ $(NH_4)_2SO_4$ and K_2HPO_4 respectively. Polymetallic concentrate was used as substrate giving liquid : solid ratio of 1 : 5. Aeration was provided by compressed air at the rate of 5 $l.min.^{-1}$ (where mentioned). The temperature throughout the experiments was 35± 2°C. Acidified (pH 2.0) distilled water was added to compensate loss due to evaporation.

Series of experiments were carried out to check the reproducibility of the results. Pulp density and medium used in all the experiments were identical.

Redox potential and pH were determined using Systronics 116 pH analyser. Fe^{2+} was determined titrometrically with potassium dichromate; copper and zinc were analysed by standard spectrophotometric and AAS methods (12).

3. RESULTS AND DISCUSSION

During the course of the experiments, all the STR systems were monitored for pH and redox potential. In the initial phase of first two days, pH and redox potential of the system were found to increase and decrease respectively. The acid production/ consumption profiles were determined on the basis of pH variation and requirement for addition of sulphuric acid. Initial and final pH on 10th day of bioleaching experiments and total amount of 10 % H_2SO_4 needed to stabilise the pH are summarised in Fig. 1. The pH of sterile mineral salt medium containing 20 % (w/v) pulp of polymetallic concentrate was found to increase from 2 to 2.46 after two days of contact time. This could be due to the presence of acid consuming gangue present in the concentrate. During bioleaching experiments, the final pH of the systems after 10 days of contact time showed the pH as low as 1.81 and as high as 2.20 with inoculum type 9 and 7 respectively (Fig. 1). This could be due to the biological action, which reduces the pH through sulphuric acid production from the available sulphide in substrate, and/or hydrolysis of biologically generated ferric. In the initial phase of experiment, variable amount of acid addition between 5 to 26.25 ml was needed which indicates faster chemical sulphide oxidation as compared to biological sulphuric acid producing activity; where biological activity was not in

Fig. 1. pH variation and acid consumption during bioleaching

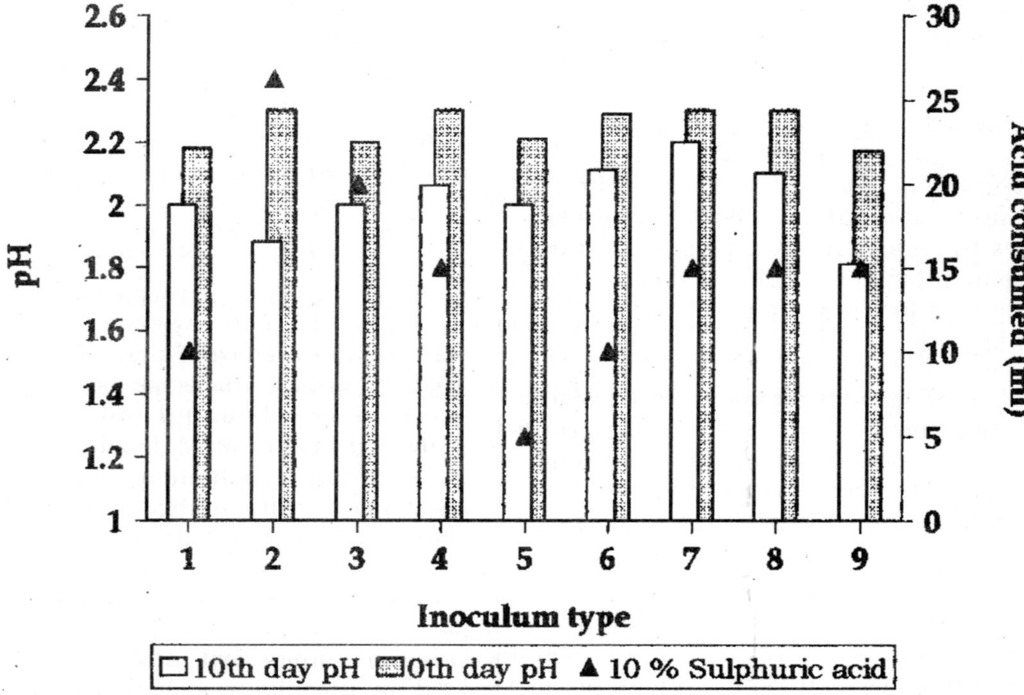

its logarithmic phase. Acid consumption was also dependent on the pattern of distribution of alkaline gangue in the concentrate particles, as well as, amount of metal sulphide oxidised during the process. Due to heterogeneous system and complexity of the process, further investigations are necessary to understand more about acid requirement behaviour.

Significant biological oxidation of sulphide mineral was confirmed in all the experiments from the observed rise in the final redox potential of the system on 10th day of experiment except where inoculum type 2 was used in leaching system (Fig. 2). Highest redox potential 432± 3 mV was seen with inoculum type 5 and 9. It can be inferred that in the presence of various types of inoculum, Fe^{2+} liberated from the concentrate due to indirect mechanism of leaching was biologically oxidised to Fe^{3+} in different proportion. The final redox potential of the system was the ultimate outcome of the overall biochemical processes.

As can be seen from the data presented in Figure 2, soluble Fe^{2+} on 10th day in the system varied significantly between 0.67 to 2.58 $g.l^{-1}$ as compared to minor variation (0.558 to 0.785 $g.l^{-1}$) in initial concentration. The final concentration of Fe^{2+} in the system increased 2 to 4 folds except with inoculum type 5.

The system with inoculum type 9 also showed second highest mV, but 1.485 $g.l^{-1}$ soluble Fe^{2+} on 10th day, which was 2.2 fold higher than the system with inoculum, type 5. Both these systems showed almost similar redox potential (432 ± 3 mV) that indicates similar Fe^{3+}/Fe^{2+} ratio on 10th day, thus system with inoculum type 9 was more biologically active as compared to system with inoculum type 5. This could be

Fig. 2. Influence of inoculum types on redox potential and ferous profile during the process.

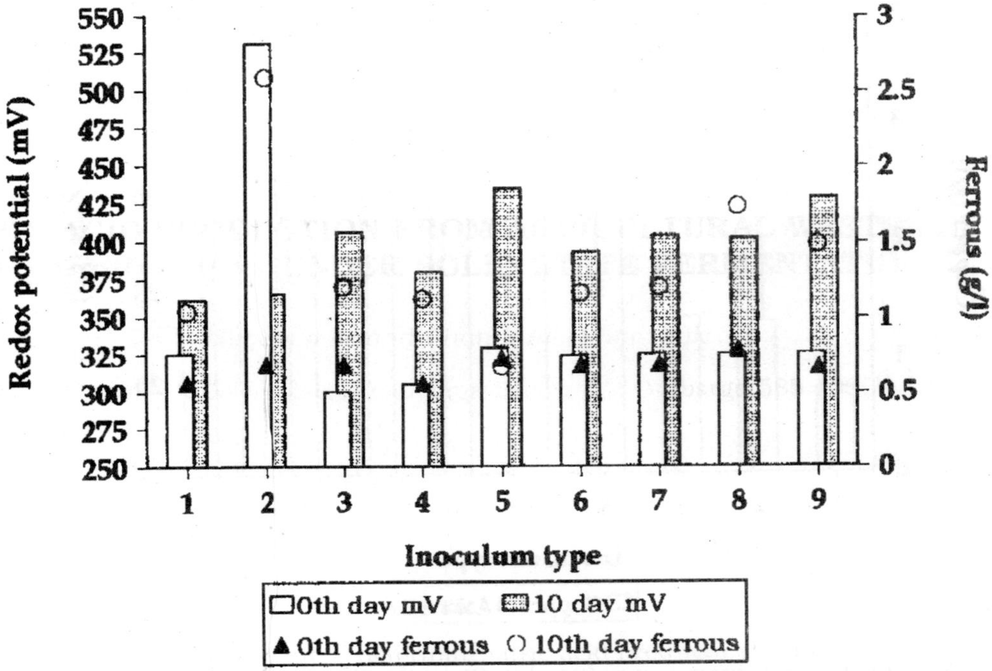

due to type and amount of inoculum and presence of baffles and aeration facilities provided. This was also supported on the basis of 1.47 and 6.575 $g.l^{-1}$ more copper and zinc extraction recorded as compared to inoculum type 5.

Data of copper and zinc dissolution in the presence of various types of inoculum on 10th day of process are shown in Fig. 3. Copper and zinc extraction range from 0.735 to 3.76 and 19.575 to 37.285 $g.l^{-1}$ with natural flora and inoculum type 9. Leachate was found to be the best type of inoculum under the experimental conditions. Amount of inoculum also showed influence on zinc extraction. Higher inoculum (20 %) resulted in 6.2 $g.l^{-1}$ more zinc extraction as compared to lower dose (10 %) of inoculum i.e. system 9 and 8 respectively.

As can be seen from the overall data, type and amount of inoculum resulted in 4 and 1.5 fold improvement in copper (0.735 and 2.962 $g.l^{-1}$) and zinc (19.575 and 29.282 $g.l^{-1}$) extraction with normal flora and inoculum type 6 respectively. Additional bioextraction experiments with improved aeration and agitation facility showed the increment of 1.3 fold for both copper and zinc extraction. Wide variation among copper and zinc extraction (4 and 1.5 fold) under the influence of inoculum types can be due to more difficult solubilization of copper by chemical or indirect mechanism from chalcopyrite than zinc from sphalerite, the minerals present in the concentrate. Thus improved inoculum addition resulted in very high increase in copper solubilization. Natural flora of the concentrate was responsible for 19.575 $g.l^{-1}$ zinc extraction and 0.735 $g.l^{-1}$ copper extraction. Thus, the increase in zinc extraction due to improved inoculum was comparatively low.

Fig. 3. Copper and Zinc extraction under the influence of various inoculum types

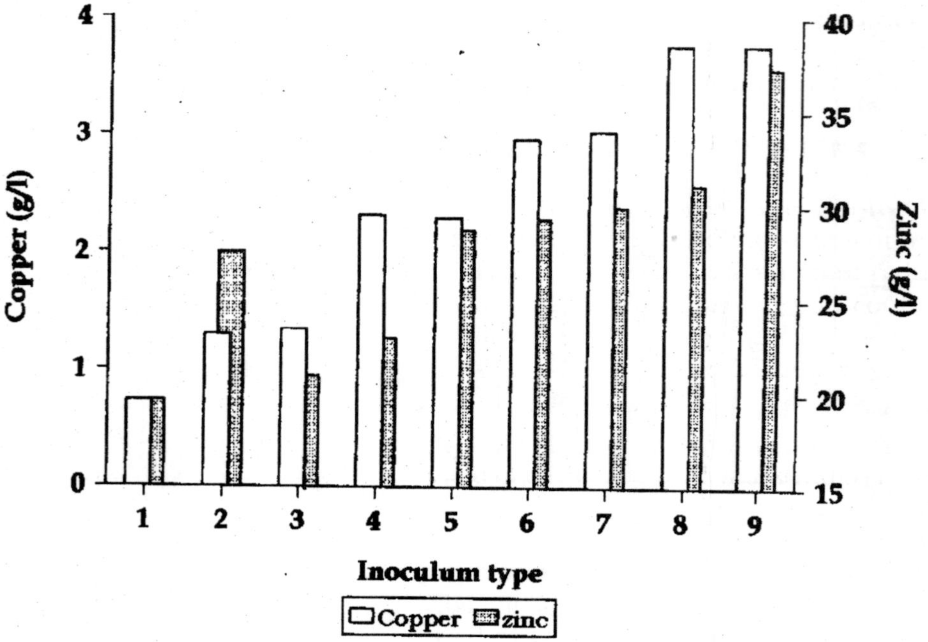

Note — Metal extraction time was reduced from 10 days to 9,7 and 8 das for copper with inoculum type 5,7 and 9 respectively and for Zinc, 9 days for inoculum types 6 and 9.

In the presence of identical inoculum, the design modification and/or improvement in aeration showed similar influence on copper and zinc extraction (1.3 fold rise) due to obvious reasons. More than 20% inoculum resulted in lower metal extraction (data not shown).

Addition of wet slurry as an inoculum was not found to be competent with leachate as inoculum. This could be due to less available suspended *Thiobacillus* cells responsible for ferrous oxidation in the supernatant. Addition of slurry also increases pulp in the system, which can adversely influence the gas mass transfer and thus lowers metal extraction.

The overall bioleaching profile in the presence of inoculum type 2 was mainly dominated by chemical reaction i.e. indirect leaching mechanism due to the presence of nearly 29 to 30 g of Fe^{3+} in the system in the form of inoculum (3 litre, 1 % Fe^{2+} grown culture having more than 99 % Fe^{3+}). Fe^{3+} ion immediately reacted with metal sulphide and oxidised copper and zinc with following reaction :

$$MS + Fe_2(SO_4)_3 \xrightarrow[\text{mechanism}]{\text{indirect}} MSO_4 + 2FeSO_4$$

This process resulted in lowering of the redox potential to as low as 308 mV from initial redox potential of 531 mV within 24 hours of contact time which yielded 23.35 $g.l^{-1}$ of zinc extraction. During this time, the soluble Fe^{2+} content of the system reached to 3.015 $g.l^{-1}$ from initial Fe^{2+} of 0.658 $g.l^{-1}$. Both these observations support the chemical leaching process. Minor biological activity was recorded thereafter, which ultimately gave the final copper and zinc extraction of 1.3 and 27.47 $g.l^{-1}$

respectively. There was gradual minor increase in redox potential during the later part of the leaching process, which also indicates weak biological activity. The observed lowest pH recorded during leaching process with this inoculum type was mainly due to precipitation of Fe^{3+} as well as the addition of sulphuric acid needed for the stabilisation of pH in the initial phase of the process.

Depending upon the variation in inoculum and experimental design, the reaction time as short as 8 days and as long as 27 days was recorded for the best metal extraction (Table 2). The variation of contact time indicates the suitability of the inoculum as well as difference in their metabolic activity for metal extraction due to availability of cells either in the form of attached biomass or suspended cells and other influencing factors in the leach system.

On the basis of data, leachate was found to be the best inoculum as compared to slurry and supernatant. Inoculum showed the following decreasing order of suitability in STR bioleaching experiments.

Leachate > Supernatant > Slurry > 1% Fe^{2+} grown consortium

Table 2. Time required for maximum metal extraction.

Inoculum type	Copper		Zinc	
	%	days	%	days
1	45.2	16	36.8	16
2	29.4	11	45.5	10
3	46.8	14	34.5	10
4	68.0	17	52.8	23
5	45.8	9	73.6	27
6	75.0	15	72.9	23
7	75.2	15	71.9	23
8	75.2	10	62.9	12
9	75.2	8	72.9	12

Copper extraction was more influenced by the type of inoculum used. Aeration and improved mixing showed similar improvement in copper and zinc extraction.

REFERENCES

1. Karavaiko GI, Kuznetsov SI & Golomzik AI, Nauka, Moscow, (1972) 248.
2. Atkins AS, In: Metallurgical applications of bacterial leaching and related microbiological phenomena (Murr LE, Torma AE & Brierly JA eds) Academic Press, NY (1978) p 403.
3. Detz CM & Barvinchak G, Am Min Cong J, 65 (7) (1979) 75.
4. Yunker SB & Radovich JM, Biotechnol Bioengg Symp Ser, 15 (1985) 307.
5. Dave SR, Natarajan KA & Bhat JV, Trans Ind Inst Metal, 32 (2) (1981) 161.
6. Silver M & Torma AE, Can J Microbiol, 20 (1974) 141.
7. McGoran CJM, Duncan DW & Walden CC, Can J Mirobiol, 15 (1969) 135.
8. Roy P & Mishra AK, Ind J Exp Biol, 19 (1981) 728.
9. Dispirto AA, Dugan PR & Tuovinen OH, Biotechnol Bioengg, 25 (1983) 1163.
10. Silverman MP & Lundgren DG, J Bacteriol, 77 (1959) 642.
11. Dave SR, Ph.D. thesis. The University of Mysore, Mysore, Karnataka. (1980).
12. Vogel AI, In: A Textbook of quantitative inorganic analysis, 3rd ed, ELBS and Longman, London (1962).

CHAPTER - 41

ROLE OF SYMBIOTIC MICRO-ORGANISMS IN SUSTAINABLE PLANT GROWTH ON HEAVY METAL CONTAMINATED INDUSTRIAL SITES

T.M. Chaudhry and A.G. Khan

Dept. of Biological Sciences, University of Western Sydney, Macarthur
P.O.Box 555, Campbelltown, NSW 2560, Australia

ABSTRACT

Pollution of the environment with toxic materials from fossil burning, mining and smelting of metalliferous ores, disposal of sewage, fertilizers and pesticides, etc. has increased dramatically since the onset of the industrial revolution. Of most commonly used methods of cleaning such contaminated sites is excavation and burial. Recently, however, the role of plants (phytoextraction) and their associated microbes for environmental clean-up has been pursed. But, knowledge of suitable plants is lacking and we know very little about the role symbiotic bacteria and fungi adapted to heavy metals play in phytoremediation. This paper reports the occurrence of plants, and their symbiotic associations with soil microbes, at heavy metal contaminated abandoned mine sites in New South Wales, Australia.

Knowledge of suitable plants that might be used for phytoremediation of heavy metal contaminated soil is lacking. Furthermore, since metal uptake and tolerance depend on both plant and soil factors, we also require information on plant-soil interactions. The plants growing on contaminated soils are normally slow growing producing a small biomass but they accumulate heavy metals in their tissues lowering the quantity and controlling the migration of heavy metals through the soil profile. Symbiotic bacteria and fungi adapted to heavy metal can be utilised to enhance the growth of such plants.

At the contaminated site species richness was poor. Roots of all the plants surveyed possessed mycorrhizal infections. On the mine tailings, species of *Baeckea, Poa, Lomandra, Acacia, Eucalyptus* and cultivated *Pinus* were the most abundant plants.

Roots of *A. melanoxylon* possessed dual symbioses, i.e., arbuscular mycorrhizae and nitrogen fixing rhizobial nodules, whereas roots of *E. dalrympleana* possessed both ectomycorrhizae and arbuscular mycorrhizae. Significance of these plants and their symbionts in phytoremediation and preserving the biodiversity and physical structure of the contaminated soils is also discussed.

1. INTRODUCTION

Heavy metal soil pollution by industrial wastes is one of the most severe ecological problems in the world. There are several chemical and physical processes for cleaning up contaminated soils (1,2). However, all of them are labour intensive and costly, and can hardly be applied to thousands of hectares of land contaminated with inorganic heavy metals (3) Most common on site techniques to decontaminate soil in Australia involve either, on site management, or, soil excavation and landfill (4-6), which merely shifts the contaminant problem elsewhere (5).

Heavy metals occur in soils mostly as inorganic compounds (7), thermal destruction or volatization, commonly applied to organic compounds, cannot be applied to decontaminate heavy metal contaminated soils (8). As a consequence, whilst a number of on-site remediation technologies are available for organic contaminants, very few *in-situ* methods are available for removal of heavy metal soil contaminants (9). Recently, Mench, et al., (10) suggested immobilisation method for heavy metal decontamination, which can be achieved by complexing the contaminants (11) or through increasing the soil pH by liming. Soil washing, another alternative to off-site burial method (12), produces a residue which, as with organic compounds, requires further treatment (13). Furthermore, the cost of soil washing and the technology involved also makes this method unpopular. Microbial bioremediation technology, well known for decontamination of organics (14), is not available for large-scale biodegradation of inorganic heavy metals. This could be, as pointed out by Burns, et al. (3), due to cheaper alternatives as landfill available.

New technologies are needed to decontaminate vast land as removing and replacing contaminated soils or soil washing in acids, are only practical for a small area. Furthermore, these technologies render the soil biologically dead and useless as a medium for plant growth as they remove all flora, fauna and microbes including useful nitrogen fixing rhizobia and P-enhancing mycorrhizal fungi in the process of decontamination. They, also, can not manufacture new topsoil to replace the old. Phytoremediation, an emerging new technology for soil remediation, which employs plants, offers an environmentally safe, potentially cheap and visually attractive alternative method to decontaminate soils which are polluted with non-volatile and immobile inorganic heavy metals (15). Some plants survive in toxic soil either by avoiding metals (hyper-tolerant) or uptaking and storing toxins in their tissues (hyperaccumula tors) (16-18), and, cultivating such plants on contaminated soils is an alternative to hauling and dumping contaminated soils (19-20).

Metal accumulator plants are usually small and to make phytoremediation a viable technology, either, find fast growing as yet undiscovered plants with extensive root system, or, engineer common plants with as yet unidentified hyperaccumulation genes. In addition to changing plant, soil also need to be amended to enhance uptake by using fertilisers (21), chelating agents (22-23) or microbes, especially symbiotic mycorrhizal fungi and nitrogen fixing bacteria.

Universal and ubiquitous symbiotic mycorrhizal associations are known to benefit plant nutrition, growth and survival due to exploration of soil for nutrients, including heavy metals (24), and pioneer plants growing on nutrient deficient soil and wastelands are often mycorrhizal (25-27). Mycorrhizal plants are partially protected against heavy metal toxicity as it is believed that heavy metals are bound by carboxyl group in hemicelluloses of mycorrhizal fungal hyphae.

The prospect of such fungi existing in nature has important implications for phytoremediation, because mycorrhizal fungi introduced in combination with their host plant at a contaminated site may survive and have sustained degradative properties, feature often not characteristics of introduced bacteria (28). They are also known to increase plant's tolerance to heavy metals (29) and nodulation of legumes equipped with nitrogen fixing rhizobial root nodules (30). These symbiotic bacteria and fungi present in contaminated soils, selected over time with greater resistance to these heavy metals (31) and adapted to them, may have evolved a heavy metal tolerance and that they may have the potential to play an important role in phytoremediation by enhancing growth and biomass of hyper-accumulator plants on contaminated sites (29). But before adopting this plant-microbe system as an ecologically sound approach to long term remediation and ecological restoration of heavy metal contaminated sites, there is an urgent need for primary reconnaissance and screening of plants growing on contaminated sites (20), and their associated symbionts such as nitrogen fixing bacteria and mycorrhizal fungi.

In view of the potential importance of symbioses in phytoremediation, the objectives of this study were to survey the distribution of plants growing on a metal-contaminated abandoned mine site and nearby non-polluted reference site, and to study their symbiotic associations. The use of these plants and their symbionts in the phytoremediation of such heavy metal contaminated and abandoned mine spoil heaps will also be discussed.

2. STUDY AREA AND METHODS

The area characterised in this study is the metal contaminated abandoned mine located at Sunny Corner, approximately 200 km. west of Sydney, Australia. Before the opening of the silver mining in 1880's, the whole area was bushland, thickly wooded by *Eucalyptus*, *Exocarpus*, *Banksia* and *Acacia* species. Numerous spoil heaps from the old mine mineral sulphide load consisted of minerals, mostly silver-lead-zinc (galena), mixed with copper and iron pyrites (32). Numerous spoil heaps around the complex of mine working were studied for vegetation and their symbiotic associations. Plants growing on these spoil heaps and their rhizosphere soils were collected from : (1) Site 1 which was located at the mine origin, close to the exit from a collapsed adit and source of heavy metal; (2) Site 2 which was 3km downstream, an area that was also disturbed, having previously experienced major workings containing no plant growth; (3) site 3 which was adjacent to site 2, but on the Daylight Creek embankment; and (4) site 4 (reference site) which was a further 3km downstream where erosion was much reduced.

The vegetation cover on the contaminated and the reference sites was studied by the quadrate method (33), and the frequency of occurrence of plant species was determined (34). The dominant plant species growing on the above sites were collected by sampling along a transact. The leaves, flowers, and fruits, where available, were also collected, dried and mounted on herbarium sheets for identification.

The roots and rhizosphere soil samples from each plant collected were also taken upto a depth of 15 - 20 cm. Roots were stored in 70% ethanol and then brought to the laboratory to be stained for mycorrhizal infection by clearing them with KOH and staining with trypan blue (35). The arbuscular mycorrhizal (AM) fungal infection in the roots collected was recorded as percentage of mycorrhizal colonisation using the gridline intersect method (36). The presence or absences of ectomycorrhizae and root nodules were also recorded for each plant species collected. Rhizospheres were analysed for the number of arbuscular mycorrhizal fungal spores per 50g by a wet-sieving and centrifugation technique (35).

3. RESULTS

3.1. Vegetation

The population density and frequency of plant species data are summarised in Table 1. Only plants found growing at site 1, the mine origin close to the exit from the collapsed adit, were shrubs *Baeckea utilis* or trees *Acacia melanoxylon* and *Eucalyptus dalrympleana*, *Poa labillardieri*, a native grass, was also very abundantly growing on this site. No plant growth was recorded for site 2 which was 3km downstream, in an area that was also disturbed, having previously experienced major workings. However, site 3 which was adjacent to site 2, but on the embankment of Daylight Creek, was dominated (76%) by myrtaceous

Table 1. Population density and Frequency of occurrence of plant species growing on Sunny Corner sites.

Species Frequency	Family	Population Density (No/sq.m)	Frequency (%)	Class
Site - 1				
Acacia melanoxylon R. Br.	*Fabaceae*	2.0	24.0	O
Eucalyptus dalrympleana Maiden	*Myrtaceae*	1.0	20.0	R
Pinus radiata D. Don	*Pinaceae*	3.0	32.0	O
Baeckea utilis F. Muell. ex Miq.	*Myrtaceae*	9.0	56.0	F
Lomandra longifolia Labill	*Lomandraceae*	3.0	28.0	O
Poa labillardieri Steud.	*Poaceae*	14.0	80.0	A
Site - 2				
No vegetation				
Site - 3				
Baeckea utilis F. Muell. ex Miq.	*Myrtaceae*	12.0	76.0	A
Poa labillardieri Steud.	*Poaceae*	10.0	68.0	A
Reference site				
Acacia dealbata Link	*Fabaceae*	2.0	28.0	O
Eucalyptus dives Schauer	*Myrtaceae*	2.0	24.0	O
Echium plantagineum L.	*Boraginaceae*	5.0	48.0	F
Centaurium tenuiflorum (Hoffm. & Link)	*Gentianaceae*	6.0	64.0	A
Verbascum virgatum Stokes	*Scrophulariaceae*	8.0	60.0	F
Poa labillardieri Steud.	*Poaceae*	16.0	84.0	VA

R= Rare (0-20%), O= Occasional (21-40%), F= Frequent (41-60%), A= Abundant (61-80%), VA= Very Abundant (81-100%), (Source : 34)

shrubs of *B. utilis*. This shrub was absent from all the other polluted sites studied, including the non-polluted reference site. The reference site, ie site 4, was inhabited by various plant species, none of which were represented on the contaminated sites , except the native grass *Poa labillardieri*. The vegetation of the reference site was more diverse and many endemic trees such as *Eucalyptus dives*, *Acacia dealbata*, *Exocarpus* and *Banksia spp.* were present (Table 1). The frequently growing herbs on this site included *Echium plantagineum* (Boraginaceae) and *Verbascum virgatum* (Scrophulariaceae).

3.2. Symbiotic Associations

The results of mycorrhizal status of the plants growing on various sites are summarized in Tables 2. The root of all the plants examined were infected with arbuscular mycorrhizal (AM) endophytes, except those of *Verbascum virgatum*, a Scruphularious plant. AM fungal colonization was higher in the roots of plants growing on the reference site than those on the contaminated sites, percentage colonization values ranging from 18-40% and 41-65% respectively (Table 2). Root cortices of arbuscular mycorrhizal *B. utilis*, a plant common to sites 1 and 3, both heavy metal contaminated, did not contain any fungal vesicles. However, many appressoria on the root surfaces and intracellular coils in the cortical cells were recorded for *B. utilis*. Roots of *Eucalyptus spp.*, growing on the contaminated as well as reference sites, possessed both ectomycorrhizae (ECM) as well as arbuscular endophytic mycorrhizae (Table 3). There was no significant difference in arbuscular mycorrhizal coloniza-

Table 2. AM- Colonization in plant species growing on Sunny Corner sites.

Species	Family	AM-Colonization			
		Hyphae (%)	Vesicles (%)	Arbuscules (%)	Total Colonization (%)
Site – 1					
Acacia melanoxylon R. Br.	*Fabaceae*	12.0	11.0	8.0	31.0
Eucalyptus dalrympleana Maiden	*Myrtaceae*	21.0	10.0	9.0	40.0
Baeckea utilis F. Muell. ex Miq.	*Myrtaceae*	13.0	nd	6.0	19.0
Poa labillardieri Steud.	*Poaceae*	15.0	14.0	10.0	39.0
Site – 2					
No vegetation					
Site – 3					
Baeckea utilis F. Muell. ex Miq.	*Myrtaceae*	11.0	nd	7.0	18.0
Poa labillardieri Steud.	*Poaceae*	13.0	10.0	9.0	32.0
Reference site					
Acacia dealbata Link	*Fabaceae*	36.0	13.0	8.0	57.0
Eucalyptus dives Schauer	*Myrtaceae*	25.0	8.0	11.0	44.0
Echium plantagineum L.	*Boraginaceae*	16.0	15.0	12.0	43.0
Centaurium tenuiflorum (Hoffm. & Link)	*Gentianaceae*	16.0	16.0	9.0	41.0
Verbascum virgatum Stokes	*Scrophulariaceae*	–	–	–	nd
Poa labillardieri Steud.	*Poaceae*	34.0	19.0	12.0	65.0

nd = not detected.

tion of roots of *E. dalrympleana* (40%), growing on the contaminated site 1, as compared to those of *E. dives* (44%) growing on the reference site 4 (Table 2). Root cortical cells of *E. dives* contained numerous degerative arbuscules as well as a network of arterial hyphae and a sheath or mantle of ectomycorrhizal fungus around their tips. Dual symbioses were also recorded for the roots of leguminous trees, i.e., *A. dealbata* on the reference site and *A. melanoxylon* on the contaminated mine site, i.e., their roots were equipped with N-fixing rhizobial nodules as well as arbuscular mycorrhizae (Table 3).

The rhizospheres of plants from the contaminated sites were found to contain a diverse range of AM fungal spores, the dominant types were *Glomus* and *Gigaspora* strains. Soil from the reference site harboured various *Glomus* spores, but no *Gigaspora* spores were present.

4. DISCUSSION

Our observations confirm the generally held view that majority of plants growing under natural conditions have mycorrhizae (24). Non-mycorrhizal status of *V. virgatum* roots in the present study is consistence with Brundrett, (37), who has listed Australian Scrophulariaceae plants as predominantly non-mycorrhizal. Our observations are also consistent with many earlier researchers who have reported AM fungal infection in plants growing on heavy metal contaminated sites (17, 29, 38-40). These observations, however, are in contrast with those of Harris and Jurgensen, (41) who reported lack of AM development in the roots of two-year-old willow and popular planted as cuttings in copper and iron ore mining tailing. These authors also found that none of the root samples from 20 trees in the copper tailing were mycorrhizal. AM fungal ecotypes from heavy metal contaminated sites were found to be tolerant to Zn and Cd than those from non-contaminated sites (42). Wang, et al., (43) also found that AM fungal spores collected from heavy metal polluted soils were more tolerant to heavy metals than those isolated from lower concentration polluted soils.

The concurrent occurrence of AM and ECM fungal endophytes in the roots of *Eucalyptus* spp. indicates a lack of antagonism once they have established themselves within the root system and supports similar observations by other researchers in Australia (44-49). The implications of this phenomena for ecosystem, as stated by Harley, (50), are "great and yet unconsidered". The occurrence of dual symbiosis of AM fungal infection and root nodulating *Rhizobia* in *A. melanoxylon* growing on the heavy metal contaminated site 1 in the present study

Table 3. Screening of nitrogen fixing and mycorrhizal tree species of the Sunny Corner sites.

Species	Family	Ectomy-corrhizae	Endomy-corrhizae	Root-nodulating Rhizobia
Acacia melanoxylon R. Br.	*Fabaceae*	−	+	+
Acacia dealbata Link	*Fabaceae*	−	+	+
Eucalyptus dalrympleana Maiden	*Myrtaceae*	+	+	−
Eucalyptus dives Schauer	*Myrtaceae*	+	+	−
Pinus radiata D. Don	*Pinaceae*	+	+	−

would also encourage mineral nutrition and increased growth (30). The reported dual symbiotic associations in the trees studied further suggests a selective advantage for these plants as pioneer species on heavy metal contaminated soils and that they may be responsible for the successful colonization of such habitats. These plant species can be manipulated as pioneering species in order to phytoremediate heavy metal contaminated soils. By growing these trees with their rhizospheres populated with selected microorganisms during the process of phytoremediation it should be possible to enhance uptake of heavy metals from soil. The potential of the trees with symbiotic associations in phytoremediation is further enhanced due to their deep rooting and perennial nature. Recently Khan and Bari, (51), reported dual symbiosis in roots of trees of *Populus euroamericana*, *A. arabica* and *Delbergia sissoo*, growing on chromium polluted tannery effluent. We recommend that these trees and their associated microflora (mycorrhizal fungi and nitrogen fixing bacteria) that are adapted to the heavy metal contaminated soils be exploited for phytoremediation and bioremediation purposes. By populating the rhizospheres of these plants with selected microorganisms during the process of phytoremediation it should be possible to enhance uptake of heavy metals from soil or water.

Recently, chelating agents have been used to enhance the bioavailability of heavy metals for plant uptake (22-23). Work is underway to screen plants for their ability to phytoremediate by increasing metal bioavailability with chelating agents and increasing uptake of the available metals by inoculating plants with appropriate microbial symbionts. Studies with mycorrhizal fungi have focussed on their ability to enhance nutrient uptake in a nutrient-deficient soil and have ignored the full role they may play in phytoremediation. The prospect of symbionts existing in heavy metal contaminated soils has important implications for phytoremediation because mycorrhizal fungi introduced in combination with their host plants at a contaminated site may survive better often not characteristic of introduced bacteria for bioremediation. An important benefit of mycorrhizal infection has been observed in sites with heavy metals, such as mine spoils, and in sites containing high levels of toxic materials (27, 52). Plants with mycorrhizal infections appear to be partially protected against the toxicity of the heavy metal. This ability of the fungus provides the host plant with added protection against toxins and allow the plant to grow in areas otherwise hazardous to the plant. Wang and Chao, (53) reported a decrease in the toxicity of heavy metals in soybean inoculated with AM fungi and planted in soil polluted by heavy metals. However, the given data shows that the mycorrhizal population response to heavy metal toxicity as well as the biochemical and molecular basis of tolerance is still poorly understood (54).

ACKNOWLEDGMENT

This research was financially supported by the Research Committee of the University of Western Sydney Macarthur (research grant K951075) and the Faculty of Informatics, Science and Technology during our investigation.

REFERENCES

1. McEldowney, S., Hardman, D.J. and Waite, S., In: Pollution, Ecology and Biotreatment, McEldowney, S, Hardman, DJ and Waite, S (eds.), Longman Publishers Pty. Ltd., Singapore, (1993) pp. 48-58.

2. Ellis, B., In: Contaminated Land Treatment Technologies, Society of Chemical Industry, Rees, JF (ed), Elsevier Applied Science, London, (1992) pp. 30-46.

3. Burns, RG, Rogers, S & McGhee, I, In: Contaminants and the Soil Environment in the Australasia-Pacific Region, Naidu, R, Kookuna, RS, Oliver, DP, Rogers, S and McLaughlin, MJ (eds.) Kluwer Academic Publishers, London, (1996) pp. 125-181.

4. Parker, R., Environmental restoration technologies. EMIAA Yearbook, pp. (1994) 169-171.

5. Smith, B., Waste Manag. Env., 4: (1993) 24-30.

6. Shoebridge, L., In: The Health Risk Assessment and Management of Contaminated Sites, Langley, A and van Alpen, M. (eds.). Proc. 2nd National Workshop on the Health Risk Assessment and Management of Contaminated Sites, 5-7 April, (1993), Canberra. South Australian Health Commission, Adelaide, South Australia.

7. Alloway, BJ, In: Heavy metals in Soils, Alloway BJ (ed.) Blackie, Glasgow, (1990) pp. 7-28.

8. Kroopnick, P.M., In: Remediation of Hazardous Waste Contaminated Soils, Wise, DL and Trantolo, DJ (eds.), Marcel Dekker, Inc., New York, (1994) pp. 779-790.

9. Peters, R.W. and Shem, L., App. Microbiol. Biotechnol., 40: (1992) 926-932.

10. Mench, M.J., Didier, V.L., Loeffler, M., Gomez, A. & Masson, P., J. Environ. Qual., 23: (1994) 58-63.

11. Wills, B, Mineral Processing Technology. 4th Ed., Pergaman Press, Oxford. (1988)

12. Elliott, H.A., Linn, J.H. and Shields, G.A., Haz. Waste Haz. Mater., 6: (1989) 223-228.

13. Dennis, R.M., Dworkin, D. and Zupko, A.J., In: Remediation of Hazardous Waste Contaminated Soils, Wise, DL and Trantolo, DJ (eds), Marcel Dekker, Inc., New York, (1994) pp. 745-778.

14. Flathman, P.E., Jerger, D.E. and Exner, J.H. (Editors) Bioremediation Field Experiences. CRC, Press, Boca Raton, Florida, (1994) 548 pp.

15. Markert, B, Plants as biomonitors for heavy metal pollution of the terrestrial environment. VCH, Weinheim, (1994).

16. Brooks RR, Biological methods of prospecting for minerals. John Wiley & Sons, Inc., New York, (1983).

17. Khan, A.G., Chaudhry, T.M., Hayes, W. J., Khoo, C.S., Hill, L., Fernandez, R. and Gallardo, P., Water, Air and Soil Pollution. 104: (1998) 389-402.

18. Chaudhry, T.M., Hayes, W.J., Khan, A.G. and Khoo, C. Australasian J. Ecotox., 4: (1998) 37-51.

19. Baker, AJM, McGrath, SP, Sidoli, CMD & Reeves, RD, Resources, Conservation and Recycling, 11: (1994) 41-49.

20. Kumar, P.B.A.N., Dushenkov, V., Motto, H. and Raskin, I., Environ. Sci. Technol., 29: (1995) 1232-1238.

21. Shetty, K.G., Hetrick, B.A.D., and Schwab, A.P., Environ. Pollut., 88: (1995)

307-314.

22. Blaylock, MJ, Salt, DE, Dushenkov, S, Zakharova, O, Gussman, C, Kapulnik, Y, Ensley, BD & Raskin, I, Environ Sci Technol, 31: (1997) 860-865.
23. Haung, J.W., Chen, J., Berti, W.R. and Cunningham, S.D., Environ. Sci. Technol., 31: (1997) 800-805.
24. Smith, S.E. & Read, D.J., Mycorrhizal Symbiosis, 2nd Ed., Academic Press, London, (1997) pp. 11-32.
25. Khan, A.G., New Phytol. 87: (1981) 363-370.
26. Garrett, S.D. (Editor), Soil Fungi and Soil Fertility, 2nd Ed., Pergamon Press, Oxford, (1981) 150 pp.
27. Paul, E.A. and Clark, F.E., Soil microbiology and Biochemistry. Academic Press Inc., San Diego, Canada, (1989) 273 pp.
28. Donnelly, P.K. and Fletcher, J.S., In:Bioremediation through Rhizosphere Technology, Anderson, TA and Coats, JR (eds) American Chemical Society, Washington DC., (1994) pp. 93-99.
29. Shetty, K.G., Hetrick, B.A.D., Figge, D.A.H. & Schwab, A.P., Environ. Pollut., 86: (1994) 181-188.
30. Bethlenfalvay, GJ, In: Methods in Microbiology, Norris, J.R., Read, D.J. and Verma, A.K. (Eds.) Academic Press, London, (1992) pp. 375-389.
31. Griffioen, W.A.J., Ietswaart, J.H. and Ernst, W.H.O., Plant and Soil, 158: (1994) 83-89.
32. Chapman, B.M., Jones, D.R. and Jung, R.F., Geochemica Cosmochemica Acta, 47: (1983) 1957-1973.
33. Krebs, C.J., In: Ecology: The Experimental Analysis of Distribution and Abundance, Krebs, CJ (ed.), 3rd Ed., Harper & Row, Publishers, New York, (1985) pp. 157-172.
34. Colinvaux, P.A., In: Introduction to Ecology, Colinvaux, PA (ed), John Wiley & Sons, Inc., New York, (1973) pp. 509-520.
35. Brundrett, M, Bougher, N, Dell, B, Grove, T & Malajczuk, N, ACIAR, Canberra, (1996) pp. 155-186.
36. Giovannett, M. and Mosse, B., New Phytol., 84: (1980) 489-500.
37. Brundrett, M, Microbiology Australia, 17: (1996) 12-13.
38. Diz, G. and Honrubia, M., Agric. Ecosys. Environ., 29: (1990) 85-89.
39. Gildon, A. and Tinker, P.B., New Phytol., 95: (1983) 247-261.
40. Ietswaart, J.H., Griffioen, W.A.J. and Ernst, W.H.O., Plant and Soil, 139: (1992) 67-73.
41. Harris, M.M. and Jurgenson, M.F., Plant and Soil, 47: (1977) 509-517.
42. Leyval, C., Singh, B.R. and Joner, E.J., Water, Air & Soil Pollut., 84: (1995) 203-216.
43. Wang, Y.P., Wu, C.G., Huang, Y.L. and Chao, C.C., J. Agri. Associ. China, 0: (1995) 55-69.
44. Sward, R.J., Australian Jr. Bot., 26: (1978) 253-264.
45. Malajczuk, N., Reddell, P. and Brundrett, M., In: Mycorrhizae and Plant

Health, Pfleger, FL & Linderman, RG (eds.), APS, Press, Minnesota, (1981) pp. 83-100.

46. Lapeyrie, F.F. and Chilvers, G.A., New Phytol., 100: (1985) 93-104.
47. Khan, A.G. Mycorrhiza, 3: (1993) 31-38.
48. Newman, E.I. & Reddell, P., New Phytol., 106: (1987) 745-751.
49. Warcup, J.H., Mycological Res., 95: (1991) 656-659.
50. Harley, J.L, Mycol. Res., 92: (1989) 129-139.
51. Khan, A.G. and Bari, A., Abstracts, Development and Environmental Impact Conference, 21-23 September, (1997), Riyadh, pp 119.
52. Leake, J.R., In: Mycorrhizae in the next Decade-Practical Applications and Research Properties, Sylvia, DM, Hung, LL and Graham, JH (eds.), Proc. 7th North American Conference on Mycorrhizae, 3-8 May, 1987, Gainesville, Florida, 332 pp.
53. Wang, Y.P. & Chao, C.C., J. Agri. Associ. China New Series, 157 (1992): 6-20.
54. Turnau, K., Wiadomosci Botaniczyne, 37: (1993) 43-58.

CHAPTER - 42

POTENTIALS OF *Pseudomonas pickettii* STRAIN US321 IN BIOREMEDIATION

Upma Gulati and Sheela Srivastava

Department of Genetics, University of Delhi, South Campus
Benito Juarez Road, New Delhi-110 021, INDIA

ABSTRACT

Pseudomonas pickettii strain US321 a Gram negative, soil isolate shows resistance against copper, zinc, cadmium and to antibiotic like ampicillin, chloramphenicol and erythromycin. The strain not only could effectively removes copper but also copper and zinc from the copper-zinc mixed salt solutions as well as the effluent samples collected from Khetri Copper Complex (KCC), Rajasthan, India. Besides the removal of metals, other parameters like pH and cell viability were also taken into consideration. Removal of metal from soluble metal fraction of chalcopyrite ore and ore tailings demonstrated its potentials in bioremediation. The simultaneous and preferential recovery of both the metals through desorbing agent treatment has been successful. The traits mentioned above justify the application of this bacterium in environmental decontamination and metal biorecovery processes.

1. INTRODUCTION

Industrial activities and deliberate and accidental discharges are the major causes due to which microroganisms are increasingly exposed to toxic levels of metal pollutants and may have to acquire resistance to these metals as a survival strategy. The fundamental importance of a mechanism may be determined by the ability to detoxify or compartmentalize the metal within the cell either by adsorption or extracellular precipitation (1, 2). Certain bacterial species exhibit the property of metal sequestration and thus offer a potential alternative to existing methods of environmental remediation and recovery of valuable metals from industrial wastes (1, 2, 3, 4, 5).

In the present investigation, *Pseudomonas pickettii* strain US321 which exhibit resistance towards copper (Cu), zinc (Zn) and cadmium (Cd) was used to remove metals from mine effluent as well as ore and ore tailings suspensions so as to establish the application of this bacterium for metal recovery and environmental bioremediation.

2. MATERIALS AND METHODS

(a) Organism, Media and Growth Conditions

Pseudomonas pickettii strain US321 isolated from garden soil has been described earlier (6). The strain was maintained on $CuCl_2.2H_2O$ (2 mM) containing unbuffered gluconate mineral salt medium (GMM), modified from Mergeay et al. (7). The trace element solution, SL7, in this medium contained 0.02 μg/ml of copper sulphate besides other metal salts. The pH was adjusted to 7.0. Sodium gluconate (0.4% w/v) was added as a carbon source. Soidified media contained 15 g/L of agar. Cultures were incubated at 30°C. Liquid cultures were raised in GMM on a Controlled Environment Incubator Shaker (New Brunswick Sci. Co. Inc. N.J., USA) at 200 rpm and 30°C followed by subculturing with 10% of stationary phase culture for the final experiment.

(b) Simultaneous Resistance

Maximum tolerable concentration (MTC) of different metals was determined by streaking the cells on GMM - agar, containing various concentrations of metal salt solution. The concentration beyond which no growth was observed was selected as MTC for that metal. To check simultaneous/multiple resistance both Cu and Zn were added and the combined effect of metals was observed.

(c) Metal Removal from Effluent/ Ore/ Ore tailings

The effluent, ore and ore-tailings were collected from Khetri Copper Mines (KCC), Rajasthan, India. The exponential phase cells grown in GMM were harvested, washed and transferred to different effluent concentrations (0-100%) supplemented with sodium gluconate (0.4% w/v) under sterilized conditions. One set under unautoclaved condition, i.e., with its natural flora, was also employed. Similarly, the ore/ore-tailings were suspended in GMM aseptically at 10% and 20% pulp density, respectively and shaken at 200 rpm, 30°C for ten days. On 11th day, the supernatant, separated by centrifugation (6000 x g/10 min), and supplemented with sodium gluconate (0.4% w/v) was inoculated with the exponentially grown cultures.

The samples were withdrawn at two days interval and three parameters were analysed : (1) pH, (2) viable counts, and (3) Cu content. In case of effluent, cells were grown for 24h and 48h and suitable dilutions in all the cases were plated on GMM agar to estimate the CFUs/ml. The culture supernatant was used for estimation of metal content and pH.

To monitor the copper retained in the solution, 4 mL of culture filtrate was collected by centrifugation, filter sterilized and acidified by adding 1 mL of HNO_3-$HClO_4$ (6:1). Metal content was represented as ug of metal present in 1 mL of filtrate (μg/mL). As a control 2 mM copper chloride and 2 mM zinc sulphate solutions were analysed and calculations were done accordingly.

3. RESULTS AND DISCUSSION

Interaction of heavy metals with microbes can be helpful in metal removal

Table 1. Comparison of viability (CFUs/mL) in gluconate minimal medium (GMM) and effluent samples

Culture media	CFUs/mL			
	0 day	2nd day	4th day	6th day
GMM	7 x 10^4	1 x 10^4	5 x 10^{10}	1 x 109
Mine water (Autoclaved inoculated)	4 x 10^4	2 x 10^5	3 x 10^{10}	5 x 10^{10}

and of commercial relevance because the removal of potentially toxic metals and radionuclides from the effluents and waste waters can lead not only to the environmental decontamination, but also to the recovery of valuable metals (1,5,8,9). In order to demonstrate the applicability of *Psendomonas* sp. Strain US321 under natural conditions, effluent samples collected from the Khetri Copper Complex, Rajasthan, India were used. The mine water sample showed a copper content of 0.614 ppm, Zinc 0.012 ppm, and a pH of 6.35 when estimated. The effluent sample when diluted (1:9) with the medium supplemented with sodium gluconate as a carbon source, allowed the growth of bacterium similar to the control (Table 1) grown on GMM. Two different sets were prepared for estimation of both Cu and Zn removal : (1) unautoclaved, inoculated; (2) autoclaved, inoculated. Although the metal

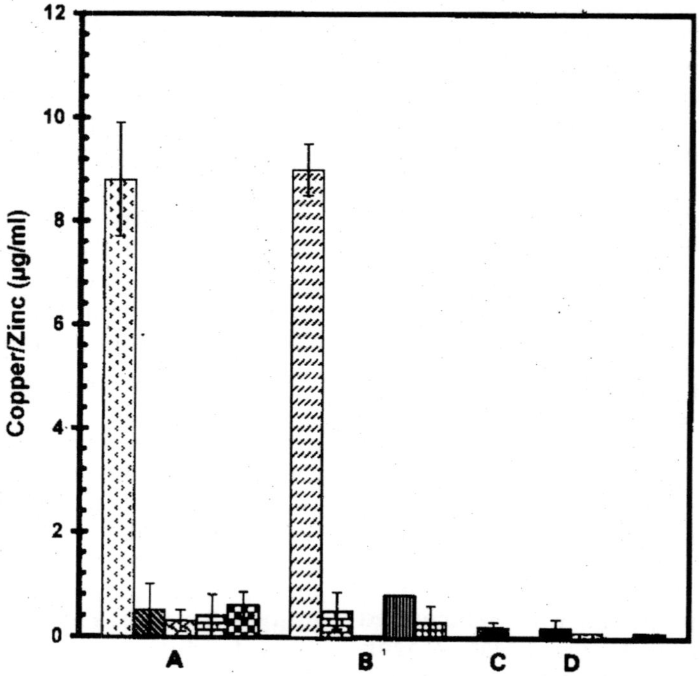

Figure 1. Copper and Zinc removal from effluent samples with the help of *Pseudomonas pickettii* strain US321.

A,C = Unautoclaved, inoculated (Cu, Zn), B,D = Autoclaved, inoculated (Cu, Zn),

0 day , 2nd day 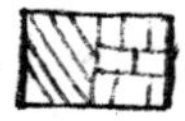, 4th day , 6th day , 8th day 

content in the effluent was very low, metals have the tendency to build-up their concentration due to their non-biodegradable nature. Thus, removal of even low concentration of the metal ions will restore the health of the ecosystem. Strain US321 could efficiently remove both Cu and Zn as shown in Fig. 1, and thus, could lead to environmental cleanup by checking the continuous deposition of these metals. It was also observed that when inoculated with the natural flora, this bacterium could remove comparable level of metals. Thus, strain US321 not only can successfully compete with the other microorganisms but can also be exploited under natural conditions. From the mine effluent, cells were able to remove ~80% of available Cu after 48h.

ACKNOWLEDGEMENT

The grant to U.G. from CSIR is gratefully acknowledged.

REFERENCES

1. Gadd, GM & While C, TIBTECH, 11 (1993) 353.
2. Ji G & Silver S, J. Indus. Microbiol, 14 (1995) 61.
3. Hutchins SR, Davidson MS, Brierley JA & Birerley CL, Annu. Rev. Microbiol, 40 (1986) 311.
4. Gadd GM, Encyclopedia of Microbiology, 2 (1992) 351.
5. Bhagat Renu & Srivastava Sheela, In : Environmental Biotechnology : Principles and Applications. (Moo-Young M, Anderson WA & Chakrabarty AM eds), Kluwer Academic Publications, The Netherlands, (1995) 158.
6. Gilotra U & Srivastava S, Curr. Microbiol, 34 (1997) 378.
7. Mergeay M, Nies D, Schlegel HG, Gents J, Charles P & Van Gusegem F, J Bacteriol, 162 (1985) 328.
8. Sharma P & Verma A, Indian J Microbiol, 31(1) (1991), 1.
9. Lakshmanan VI, Christison J, Knapp RA, Scharer JM & Sanmuga-sunderam V, In: Proc Second Annual General Meeting of Biominet (McGready RGL ed.), CANMET special publications SP85-6, Canadian Government Publishing Centre, Ottawa, (1986).

CHAPTER - 43

BIOREMEDIATION OF CADMIUM BY METAL RESISTANT STRAIN OF CYANOBACTERIA

S.K. Verma Shuchi Jain, V. Raveender and P.I. Rosanto*

Biological Sciences Group, Birla Institute of Technology and Science, Pilani (Raj) - 333 031 India

ABSTRACT

The Biosorption of cadmium by parent, Met-R1 and Met-R2 strains of the cyanobacterium *Nostoc calcicola* was studied. The two mutant strains used in the study showed equal growth in metal-free control and in the medium containing 2.0 mM Cd, whereas, the same concentration was highly toxic to parent culture. A comparison of cadmium removal by dead and viable cells under free and immobilized conditions showed superiority of immobilized viable cells over dead cells. The Met-R2 strain showed a significantly higher biosorption of cadmium than Met-R2 strain. The suggested reasons for this difference are, low secretion of exopeptide, higher adsorption at the cell surface and higher intracellular accumulation of cadmium. Efficient Cd removal from aqueous solution over a range of metal concentration and the availability of biosorbent for repeated use are the suggested advantages of this technique.

1. INTRODUCTION

Presence of toxic metal ions in aqueous solution due to increased industrio-urban activity is one of the challenging problems of environment. The present common practice for the removal of soluble metal ions includes simple chemical precipitation, ion exchange resins, solvent extraction and electro winning. Low bulk metal concentration, ion specificity, narrow range and poor stalling colloidal properties may limit the effectiveness of these conventional processes. In the last decade, considerable research have been carried out on the material of biological origin for metal sorption. (1-4). The phenomenon of retention of cation from solution by

* Author for correspondence

biological material has been termed as 'biosorption'. The superiority of biosorbent over conventional material is well established, owing to their wide working pH range and non-specific metal binding. The mechanism of biosorption is little understood and the information available is largely qualitative, fragmented and rather limited. Removal of metal ions in general, by microbes include cellular adsorption and carrier mediated cellular uptake (5,6). Both these processes contribute significantly in removal of metals from aqueous solution. Wild type strains being sensitive towards elevated metal concentrations, may not be utilizing the potentials of later process as their metal-sensitive proteins are inactivated under such conditions. These limitations of metal toxicity of biosorbent provide the need to improve genetic potentials of biosorbent, which can function efficiently under field conditions. In this regard we have developed single and multiple metal resistant cyanabacterial strains (7,8).

In the present paper we report our findings with regard to the characterization and efficiency of metal resistant strains of the cyanobacterium *Nostoc calcicola*, towards the development of a suitable cadmium biosorbent.

2. MATERIALS AND METHODS

The cultures used in this study were filamentous cyanobacterium *Nostoc calcicola* and its two mutant strains Met-R1 and Met-R2 developed and characterized in our laboratory (7,8). Cultures were gown in Allen Arnon's medium at 25 1°C under continuous illumination of 75-80 E/m^2.

(a) Preparation of immobilized beads

The exponential phase cyanabacterial cells (400g proteins ml^{-1}) obtained by centrifugation and repeated washings, were suspended in 5% (w/v) solution of alginic acid (Sodium salt, Fluka) prepared in the growth medium as above. The mixture was pumped out drop wise in 0.2M $CaCl_2$ solution aseptically in a Laminar flow cabinet. The beads thus formed of cyanabacterial cells to a time of 55 ug protein bead^{-1} were subsequently harvested, washed several times with sterile triple glass distilled water and resuspended in a 20 ml. volume of growth medium contained a 250 ml cotton stoppered culture flasks. The beads thus formed were used in subsequent experiment.

(b) Cadmium removal experiment

Batch experiments were conducted with 100 ml. culture of free cells or about 50 immobilized beads, gently agitated under constant illumination at 75 E/m2 and at a temperature of 25 + 1°C. About 60 hr after inoculation, the cultures were dosed with cadmium to achieve a desired concentration (1.0-5.0 mM). Samples were taken at pre-determined time intervals for the estimation of cell dry weight, the metal concentrations at the cell surface or cell interior and the residual metal concentrations in the solution.

The surface adsorption of cadmium was assayed using EDTA (10 mM.) washable fraction as described earlier (1,10). All the samples collected were immediately filtered through pre-washed and pre-weighed membrane filters (Millipore 0.45 m) and the cell dry weights were determined after drying the filter at 103°C for 24 hrs. The dried cells were digested in concentrated nitric acid after having dissolved the filter in acetone. The analysis of the cellular metal concentrations in solution were carried out using an Atomic Absorption Spectrophotometer (Varian model 1275).

3. RESULTS AND DISCUSSION

The removal of metal ions from aqueous

Fig. 1. Growth pattern of parent and two mutants; Met - R1 and Met - R2 of *Nostoc calcicola* exposed to different concntration of cadmium for 200 hrs.

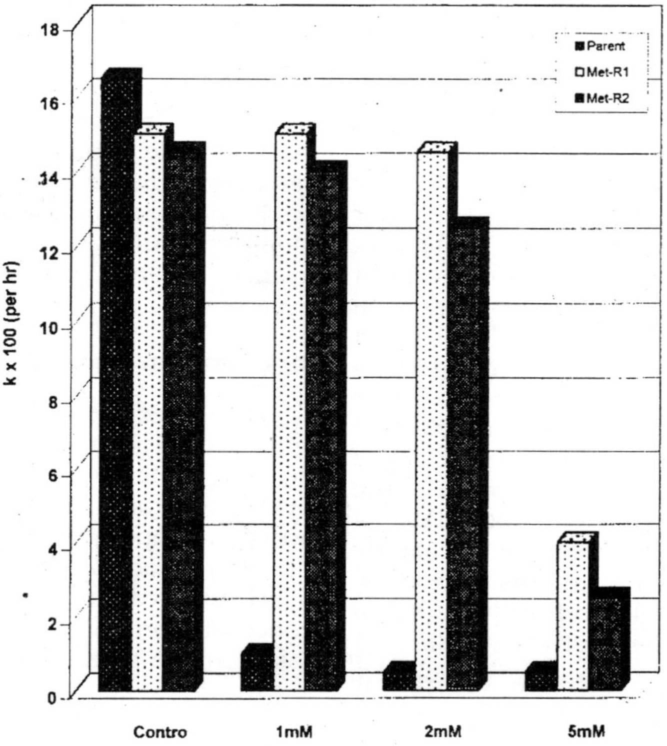

solution is shown to be a biphasic phenomenon involving an energy-independent bivalent metal ion exchange with counter ions present in cell wall (11) and energy-dependent active uptake (1). The relative contribution of these two processes in metal removal can be demonstrated by comparing the sorption of metal ion by living and dead biomass as only viable cells can demonstrate energy-dependent phenomenon. To study the behaviour of dead and viable biomass, we selected parent and two mutant strains of *Nostoc calcicola.* A 3.0 M of Cd. was found to be completely lethal for parent whereas both the mutants could grow normally upto 2 mM of cadmium (Fig. 1). A comparison of dead and viable biomass exposed to 2.0 mM Cd-solution for 200 hrs shows a clear superiority of viable biomass over the dead ones (Fig. 2). Also a comparison between two mutant strains with equal metal tolerance (2.0 mM) showed different behaviour, Met-R2 showing three hundred percent more cadmium removal than the Met-R1 strain. These data clearly indicate the role of some metal resistant proteins responsible for the viability of cells in the metal removal.

The immobilization of cells is known to inhance various metabolic processes like photosynthesis O_2 evolution (12) H_2 production in marine photosynthetic bacteria (13, 14) and Cu uptake and accumulation (10). A comparison of Cd removal by free cells and immobilized beads prepared from three strains and exposed over a period of 200 hrs. in 2.0 mM $CdCl_2$

Fig. 2. Cd removal by dead and viable biomass of parent; Met –R1, Met –R2 strains of *N. calcicola* exposed to 2mM Cd solution for 200 hrs.

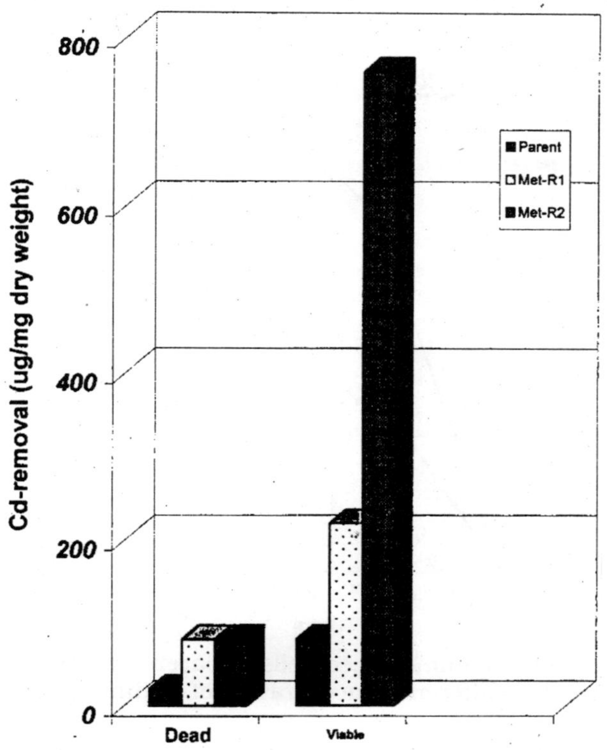

solution showed almost 1.5 fold increase in Cd-removal in all the cases (Fig. 3) thus, clearly establishing the role of immobilization in Cd-removal by the both metal-sensitive and metal-resistant strains. The immobilized beads were used in further studies of cadmium removal.

The various mechanisms suggested for metal tolerance in microbes include as cellular exclusion of metals by the secretion of exopeptides (15) or metal transport across the membrane (8,16), extrusion of metal (7) and immobilization at cell interior (17, 18). The Met-R1 strain falling in first category, may not allow sufficient Cd-accumulation due to retarded Cd transport. The microbes are known to release small peptides and proteins extracellularly, when subjected to metal stress. These peptides can bind the metal in the solution thus reducing their bioavailability and hence toxicity (15). A comparison of exopeptide secretion by 3 strains with increasing Cd concentration showed an enhancement in case of Met-R1 strain till 2.0 mM as compared to other two strains which did not show a significant increase under similar conditions (Fig. 4).

The metal-ion specific cellular adsorption of Cd was found to be maximum in case of Met- R2 (250 g Cd/mg dry wt.) contributing about 30% of the total Cd-removed from the solution after 200 hr. exposure (Fig. 5). The same process contributed only 16% in case of Met-R1 strain metal removal process. The major process in this strain being exopeptides released from the cells but trapped in alginate beads (80%).

Fig. 3. Cadmium removal by free and immobilized cells of Met-R2 strain of *Nostoc calcicola* exposed to 2 mM of cadmium for 200 hr.

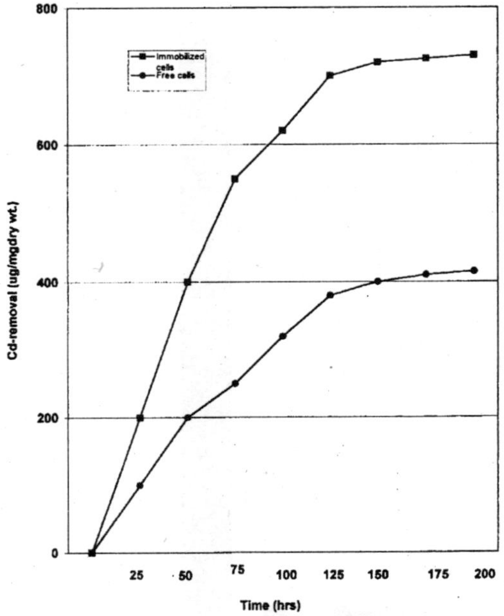

Fig. 4. Exopeptide secretion by immobilized cells of parent and two mutants of *Nostoc calcicola* exposed to different concentration of Cadmium for 200 hrs.

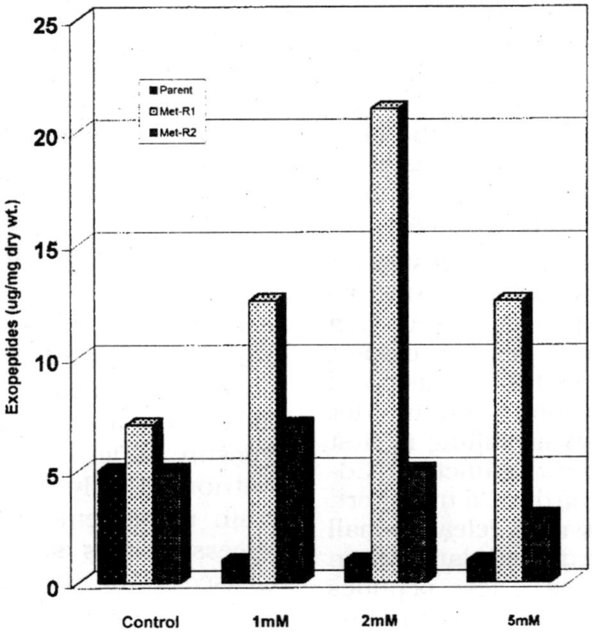

Fig 5. A comparison of extracellular chelation, cellular adsorption, cellular accumulation and total removal of cadmium by immobilized beads prepared from parent, Met-R1 strain of *Nostoc calcicola* and exposed to 2 mM cadmium for 200 hr.

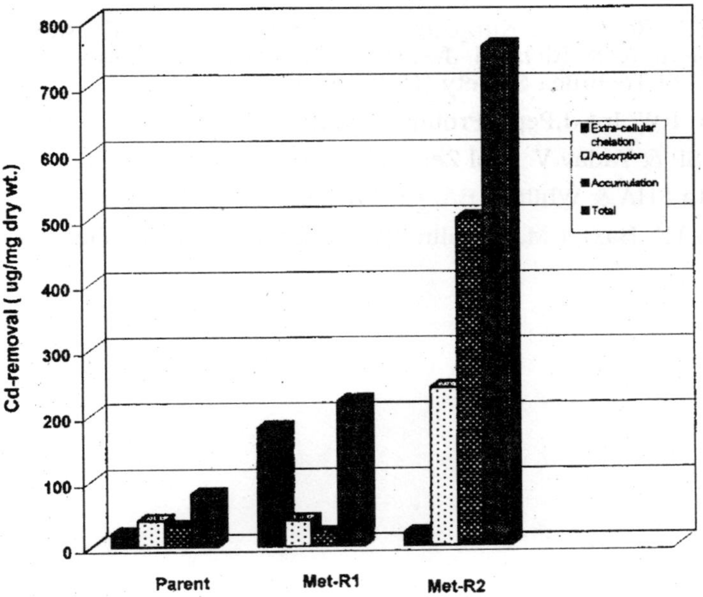

The apparent superiority of immobilized cells of Met-R2 strain in Cd-removal over the parent or Met-R1 cells suggests that such system could be successfully applied to scavenge Cd in repeated cycles without loss of cells.

ACKNOWLEDGEMENTS

The authors are thankful to the Director, BITS for providing laboratory facilities, and to the UGC for financial assistance.

REFERENCES

1. Ting YP, Lowson F & Prince IG, Biotech Bioeng 37 (1991) 445.
2. Kuyucak N & Volesky B, Biotech. Lett 10 (1988) 137.
3. Volesky B & Prasetyo I, Biotech. Bioeng 43 (1994) 1010
4. Uberol M & Shadman F, Environ Sci Technol 25 (1991) 1285
5. Verma SK & Singh SP, Cur Microbiol 21 (1990) 33
6. Verma SK & Singh SP, Biol Met 4 (1991) 192
7. Verma SK & Singh HN, FEMS Microbiol Lett 84 (1991) 291
8. Verma SK & Singh SP, Bull Environ Contain Toxicol 54 (1995) 614
9. Allen MB & Arnon DI, Plant Physiol 30 (1955) 366
10. Singh SP, Verma SK, Singh RK & Pandey PK, FEMS Microbiol Lett. 60 (1989) 193
11. Hoyle BD & Beveridge TJ, Can J Microbiol 30 (1984) 204

12. Bailliez C, Largeau C, Berkaloff C & Casadevel E, Aplpl Microbiol Technol 23 (1986) 361

13. Matsunga T, Mitsui A, Proc 4th Symp. Biotechnology in Energy Production and Conservation, Catlinburg Tennessee (1982) pp. 32

14. Mitsui A, Rosner D, Kunazawa S, Barciela S, Pilips E, Ramchandran S, Takahashi A & Richard J. Proc. 22nd Space Congress (ed) Conavera Council of Technica Society (1985) pp.11.7

15. Olafson RW, Int J.Pept. Protein Res. 24 (1984) 303

16. Singh SP & Yadav V Biol Zentralbl 105 (1986) 539

17. Shehata FHA & Whilton BA, Br. Phycol J 17 (1982) 5

18. Jensen TE, Baxter M. Rachlin JW & Jani V. Environ Pollut. 27 (1982) 119

CHAPTER - 44

MICROBIAL DEGRADATION OF CRUDE OIL AND SOME PETROCHEMICAL POLLUTANTS

Anjana Desai , Hanson K.G., Arpana Acharya and Dharmendra Patel

Department of Microbiology & Biotechnology Centre,
Faculty of Science, M.S. University of Baroda, Baroda 390 002.

1. INTRODUCTION

Gale (1) best summarized the concept of microbial infallibility by suggesting that there existed in nature a microorganism capable of metabolizing any conceivable compound that the organic chemist might choose to synthesize. The opposing doctrine of molecular recalcitrance was advanced by Alexander (2) who noted that many synthetic chemicals persisted in soil. It is estimated that nearly 60,000 tons of hydrocarbon oil enters the environment each year (3). However, we do not have information on the amount of petrochemicals entering the environment.

Environmental damage due to oil spill in the past and recent times has focussed on the need for environment friendly strategies for remediation of the contaminated sites. Based on the fact that bioremediation was the major process for removal of non-volatile oil components from the environment, it has been projected to play an important role in the environment clean up (4). The ability of microorganisms to degrade hydrocarbons has been known since 1895 with the report of paraffin utilization. Thus it is a group of microorganisms rather than a single organism, that would be required, to degrade a mixture of hydrocarbons like crude oil, kerosene, or a waste water from related industries.

Phthalic acids and their esters are one of the major petrochemicals produced and consumed on a large scale for the manufacturing of synthetic fibres, plastics, paints etc. The US Environmental Protection Agency has listed these among the priority pollutants and their presence is a matter of serious environmental concern (5,6). There are reports on the efficient degradation of phthalate esters by microbes (7-9) however, the degradation of dimethyl-terephthalate (DMT) is not widely documented. Nitriles are another group of notoriously poisonous compounds having widespread industrial use as solvents, in manufacture of plastics, synthetic rubber, pharmaceuticals, herbicides and other chemicals (10). Nitrile-wastes dispersed throughout the biosphere are an important form of environmental pollution. Enzymes capable of metabolizing nitriles have evolved in many microorganisms; apparently due to wide spread occurrence of natural nitriles (11).

The present paper describes biodegradation studies with respect to crude oil and two major petrochemical based pollutants, viz., phthalate and phthalate esters; and nitriles. Biological treatment coupled with physicochemical treatment is the most widely used and cost effective pollution control technology. Number of advancements in the "engineering hardware" have been reported, However, only recently has attention been paid to the "Biological software", as a result, the specific tailor made microbial culture products are now available in developed countries in recent times for wastewater treatment and bioremediation purpose.

2. METHODS AND MATERIALS

Degradation of Bombay High Crude Oil (BHCO) during growth of *Acinetobacter* sp. A3 in shake flask experiments was carried out by inoculating 1ml stationary phase culture (10^7cells) into 50ml Bushnell-Hass Medium (BHM) (12) with 0.5% BHCO, on a rotary shaker (180rpm; 28 ± 2°C) for the required time. Uninoculated flasks containing medium and BHCO served as control for estimating environmental loss. Residual crude oil was extracted with chloroform and estimated by gas chromatography as described earlier (13). Assessment of BHCO biodegradation in soil was done as per the method described by Hanson *et al.* (14). Protoplast preparation, fusion and hydrocarbon degradation studies by parent cultures and fusant were carried out as described earlier (15). Growth, degradation of Dimethylterephthalate (DMT) and estimation of esterase activity were done as per the method described (16). Acetonitrile degradation and estimations of nitrile hydratase and amidase activities were done as per the methods described (17).

3. RESULTS AND DISCUSSION

Many hydrocarbon degrading microorganisms were obtained from various hydrocarbon rich environmental niches by enrichment techniques. Amongst these, one of the isolate *Acinetobacter* sp. A3, showed relatively higher and faster crude oil degradation and hence was selected for further studies.

Crude oil degradation by *Acinetobacter* sp. A3 was monitored by allowing cells to grow in BHM containing 0.5% BHCO as sole carbon source. Quantitative estimates indicated that of about 70% BHCO degraded in 120h, 60% was biodegraded and approx. 10% was lost due to weathering (Fig : 1). Though majority of BHCO was degraded by 24h, there was a steady increase in the growth of *Acinetobacter* sp. A3 till 72h probably supported by the breakdown intermediates, which were not detected. However by end of 72h, growth and degradation remained almost constant showing that all the crude oil components which could be attacked by *Acinetobacter* sp. A3 have been completely metabolized. Fractionation of the residual BHCO into discrete chemical classes following

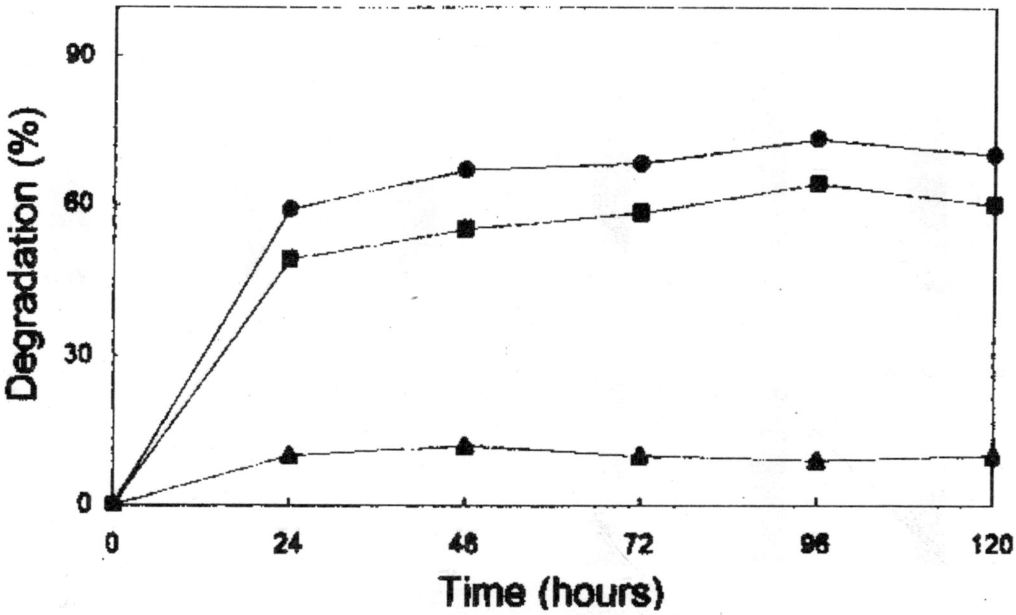

Fig. 1. Growth of *Acinetobacter* sp. A3 on BHCO. Biodegradation (■), Weathering (▲) and total degradation (•)

degradation by *Acinetobacter* sp. A3 revealed that the saturate fraction was most rapidly and extensively degraded with biodegradation being primarily responsible for it. Biodegradation of resins, however, was slow initially, but once the majority of saturates had been degraded (between 24 and 72 h), the resin fraction was attacked and over 60% was degraded by 120h. Such preferential microbial utilization of saturates over other fractions in North Central and Kuwait crude oils have been documented (18,19). On the other hand weathering accounted for the loss of the aromatic fractions, a majority of which was lost by 24h, which probably negated its toxic effects and hence did not interfere with growth and subsequent BHCO degradation by *Acinetobacter* sp. A3. The asphaltenes being high molecular weight, highly recalcitrant compounds (20) remained unaffected by weathering or biodegradation within the time period tested (Fig. 2).

Based on the fact that biodegradation is the major process for removal of non-volatile petroleum hydrocarbons from the environment, bioremediation has been projected to play an important role in environmental cleanup of petroleum pollutants. In soil, adsorption of oil onto soil particles, percolation into deeper layers where O_2 becomes limited, and toxicity of the oil to the indigenous microorganisms, retard biodegradation of oil. Bioaugmentation with *Acinetobacter* sp. A3 to BHCO contaminated soil revealed that the cells, not only remained viable but also proliferated such that the cell count increased by about 4.5 times over 30 days incubation period (Table 1).

On the other hand, crude oil exerted an inhibitory effect on the majority of the indigenous microorganisms (Table 1). Though some of the indigenous microorganisms remained viable as seen by the stabilization of the colony count after 10 days incubation, they did not contribute to the degradation process as evidenced by no significant change in the amount of BHCO degraded. In addition, BHCO contaminated

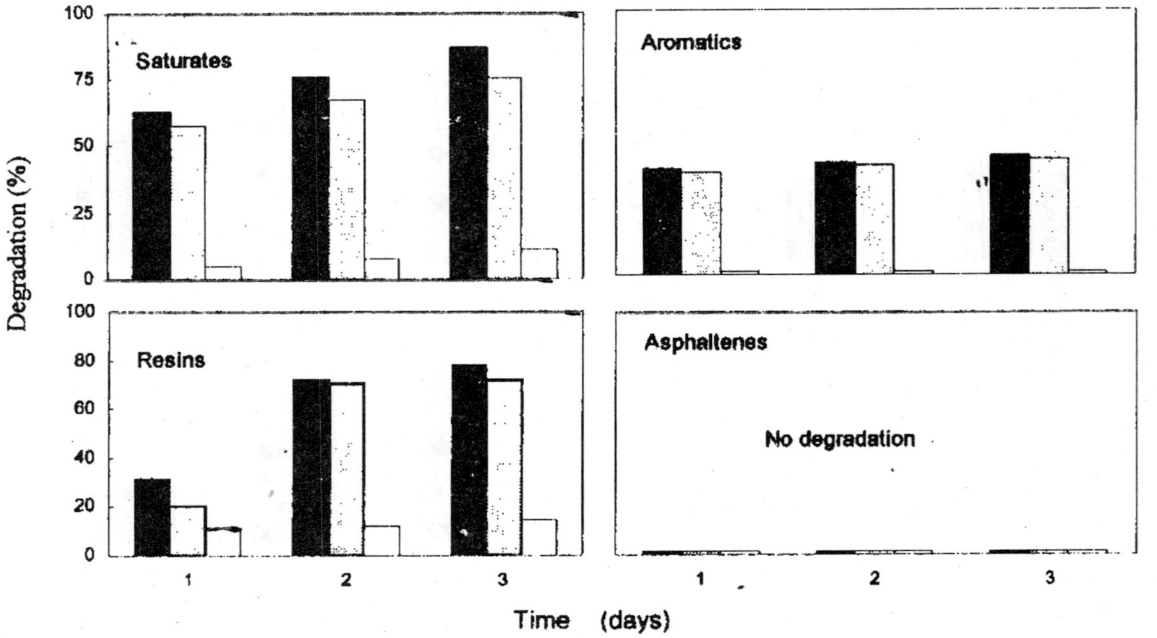

Fig. 2. Degradation of BHCO components during growth of *Acinetobacter* sp. A3. Total degrdaation (▇), biodegradation, (▨), weathering (□)

soil bioaugmented with *Acinetobacter* sp. A3 for 30 days prior to planting Mung beans, supported about 80% seed germination as opposed to 35% germination in the untreated soil. Treatment for shorter time intervals also supported better germination than the untreated soil incubated for the same period (Table 2). The better rate of seed germination in the treated soil could be specifically attributed to the decrease in phytotoxicity of crude oil owing to its enhanced degradation.

Attempts to increase the substrate range of *Acinetobacter* sp. A3 to include aromatics as well, by the transfer of TOL plasmid (pWWO) was not successful since the transconjugants obtained by mating *E. coli* containing either pRK2013 or pSUP5011 as the helper vector and *P. putida* KT2442 (pWWO) were unstable. Genetic competence for transformation

Table 1. Effect of *Acinetobacter* sp. A3 bioaugmentation on degradation of BHCO in soil

Incubation Period (days)	Fold increase / decrease of CFU*		BHCO degradation (%)	
	Soil + BHCO	Soil + BHCO + *Acinetobacter* sp. A3	Soil + BHCO	Soil + BHCO + *Acinetobacter* sp. A3
5	-3.51	1.6	3.49	10.53
10	-2.74	2.48	11.61	25.34
20	-1.62	4.4	19.16	37.11
30	-2.38	4.48	20.06	49.88

* Negative sign indicates fold decrease of CFU (Colony forming unit).

Table 2. Effect of *Acinetobacter* sp. A3 augmentation on germination of Mung (*Phaeseolus aureus*) beans in BHCO(5%) - contaminated (5% soil)

Incubation period (days)	Number of seeds germinated (mean ± SD)		
	Soil	Soil + BHCO	Soil + BHCO + *Acinetobacter* sp. A3
5	9.8 ± 0.44	0.0	2.6 ± 1.44
10	9.8 ± 0.44	0.6 ± 0.54	5.3 ± 0.83
20	10.0 ± 0.89	2.4 ± 0.89	6.4 ± 0.54
30	9.8 ± 0.44	3.6 ± 0.54	8.2 ± 0.83

seems to be a rare phenomenon in the genus *Acinetobacter* (21) and hence transformation of *Acinetobacter* strains with DNA from bacteria not of this genus does not occur at all, while very few Acinetobacter strains such as BD413 (22) and NCIB 8250 (23) can be made competent for transformation with DNA from members of the same genus. Other than mutagenesis and cloning techniques, protoplast fusion has been recently recognized as very useful tool for developing microorganisms with desired properties (24).

As mentioned earlier *Acinetobacter* sp. A3 was capable of degrading only alkanes and resins in BHCO, which limited total BHCO degradation. *Pseudomonas putida* DP99, on the other hand was capable of degrading polycyclic aromatic hydrocarbons such as naphthalene. Protoplast fusion between *Acinetobacter* sp. A3 and *P.putida* DP99 was thus envisaged to produce fusants (hybrids) with ability to degrade both the classes of hydrocarbons viz.saturates and aromatics. For the purpose of this study, tetradecane and naphthalene were selected as model alkane and aromatic hydrocarbons. PEG-6000 mediated protoplast fusion between *Acinetobacter* sp. A3 and *P.putida* DP99 resulted in a number of fusants capable of degrading both naphthalene and tetradecane. Fusant F4/ 13 selected at random was capable of degrading both the hydrocarbons when supplied individually as opposed to its parent strains, which could degrade just one of the hydrocarbons (Fig. 3). Whereas the extent of naphthalene degradation by fusant F4/13 was similar to that of *P. putida* DP99 (99%) by the end of five days, it brought about greater

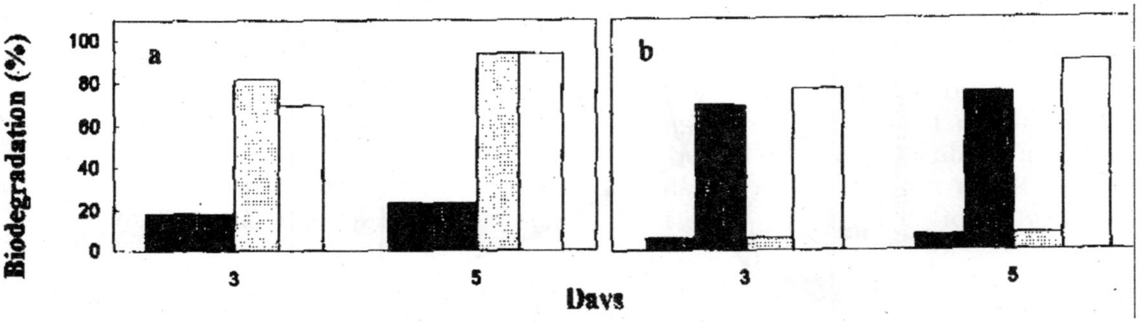

Fig. 3. Biodegradation of (a) naphthalene and (b) tetradecane by *Acinetobacter* sp., A3 ( ), *P. putida* DP99 ( ), fusant F4/13 ( ) and weathering ( )

degradation of tetradecane (82%) than *Acinetobacter* sp. A3 (77%) in the same period (Fig. 3). Thus the fusion resulted in the formation of hybrids possessing hydrocarbon degradative ability equal to or even better than its parent strains and the capability to degrade both the classes of hydrocarbons. On examining the BHCO degradation potential, fusant F4/13 exhibited enhanced BHCO degradation compared with *Acinetobacter* sp. A3 by the end of five days. *P. putida* DP99 did not bring about any significant degradation (Table 3). At first glance, enhancement of BHCO degradation by 2% by the fusant F4/13 does not appear significant. But this enhancement was due to the degradation of the aromatic fraction (Table 4). Since the aromatic content of BHCO was low, the enhancement in overall degradation of BHCO was thus small. The fusant, thus, will find application in treatment of crude oil with higher aromatic content, due to its ability to degrade aromatics in addition to the saturate and resin fractions. The fusant also retained its hydrocarbonoclastic properties on repeated transfers in hydrocarbon containing media over a period of ten months thus confirming its genetic stability.

Table 3. BHCO degradation by *Acinetobacter* sp. A3, *P. putida* DP99 and Fusant F4/13

Organism	Time (days)	BHCO degradation (%)
Acinetobacter sp. A3	3	32.7
Fusant F4/13	3	27.6
P. putida DP99	3	9.8
Weathering*	3	8.6
Acinetobacter sp. A3	5	37.5
Fusant F4/13	5	39.5
P. putida DP99	5	10.4
Weathering*	5	8.8

*BHCO loss from uninoculated flasks during incubation.

Studies were also conducted towards isolating and studying degradative kinetics of two other petrochemical pollutants viz, dimethylterephthalate and nitriles. *Comamonas acidovorans* D4, isolated from petrochemical wastewater utilized 0.5% dimethylterephthalate, phthalates and phthalate esters as sole carbon source without growth supplements. Maximum DMT degradation (66%) was observed in the first 24h of growth reaching to completion by 72h (Fig. 4). The biodegradation of

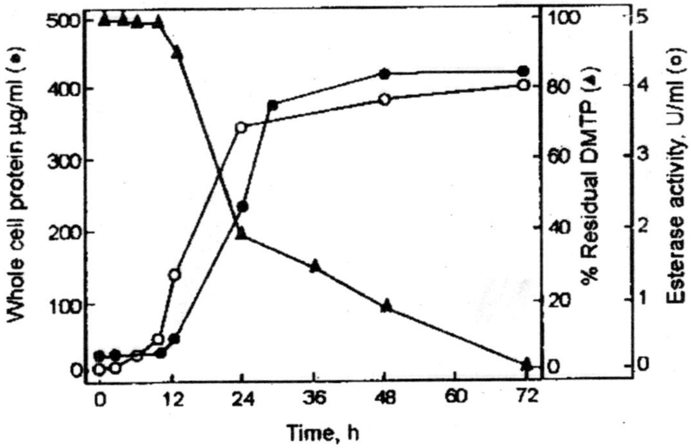

Fig. 4. Time dependent growth, DMT utilization and esterase activity in *Comamonas acidovorans* D-4.

Table 4. Degradation of BHCO fractions by *Acinetobacter* sp. A3 and Fusant F4/13

Organism	Time (days)	BHCO degradation (%)			
		Saturates	Aromatics	Resins	Asphaltenes
Acinetobacter sp. A3	3	34	26.7	36.6	0
Fusant F4/13	3	25.2	36.4	2	0
Weathering*	3	2.8	25.1	1.5	0
Acinetobacter sp. A3	5	40.6	27.2	36.8	0
Fusant F4/13	5	43	51.8	29.5	0
Weathering*	5	4.6	26.2	9.3	0

*Hydrocarbon loss from uninoculated flasks during incubation.

Table 5. Nitrile hydratase and amidase activities of acetonitrile grown cells of *R. erythropolis* A10 after 72 h of growth.

Substrate	Enzyme	Specific activity (µmolNH3/min/mg protein)
Isobutyronitrile	Nitrile hydratase	117
Acrylonitrile	Nitrile hydratase	69.57
Succinonitrile	Nitrile hydratase	115.7
Propionitrile	Nitrile hydratase	177.03
Acetonitrile	Nitrile hydratase	167.57
Benzonitrile	Nitrile hydratase	89.83
Acetamide	Amidase	2857
Acrylamide	Amidase	657

phthalate esters has been reported to be initiated by the enzymatic cleavage of ester bonds by esterase. A parallel increase in cell associated esterase activity with the growth and DMT utilization was observed. (Fig. 4). Partial degradation of DMT to terephthalic acid after six days of incubation by *Sclerotium rolfsii* (25) and *Aspergillus niger* (26) have been documented however, in a *Bacillus* sp. (27) complete degradation occurred, but at a very low concentration (0.01%) of DMT.

R. erythropolis A10 when grown under shake flask condition in SMM (Sterile Minimal Medium) containing 1% (v/v) (200mM) acetonitrile as the sole source of carbon and nitrogen, degraded about 70% acetonitrile within 72h (Fig. 5). A parallel increase in ammonia concentration with the growth is indicative of the utilization of acetonitrile. It has been suggested that the aromatic and unsaturated nitriles are metabolized by a single enzyme, nitrilase (28, 29), whereas simple aliphatic nitriles are metabolized by a two step process involving nitrile hydratase and amidases (30-32). The cell free extract of acetonitrile grown cells of *R. erythropolis* A10 contained both the hydrolytic enzymes which are known to participate in the sequential hydrolysis of acetonitrile. Nitrile hydratase showed relatively broad substrate specificity when tested with several nitriles except for benzonitrile and acrylonitrile which were hydrolysed with lower activity (Table 5). The ability of *R. erythroplis A10* to grow over a wide range of pH values (5.5-8.5), at a high concentrations of acetonitriles and to utilize a wide range of nitriles as sole C and N source indicates the potential of this organism for its use in treating the toxic nitrile containing wastes.

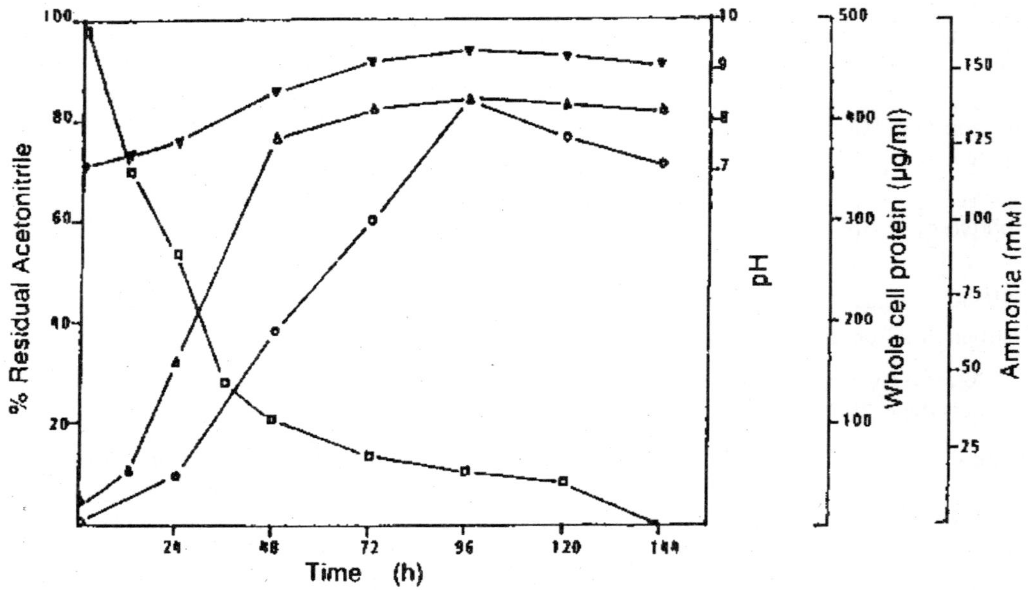

ʼig. 5. Growth and ammonia formation by *R. erythropolis* A10 during culture on acetonitrile at 28°C : □—Acetonitrile; Δ—Whole cell protein; O—Ammonia and ▽—pH. Initial acetonitrile concentration was 200 mM.

Inoculation of *C. acidovorans* D4 and *R. erythropolis* A10 individually and in consortium into effluents of two Baroda based petrochemical industries brought about 80 to 90 % reduction in COD within 48h of incubation, implying the potential of these organisms in biotreatment of petrochemical waste.

REFERENCES

1. Gale F.F. In: The chemical activities of bacteria, London, Academic Press. (1952)
2. Alexander M. Adv Appl Microbiol. *7* : (1965) 35-76.
3. Brown L.R. Chem Engg Progress. (1987) p35.
4. Swanell R.P.J. (1994). Nature. *368* : 396-397.
5. Gueven K.C., Reisch J., Kizil Z., Guevenev B. & Cevher E. Phytochemistry. *29* : (1990) 3115.
6. Woodward K.N., Smith A.M., Mavisrotti S.P. and Tomlinson N.J. Toxicol Rev. *14*:(1986) 183.
7. Dutton J.R., Venables W.A. and Saint C.P. Microbiology *14* :(1995) 1673.
8. Eaton R.W. and Ribbons D.W. J Bacteriol, *151*: (1982) 48.
9. Kurane R., Suzuki T. and Takahara Y. Agric Biol Chem, *41*: (1977) 2119.
10. Henahen J.F and Idon J.D. Chem Engg News *49* : (1971) 16.
11. Ingorsen K., Yde B., Gotfredsen S.E and Tsuchiya R. In: Cynamide compounds in biology, Ciba Foundation Symposia, 140.(Eds. D.Evered and S. Harnet), (1988) pp 16-31, Chichester, John Wiley & Sons.

12. Bushnell L.D. and Haas H.F. J Bacteriol. *41* : (1941) 653-673.
13. Hanson K.G., Kale V.C. and Desai A.J. FEMS Microbiol Lett *122* : (1994) 275-280.
14. Hanson K.G., Nigam A., Kapadia M. and Desai A.J. Ind J Exptl Biol. *35* : (1997) 191- 193.
15. Hanson K.G. and Desai A.J. Biotechnol Lett. *18* :(1996) 1369-1374.
16. Patel D.S., Desai A.J. and Desai J.D. Ind J Exptl Biol, *36* : (1998) 321-324.
17. Acharya A. and Desai A.J. World J Microbiol Biotechnol. *13* : (1997) 175-178.
18. Jobson A., Cook F.D and Westlake D.W.S. Appl Environ Microbiol. *23* : (1972) 1082-1089.
19. Boyles D.T. Biotechnol Lett. *6* : (1984) 31-36.
20. Perry J.J. In: Petroleum Microbiology, Atlas, RM (ed.), (1984) pp 61-98, New York, Macmillan Publishing Co.
21. Juni E. and Janik A. J Bacteriol. *98* : (1969) 281-288.
22. Cruz J.A, Singer J.T. and Finnerty W.R. Curr Microbiol. *3* : (1979) 129-132.
23. Ahlquist E.F., Fewson C.A., Ritchie D.A., Podmone J. and Rowell V. FEMS Microbiol Lett. *7* : (1980) 107-109.
24. Gokhle D.V., Puntambekar U.S. and Deobagkar D.N. Biotechnol Adv. *11*: (1993) 199-217.
25. Sivamurthy K., Swamy B.M. and Pujar B.G. FEMS Microbiol Lett. *79*: (1991) 375-377.
26. Ganji S.H., Karigar C.S. and Pujar B.G. Biodegradation. *6* : (1995) 61-66.
27. Sivamurthy,K and Pujar, B.G. J Ferment Bioeng., *68* : (1989) 375-377.
28. Bandopadhyay A.K., Nagasawa T., Asano Y., Fujishiro K., Tani Y. and Yamada H. Appl Environ Microbiol. *51* : (1986) 302-306.
29. Nagasawa T., Kobayashi M. and Yamada H. Arch Microbiol. *150* : (1988) 89-94.
30. Arnand A., Galzy P. and Jallaglas J.C. Agric Biol Chem. *41* : (1977) 2183-3191.
31. Asano Y., Fujishiro K., Tani Y. and Yamada H. Agric Biol Chem. *46* : (1982) 1165-1174.
32. Asano Y., Yachibana M., Tani Y. and Yamada H. Agric Biol Chem. *46* : (1982) 1175-1181.

CHAPTER - 45

HERBICIDES INDUCED *In vitro* PROTEIN DEGRADATION, MUTAGENICITY AND CYTOTOXICITY IN BACTERIA.

Richa Jaiswal , Saba Siddiqui and Javed Musarrat

Department of Microbiology, Institute of Agriculture, Aligarh Muslim University, Aligarh 202002, India

ABSTRACT

The paraquat (1,1'- dimethyl-4-4' bipyridilium dichloride) and 2,4-D (2,4-dichlorophenoxy acetic acid) herbicides were quantitatively assessed for their mutagenic and cytotoxic activities. Upon photosensitization, paraquat produced significant amount of superoxide anions, however, 2,4-D did not exhibit any free radical formation even after 8 hr of incubation. Considerable protein degradation occurred with the paraquat treatment both in presence of white light and dark. The herbicides also induced substantial cytotoxicity in *Rhizobium* species. The minimum inhibitory concentrations were determined to be 0.24 and 5.0 μg/ml for paraquat and 2,4-D, respectively. A dose dependent reduction in the survival of *Rhizobium leguminosarum* and *Bradyrhizobium japonicum* and lentil *Rhizobium* species was invariably observed. The herbicides in concentration range of 2.5 to 10 μg/plate produced significant number of *his*$^+$ revertants from *his*$^-$ phenotypes of the *Salmonella* strains, as an index of mutagenicity. Comparative analysis ascertained paraquat to be more cytotoxic and mutagenic herbicide compared with 2,4-D, presumably due to its tendency to photo-generate free radicals.

1. INTRODUCTION

The application of pesticides in agriculture has resulted in increased productivity, improved quality and cost reduction of a large number of agricultural products. Without pesticides use, agricultural yield of fresh produce and field crops drops by 30 to 50 percent on annual basis (1,2). Nevertheless, the benefits of pesticide application must be considered in the light of increasing concerns related to farmers' safety and public health, persistence and biodegradability. The available data are insufficient for assessing the pesticide related disorders in the general population in this country. However, in United States, approximately 2000 persons suffer from pesticide poisoning every year. In fact, acute pesticide poisoning is a major global health problem with approximately 25 million occupational poisoning cases reported per year (3). According to a recent report, pesticide use in India has doubled in the last eight years with an average consumption rate of 450 grams per hectare. Although, the consumption rate is relatively less compared with the developed countries, the hazards associated with pesticides are much more (4). This may be due to inadequate pesticide education, monitoring and use of safety guidelines. It has been suggested that the indiscriminate use of pesticide may exert toxic effects and damage cellular macromolecules, including the hereditary material (5). Thus, the genotoxic and mutagenic activities of pesticides have been the subject of extensive research. In contrast, the information related to the mechanism of interactions of pesticides with biological macromolecules (proteins and DNA) is not well-cited in literature. This vital issue has not been thoroughly investigated, though it is an important aspect of the carcinogenic evaluation process. Therefore, emphasis in this investigation is laid on the evaluation of damaging effects of paraquat and 2,4-D in bacterial systems and their *in vitro* interactions with serum albumin.

Earlier studies have also demonstrated pesticide mediated growth inhibition of indigenous nitrogen fixing *Rhizobium* species and its adverse effects on crop productivity due to retarded biological nitrogen fixation (6,7). Thus, in the present study, the biological and/or cytotoxic activities of selected herbicides were assessed by monitoring the (i) quantitative production of reactive oxygen species, (ii) damage to proteins under *in vitro* conditions, (iii) dose dependent relationship with the survival of *Rhizobium* species and (iv) mutagenicity with Ames *Salmonella typhimurium* strains to get an insight into the mechanism of pesticide induced mutagenicity and toxicity and its implication in crop productivity.

2. MATERIALS AND METHODS

(a) Quantitation of superoxide anions

The production of superoxide anions was determined spectrophotometrically utilizing nitroblue tetrazolium (NBT) reduction assay (8). In brief, in a total reaction volume of 5ml, containing 10mM Tris-HCl buffer, pH 7.4; 30 μM NBT; 1mM EDTA; 0.6% Triton X-100, the increasing concentrations (0 – 80 μM) of paraquat were added. The tubes were exposed to white light (Philips, 40W Fluorescent tube) for 6 hr at a constant distance of 8 cm from the source. The blue colour developed was read at 560nm against the reagent blank. The parallel controls without paraquat and with paraquat (incubated in dark) were always run under identical conditions.

(b) Herbicide induced degradation of proteins

In order to determine the extent of protein degradation, increasing amounts of paraquat were added to a fixed concentration of protein (BSA, 10mg/ml) and exposed to white light for 30 min. The reaction was stopped with the addition of 1mM EDTA and precipitation was achieved with 5% TCA. To the supernatant, obtained after centrifugation at 2500 rpm, 0.1 ml of 0.1% trinitrobenzene sulphonic acid (TNBS) In 0.03M borate buffer, pH 9.5 was added. The colour developed after 30-min incubation at room temperature was measured at 340 nm against a reagent blank. The amount of acid soluble amino acids released were estimated from the calibration curve of glycine, developed independently every time the protein degradation was performed by TNBS method (9).

(c) Herbicide induced cytotoxicity in *Rhizobium* species

Cells from the freshly grown exponential cultures of *Bradyrhizobium japonicum* (strain SB-120), *Rhizobium leguminosarum* (strain P-14) and *Rhizobium* species (L_1-87) from lentil were harvested and suspended in PBS, pH 7.5. The cells of SB-120 and P-14 strains were treated with 50 and 100μg/ml, while L_1-87 cells were treated with 1 and 10 μg/ml of paraquat and 2,4-D, respectively. The cells were treated with paraquat for 2 hr whereas the treatment in case of 2, 4-D was prolonged upto 6hr. At regular intervals, the aliquots were withdrawn, serially diluted and plated on yeast mannitol (YEM) agar. The plates were invariably incubated at 28^0c for 48 hr. The colonies were counted and the percent survival was determined and plotted as a function of herbicides concentrations.

(d) Ames mutagenicity assay

Standard *Salmonella typhimurium* tester strains TA97a, TA98, TA100, TA102, and TA104 were used to determine the mutagenic potential of test herbicides, paraquat and 2,4-D, following the plate incorporation assay (10). In brief, the freshly grown cells ($1x10^8$ CFU/ml) were harvested and plated separately on histidine deficient Vogel-Bonner medium (minimal glucose) plates in presence of herbicides at varying concentration range of 0-10μg/plate. The negative and positive controls, without pesticide and with a known mutagen, ENU, respectively were always run in parallel. The *his*$^+$ revertants were counted and the mutation frequency was determined.

RESULTS AND DISCUSSION

In this study, the role of paraquat and 2,4-D herbicides as putative inducers of reactive oxygen species (ROS) and their ability to degrade proteins have been quantitatively assessed. The minimum inhibitory concentrations, mutagenicity and reduction in the survival of various *Rhizobium* species have also been studied to determine the toxic and/or mutagenic potential of these herbicides. The data revealed that the photosensitized 2,4-D does not yield detectable amounts of ROS under *in vitro* conditions (results not shown). However, the paraquat undergoes single electron reduction to form a free radical that is stable in the absence of oxygen. In presence of oxygen, this radical immediately reforms a cation with a concomitant reduction of O_2 to form superoxide anion (O_2^-) (11). This prompted us to quantitate the amount of superoxide anion liberated upon *in vitro* photosensitization of paraquat. The extent of production of blue coloured formazon as a result of paraquat exposure to white light in presence of nitroblue tetrazolium was measured spectrophotometrically and the absorbance was plotted as a function of paraquat concentration at different time intervals (Fig. 1). A linear increase in the development of blue colour was noticed. Based on the absorbance data and the extinction coefficient (ε = 15,000) of formazon, the absolute amount of superoxide anions were estimated and plotted as shown in the inset of Fig. 1. The data exhibit quantifiable production of superoxide anions (40 μmoles) with as low as 10μM paraquat upon 60 min exposure to white light. Prolonged exposure upto 6 hr at a similar dose resulted in approximately 3-fold increase (119 μmoles) in the amount of superoxide anions. The non-linear increase in colour development at higher paraquat dose (40 μM) and longer photoillumination (6 hr) signifies the saturation of superoxide anions and depletion of paraquat radicals. This could

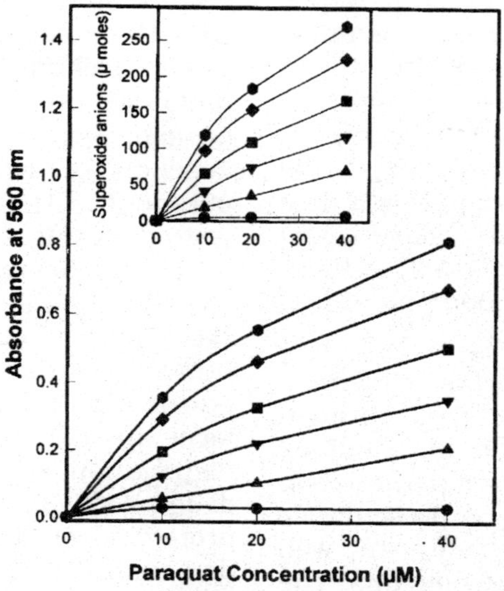

Fig. 1. Quantitative estimation of superoxide anions as a function of paraquat concentration by NBT reduction assay. The intensity of blue colour developed at varying doses was measured at 560 nm. The amount of superoxide anions produced in concentration range of 0-40 μM paraquat are shown in the inset. The values expressed are the mean of two independent experiments done in triplicate.

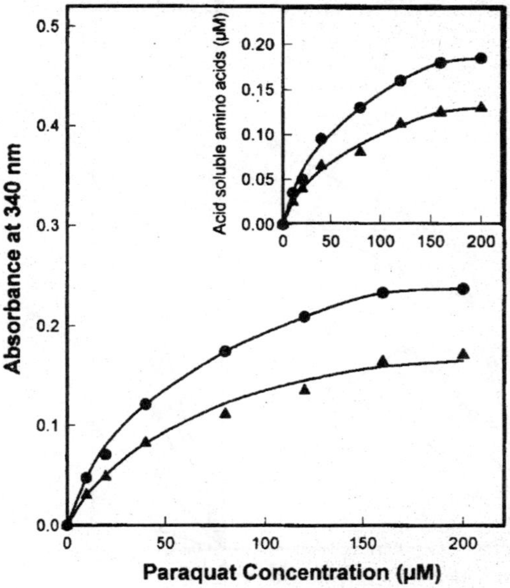

Fig. 2. Paraquat induced degradation of proteins. BSA was taken as a representative protein and treated with varying doses of paraquat. The TCA soluble amino groups, were determined by TNBS method as described in the Methods and Materials. The inset shows the amount of acid soluble amino acids released in the • - light and ▲-dark, respectively.

be due to rapid reaction between paraquat radicals and oxygen with a rate constant of 7.7×10^8 $M^{-1}S^{-1}$ (11). Indeed, no colour formation was observed in dark even after 8 hr of incubation at highest concentration of paraquat. This clearly implies that generation of ROS with paraquat is a light dependent process and may exert detrimental effects on normal cellular functioning due to macromolecular alterations.

In an attempt to develop a general model for herbicide induced protein modification and degradation, we examined the effect of sub lethal doses of paraquat on albumin, chosen as a representative protein, due to its easy availability and well characterized ligand binding domains. The results shown in Fig. 2 clearly demonstrated substantial protein degradation upon treatment with photosensitized paraquat. Interestingly, the degradation of protein was also noticed even in the absence of light at similar doses of paraquqt. This alludes that protein degradation is solely not due to formation of superoxide anions but could also be due to metal ions catalyzed reactions with the excessively bound paraquat cations on proteins. Cation binding to different sites on the albumin molecules is an important facet in the binding and physiological distribution of various metabolites and drugs (12,13). Quantitative analysis of protein degradation by chemical assay demonstrated the release of 0.18 and 0.12

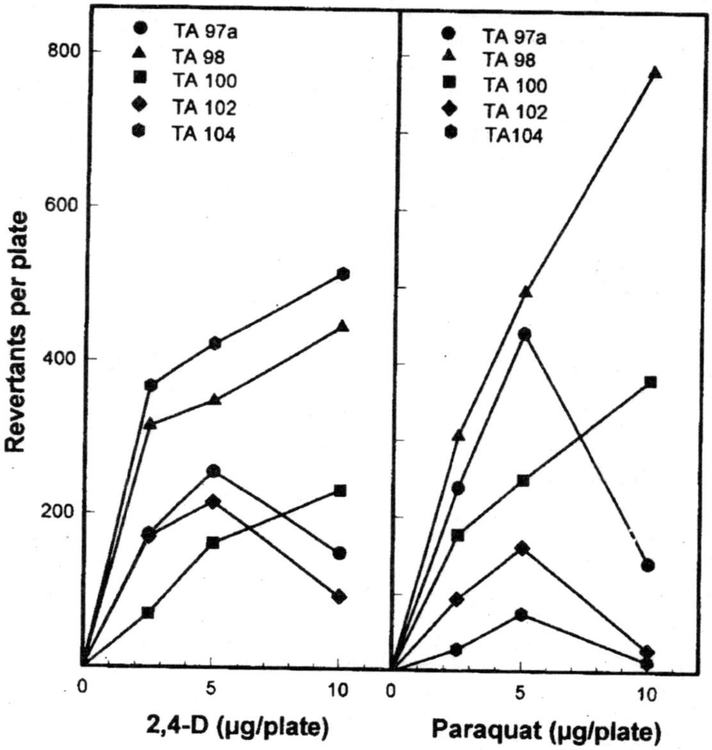

Fig. 3. Mutagenic dose response effect of 2,4-D and paraquat on the reversion of *Salmonella* tester strains. The number of actual revertants were plotted as a function of pesticides concentration after subtracting the spontaneous revertants/plate. Each data point is a mean of two independent experiments done in duplicate.

μM acid soluble amino acids in light and dark, respectively within 2 hr incubation at 200 μM paraquat. The protein degradation curve exhibited linearity upto 120 μM and plateaued at higher (180 –200 μM) paraquat concentrations. Based on these observations, it is surmised that the paraquat not only generates ROS upon photosensitization but may also bind to specific sites on proteins, augmenting the protein fragmentation process. It is generally believed that the covalent binding of chemical carcinogens to cellular macromolecules, particularly proteins and DNA, is the initial critical event between a carcinogen and the target cell. Therefore, it is important to determine the conformational changes in the agro-chemically modified target molecules to get an insight into the mechanism of pesticide induced mutagenesis and/or carcinogenesis.

Also the effect of these herbicides on various *Rhizobium* species have been

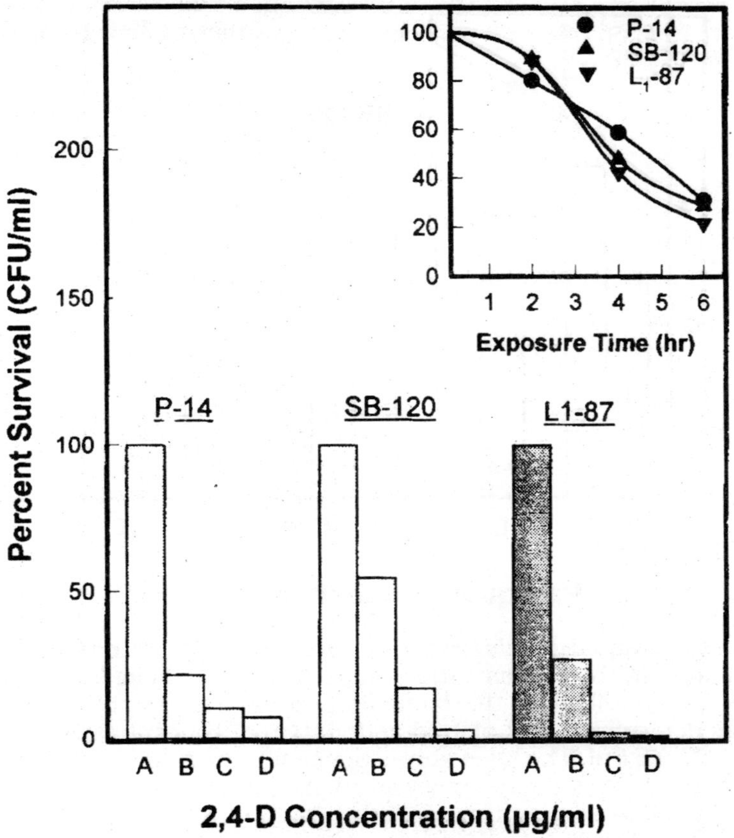

Fig. 4. Cytotoxicity of 2,4-D in *Rhizobium* species. The cells were treated with the serially diluted 2,4-D in the concentration range of 0.5 to 500 μg/ml for 6 hr, for the determination of minimum inhibitory concentration. Treated cells were plated on YEM agar as described in Methods and Materials. The per cent survival was determined and plotted as a function of concentration, where A-0.5, B-5.0, C-50 and D-500 μg/ml. The inset shows the per cent reduction in the viability of *Rhizobium* species with constant 2,4-D concentrations of 100 μg/ml for P-14 and SB-120 while 10 μg/ml for L_1-87, at different time periods.

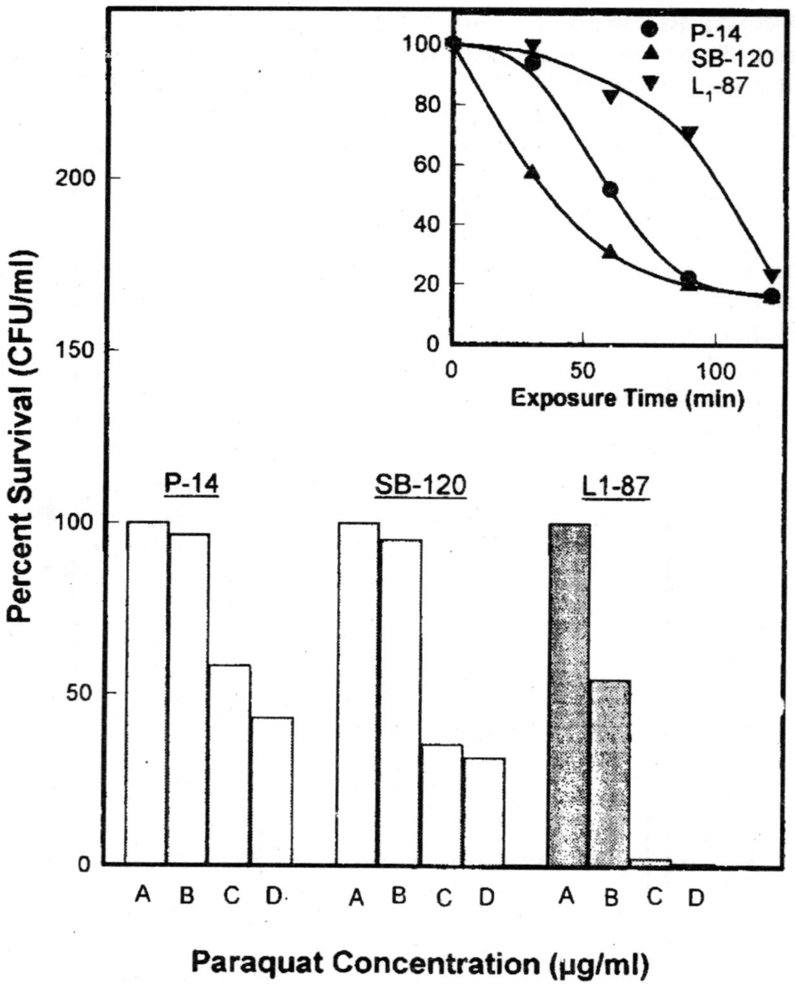

Fig. 5. Cytotoxicity of paraquat in *Rhizobium* species. The cells were treated with the serially diluted paraquat, in the concentration range of 0.024 to 24 µg/ml for 2 h. Treated cells were plated as described in the Methods section. The per cent survival in terms of CFU/ml was determined and plotted as function of concentration as A-0.024, B-0,24, C-2.4 and D-24 µg/ml. The inset shows per cent reduction in the viability of *Rhizobium* species with constant paraquat concentrations of 50 µg/ml for P-14 and SB-120 while 1 µg/ml for L_1-87, at varying time periods.

studied by measuring the bacterial survival on YEM agar plates (Fig. 3 and 4). Lower concentrations of paraquat (0.024 and 0.24 µg/ml) did not produce any toxic effects on *Rhizobium leguminosarum* and *Bradyrhizobium japonicum* strains (Fig. 3). However, at similar doses, the colony forming ability was reduced by 50% in case of *Rhizobium* species from lentil. The damage was found to be more pronounced at 10- and 100-fold higher concentrations (2.4 and 24 µg/ml) of paraquat. The decrease in the survival if *Rhizobium leguminosarum* and *Bradyrhizobium japonicum* at a fixed concentration (50 μg/ml) of paraquat was shown to be time dependent. Among all the tested strains, the lentil *Rhizobia* exhibited higher sensitivity (24% survival) with 1 μg/ml of paraquat. A significant effect of 2,4-D on the viability of all *Rhizobium* species was invariably observed (Fig. 4). The order of sensitivity of various *Rhizobium* species was found to be $L_1$87> SB-120 > P-14. A substantial loss in the colony forming abilities of SB-120 and P-14 strains was observed at 100 μg/ml of 2,4-D while a 10-fold lesser amount (10 μg/ml) of 2,4-D inflicted cytotoxicity effect of similar magnitude in $L_1$87 strain. The differential sensitivity of *Rhizobium* species for herbicides could be due to variability in the production of exopolysaccharides. The polysaccharide sheath of *Rhizobia* has been reported to act as a barrier to the entry of herbicide, thus reducing their toxicity (14). If these differential responses become evident under field conditions, inoculum-containing strains tolerant to particular pesticide can be recommended for a particular area. Hence, the development of resistance to pesticide without loosing the ability to fix nitrogen is of great agronomic importance.

The results shown in Fig. 3 revealed that much lower (5-10 μg/plate) concentrations of both paraquat and 2,4-D were adequate to induce significant number of his^+ revertants from his^- phenotypes of *Salmonella* tester strains. Earlier studies have also demonstrated the mutagenic activities of various pesticides (15). The strong mutagenic effects of paraquat could be due to the presence of several reactive alkyl groups, which have been reported to alkylate DNA (16,17). The data indicated that the reversion frequency enhances with the increasing concentrations of test herbicide upto 10 μg/plate. The bacterial population decline at higher concentrations due to the preponderance of the toxic effects of herbicides. It is, therefore, concluded that the indiscriminate and non judicious use of herbicides in agriculture not only influence the leguminous crop yield (7), but may also exert severe detrimental effects on sprayers and farm workers (2,5). The subtle danger of their large-scale use lies in the possibility that some of them damages the genetic material. Further research in this direction is warranted to better characterize and quantitate the adverse effects of agrichemicals on human health. Moreover, in-depth studies are in progress in order to probe the mechanism of pesticide DNA/ protein interactions and its impact on *Rhizobium* – legume symbiosis and its ultimate effects on crop productivity.

ACKNOWLEDGEMENTS

The authors thank Dr. Zaka-ur-Rab, Director, Institute of Agriculture for providing laboratory facilities. This work was supported in part by a grant-in-aid from the Lady Tata Memorial Trust, Mumbai.

REFERENCES

1. Zilberman D. Shmitz A. & Casterline G. *Science* **253** (1991) 518.
2. Spigiel R. W. Gourley D. R. & Holcslaw T. L. *Clin. Toxicol. Consult.* **3** (1981) **45.**
3. Jeyratnam J. *World Health Stat Q* **43** (1990) 139-144.
4. Lal O. P. *Pesticide World* **3** (1998) 43.
5. Weisenburger D.D. *Human Pathol.* **24** (1993) 571.

6. Gaur A. C. & Misra K. C. *Indian J. Microbiol.* **12** (1972) 45.
7. Rennie R. J. & Dubetz S. *Agron. J.* **76** (1984) 451.
8. Nakayama T. Kimura T. Kodama T. & Nagali C. *Carcinogenesis* **4** (1983) 765.
9. Synder S. L. & Sobocinski P. **Z.** *Anal. Biochem.* **64** (1975) 284.
10. Maron D. M. & Ames B. N. *Mutat. Res.* **113** (1983) 173.
11. Frrington J. A. Ebert M. Lard E. J. et. al. *Biochim. Biophys. Acta.* **314** (1973) 372.
12. Marx G. *Rev. Biochem. Toxicol.* **6** (1984) 95.
13. Henkin R. I. *Adv. Exp. Med. Biol.* **48** (1975) 299.
14. Grossbard E. *Occas. Symp. Br. Grass land Soc. White clover Res.* **6** (1970) 47.
15. Saxena S. Ashok B. T. & Musarrat J. *Biochemistry and Molecular Biology International* **41** (6) (1997) 1125.
16. Musarrat J. Arezina J. Uddin S. & Wani A. A. *Biochim. Biophys. Acta. (Gene Structure and Expression)* **1260** (1995) 274.
17. Musarrat J. Arezina-Wilson J. & Wani A. A. *Biochim. Biophys. Acta. (Gene Structure and Expression)* **1263** (1995) 201.

CHAPTER – 46

STUDIES ON THE INTERACTION AMONG SOIL ALGAE *In vitro*

Anil Kumar Sinha[1], *Kamini Kumari*[2] *and B.N. Verma*[3]

1. Department of Botany, K.C.T.C. College, Raxaul East Champaran - 845 305, Bihar
2. Bhagya Narayan Kanya Mahavidyalaya, Barauni, Begusarai - 851 113, Bihar
3. University Department of Botany, B.R.A. Bihar University, Muzaffarpur - 842 001, Bihar

ABSTRACT

Algae growing in complex association interact with each other through secreting extracellular metabolites. This aspect of algae is least studied. The effects of culture filtrates of ten soil algae have been studied on each other with reference to the long term toxicity and relative sensitivity in the present endeavour. Studies revealed three categories of the effect i.e. probiotic, antibiotic and neutral. It has been concluded that the culture filtrates are invariably probiotic. However, none of the culture filtrates was either exclusively probiotic or exclusively antibiotic as the same culture filtrates enhanced the growth of one test organism, it inhibited growth of others. Similarly the age of culture filtrate has not also been recorded as the deciding factor of its efficacy.

It has been concluded that the metabolites excreted by the algae have narrow range effects and also that different algae are selective in their mode or response to the metabolites present in the culture filtrate of co-existing algal species. It has been further concluded that *Ankistrodesmus fractus* is highly resistant against the growth inhibitory effect of the culture filtrates of the remaining nine taxa. On the other hand *Nostoc linckia*, *Scenedesmus arcuatus* and *Synechocystis aquatilis* are very good test organisms for bioassay experiment being sensitive to the growth inhibitory effect of culture filtrates of remaining nine texa.

1. INTRODUCTION

Most plants and microorganisms from the simplest unicellular algae to complex, higher plants produce a variety of unusual compounds collectively referred to as secondary metabolites. Thousands of these compounds have been chemically characterized and classified on the basis of their chemical groups and/or chemical properties. However, the possible role of these secondary metabolites in the ecosystem has not received due attention. Since they serve no basic metabolic function and vary in concentration and structure, even in closely related species, they were considered metabolic waste products but they have a rapid turnover rate in the cell. Such properties are hardly those of waste products. Advances in plant physiology and biochemistry have shown that many secondary metabolites facilitate interaction between plants and their environment. Examples of these are the plant hormones, auxins, gibberllines and phenolic compounds, which govern plant growth and development in response to environmental factors.

The synthesis of some secondary metabolites like phytoalexins, is induced by parasitic invasion and may enable plants to fight off infections. It has been also suggested that many secondary metabolites are allelopathic substance i.e. compounds of one organism which inhibit the growth and development of competing species. While the secondary metabolism of terrestrial plants has received the greatest attention, aquatic plants and especially algae, although are known to produce a complex array of secondary metabolites have been ignored. In the present study an attempt has been made to ascertain the role of extracellular metabolites of selected algae isolated from the paddy fields under investigation in regulating population cycle of each other in natural associations. Since the growth promoting substances produced by the algae have been thoroughly investigated in the past, emphasis has been given on their growth inhibitory compounds produced by them.

2. MATERIALS AND METHODS

(a) Selection of the Experimental Materials

23 species belonging to 19 genera of Cyanophyceae and Chlorophyceae were observed during different period in the paddy fields under investigation. Out of these, ten species viz., *Nostoc linckia, Anabaena oryzae, Microcystis aeruginosa, Oscillatoria willei, Synechocystis aquatilis, Oocystic pyriformis, Westiellopsis prolifica, Ankistrodesmus fractus, Merismopedia elegans* and *Scenedesmus arcuatus* were selected for interation studies on account of their high frequency of occurrence in the field.

(b) Isolation and purification of the test organisms

Wild strains of ten aforesaid taxa were isolated from the paddy field soils in and around Raxaul at East Champaran as per the methods described by Das *et al.*, (1).

Clonal as well as axenic cultures of the isolated taxa were raised under laboratory conditions as per method described by Kaushik (2).

(c) Maintenance of culture under laboratory conditions

Allen's modified Hughe's inorganic medium (3) was used as the basal medium for the green and non-nitrogen fixing blue-green algae under laboratory conditions. The same medium without nitrogen source was used for growing nitrogen-fixing blue-green algae. The pH of the medium was fixed at 7.5 in first case and 7.8 in the second. Cultures were exposed to a daily cycle of 14 hours of light and 10 hours of darkness under laboratory conditions. The fluorescent tubes (about 3K lux) were used as source of light. The temperature ranged between 25-31°C during the course of investigations.

Corning Erlenmeyer conical flask (250 and 1000 ml), test tubes (25 x 150 mm) and petriplates (75 mm in diameter) were used as the culture vessels. They were cleansed by treating them with chromic acid (100 gms potassium dichromate in 750 ml water + 250 ml conc. sulphuric acid) and subsequently by washing them with Teepol and again with tap water to eliminate every traces of chromic acid. Thoroughly washed glass-wares were finally rinsed with glass distilled water before taking them in use after being sterilized at 170°C in hot air even at least for one hour.

(d) Interaction studies

The bioassay experiment was conducted with the culture filtrate of all the selected ten taxa. Since the leaching of the metabolites vary with age, the culture filtrate for study was taken of two ages i.e. 15 and 25 days old to cover more metabolites leached out by the alga in question.

(e) Long term toxicity test

The two ages of the culture filtrate of each of the ten selected species was obtained in bulk and was sterilized by filtering it through sintered glass. Exponentially growing population of each of the ten test organisms were grown in the culture filtrate of remaining nine species in triplicate. The population was also inoculated in the basal medium, similarly to serve as control for comparison. The persistent effects of the culture filtrates were evaluated by scoring the growth of the alga inoculated therein and by computing the growth constant. The average of the three growth constants obtained from these replicates in each case was taken into consideration for interpretation.

(f) Growth measurements

The growth of the test organisms was measured at an interval of 24 hrs, in liquid medium through spectrophotometric method at a wave length of 663 nm by Shimadzu UV-VIS Spectrophotometer model 160. Initial O.D. (Optical Density) of 0.5 was taken constant on the day of inoculation in each case. The growth of filamentous algae was measured by pigment extraction method in Acetone (4).

(g) Growth constant

For the large groups of algae, generation time may be several days, therefore, growth can be expressed as units of increase per day (Day^{-1}). Under suitable circumstances a culture of algae grows so that the rate of addition of cells is proportional to the number present - "exponential growth". Population development follows the solution of the equation :

$$dN/dt = KN$$

Where N is number or concentration of the cells in the culture, t is time and K (the growth constant) is a quantity with dimensions of t^{-1}. The solution is :

$$K_{10} = \frac{\log (N_1/N_0)}{t_1 - t_0}$$

Where the subscripts denote values at two times and the growth constant K is the number of "logarithm-to-base 10" units of increase per day. Time is expressed as hours for unicellular organisms with shorter generation times. If growth rates (K) are thus given in logarithmic increases per hour (h^{-1}) multiply by 24 to obtain the corresponding values expressed as units of increase per day (day^{-1}). From K, division (generation) time can also be calculated as:

$$Td = 1/K \text{ days per division}$$
$$= 24/K \text{ hours per division.}$$

(h) Computation of K from two counts

If N_i and N_o are the cell concentrations at the end and beginning of a period of time t days, then K = long (N_i/N_o) 3.322/t).

(i) Relative sensitivity test

It was made as per plate diffusion method described by Collins *et al.*, (5). The petriplates were loaded with the basal medium solidified by 1% agar-agar. It was flooded with cell suspension of the desired test organisms containing about 10^6 cells/ml. The excess of suspension was removed after homogenous spreading. Holes of 8-9 mm were punched in the solidified medium with the help of sterile cork borer and the plugs were left out without disturbing the surrounding medium. The holes were filled up with the culture filtrate to be tested. These petriplates were exposed to the daily cycle of 14 hrs of light (3300 lux) followed by 10 hrs of dark period to grow the inoculated population and were watched carefully for the appearance of the inhibition zone if any, zones of inhibition wherever appeared were measured carefully.

3. OBSERVATIONS

The primary screening of the extracellular metabolites of ten wild strains of the selected algae was made by growing them in 15 and 25 days old CF (culture filtrate) of each other. The overall growth was recorded and on the basis of the data obtained, their growth constants were calculated (Table 1 and 2).

Critical analyses of the data on growth experiments with different CFs revealed three categories of the effect i.e. probiotic (growth promoting), antibiotic (growth inhibitory) and neutral. It further revealed that none of the CF was recorded exclusively any one of three category. While it was proved promoting for one test organism, it was noted growth inhibitory or neutral for others (Table 3). Growth response of each of the ten test organisms to the different CFs may be enumerated as follows :

The growth of *N. linckia* was inhibited through 15 and 25 days old CFs of all the species at hand except 15 days old CFs of *A. oryzae* and *S. arcuatus* by which it was rather triggered. The 15 days old CF of *O. pyriformis* generated neither positive nor negative effect.

15 days old CF of each of the nine taxa and 25 days old CF of 1,6,7,8 and 9 enhanced growth of *A. oryzae* but 25 days old CF of remaining four species acted adversely on its growth.

No effect could be generated on the growth of *M. aeruginosa* by 15 days old CF of 1,2,8,9 but 25 days old CF of 6 and 8 inhibited its growth. The remaining treatments proved beneficial for the growth of the species under reference.

The growth of *O. willei* was, in general, triggered except 15 days old CF of 3 and 10 as well as 25 days old CF of 2, 3 and 6 in which it was inhibited.

The growth of *S. aquatilis* in general was inhibited by the CFs of all the nine species except 15 days old CF of 1 and 3 and 25 days old CF of 9 and 10 which enhanced its growth. 15 days old CF of 7 proved neutral for the growth of the alga under reference.

15 days old CFs of all the nine species acted adversely and that of 25 days old CFs proved growth promoter for *O. pyriformis* except few instances. 15 days old CF of 8 promoted the growth instead of inhibition On the other hand 15 days old CF of 9 and 25 days CF of 3 produced no effect.

W. prolifica responded positively to 15 days old CFs of 1,2 and 3 negatively to that of 4,6,8 and 10 remained neutral in the 15 days old CFs of 5 and 9. But 25 days old CFs of all the species proved beneficial for the growth of alga under reference except 8 in which it was inhibited and 9 which remained neutral.

The younger CFs of all the nine species enhanced the growth of *A. fractus* and the

Table 1. Growth constant of different algae in 15 days old culture filtrates of each other

Test organisms	Growth constant of test organisms in culture filtrates of each other										
	Control	*Nostoc linckia*	*Anabaena oryzae*	*Microcystis oerugenosa*	*Oscillatoria willei*	*Synechocystis aquatilis*	*Oocystis pyriformis*	*Westiellopsis prolifica*	*Ankistrodesmus fractus*	*Merismopedia elegans*	*Scenedesmus arcuatus*
	C	1	2	3	4	5	6	7	8	9	10
Nostoc linckia	0.110 ±0.011	—	0.128 ±0.002	0.077 ±0.002	0.066 ±0.001	0.068 ±0.001	0.112 ±0.002	0.103 ±0.002	0.087 ±0.001	0.095 ±0.002	0.117 ±0.001
Anabaena oryzae	0.08 ±0.001	0.105 ±0.002	—	0.085 ±0.001	0.124 ±0.002	0.130 ±0.001	0.106 ±0.002	0.095 ±0.002	0.087 ±0.001	0.105 ±0.002	0.096 ±0.001
Microcystis aerugenosa	0.105 ±0.003	0.103 ±0.001	0.103 ±0.001	—	0.133 ±0.001	0.115 ±0.001	0.122 ±0.002	0.115 ±0.001	0.107 ±0.001	0.104 ±0.002	0.124 ±0.002
Oscillatoria willei	0.075 ±0.002	0.092 ±0.001	0.083 ±0.001	0.067 ±0.001	—	0.077 ±0.001	0.095 ±0.002	0.097 ±0.001	0.095 ±0.002	0.155 ±0.002	0.054 ±0.002
Synechocystis aquatilis	0.123 ±0.002	0.133 ±0.001	0.085 ±0.002	0.163 ±0.001	0.064 ±0.002	—	0.064 ±0.001	0.126 ±0.002	0.067 ±0.001	0.103 ±0.002	0.085 ±0.002
Oocystis pyriformis	0.039 ±0.002	0.034 ±0.002	0.026 ±0.002	0.015 ±0.002	0.034 ±0.002	0.025 ±0.002	—	0.024 ±0.001	0.044 ±0.001	0.041 ±0.002	0.027 ±0.001
Westiellopsis prolifica	0.063 ±0.001	0.071 ±0.001	0.106 ±0.003	0.134 ±0.002	0.053 ±0.003	0.064 ±0.002	0.053 ±0.001	—	0.055 ±0.002	0.064 ±0.002	0.034 ±0.001
Ankistrodesmus fractus	0.064 ±0.001	0.104 ±0.001	0.106 ±0.002	0.093 ±0.001	0.097 ±0.001	0.123 ±0.002	0.095 ±0.002	0.106 ±0.002	—	0.094 ±0.002	0.122 ±0.002
Merismopedia elegans	0.086 ±0.002	0.085 ±0.002	0.067 ±0.001	0.085 ±0.002	0.086 ±0.002	0.066 ±0.002	0.104 ±0.002	0.072 ±0.001	0.032 ±0.001	—	0.113 ±0.001
Scenedesmus arcuatus	0.036 ±0.002	0.0012 ±0.004	0.004 ±0.002	0.003 ±0.001	0.002 ±0.002	0.025 ±0.001	0.033 ±0.001	0.024 ±0.002	±0.045 0.002	0.025 ±0.002	—

olders inhibited its growth.

The growth of *M. elegans* remained unaffected in the younger CFs of 1,3,4 and olders of 3 and 8. It was inhibited in 15 days old CFs of 2,5,7 and 8 as well as in 25 days old CF of 7. Treatment with the remaining CFs proved beneficial.

The growth of *S. arcuatus* was inhibited in the CFs of all the species except *A. fractus* (15 days old) and *M. elegans* (25 days old) as well as *O. pyriformis* (15 days old) and *A. oryzae* (25 days old) in which it was triggered and remained unaffected respectively.

Critical analyses of the data on growth experiments (Table 1, 2) reveal that the CFs of different taxa varied in regard to the coverage of their growth inhibitory effects. As shown in Table, 4, the CFs of *M. aeruginosa, O. willei, O. pyriformis, A. fractus* and *S. arcuatus* were proved inhibitory against six test organisms in different combinations, of *A. oryzae, S. aquatilis* and *W. prolifica* against five test organisms and of *N. linckia* as well as *M. elegans* against only three test organisms.

It is further apparent from the same table that the growth inhibitory effect of the CFs of different taxa varied also with respect to their age. Thus, while CF of N. linckia was found growth inhibitory irrespective of its age, it was recorded exclusively through 15 days old CFs of *O. pyriformis, W. prolifica as well as M. elegans and exclusively through 25 days old CFs of A. oryzae* and *M. aeruginosa.* However, no definite pattern of growth inhibition was recorded in regard to the age of the culture filtrates of *O. willei, S. aquatilis* and *S. arcuatus.*

The magnitude of the growth inhibitory effect of different culture filtrates was also

Table 2. Growth constant of different algae in 25 days old culture filtrates of each other

Test organisms	Growth constant of test organisms in culture filtrates of each other										
	Control	*Nostoc linckia*	*Anabaena oryzae*	*Microcystis aeruginosa*	*Oscillatoria willei*	*Synechocystis aquatilis*	*Oocystis pyriformis*	*Westiellopsis prolifica*	*Ankistrodesmus fractus*	*Merismopedia elegans*	*Scenedesmus arcuatus*
	C	1	2	3	4	5	6	7	8	9	10
Nostoc linckia	0.11 ±0.011	—	0.103 ±0.012	0.101 ±0.002	0.058 ±0.001	0.022 ±0.001	0.027 ±0.002	0.08 ±0.012	0.107 ±0.002	0.105 ±0.01	0.064 ±0.008
Anabaena oryzae	0.08 ±0.001	0.108 ±0.003	—	0.074 ±0.002	0.063 ±0.003	0.063 ±0.002	0.106 ±0.002	0.099 ±0.001	0.276 ±0.003	0.102 ±0.003	0.056 ±0.002
Microcystis aerugenosa	0.105 ±0.003	0.101 ±0.003	0.135 ±0.003	—	0.125 ±0.002	0.125 ±0.003	0.098 ±0.003	0.116 ±0.004	0.011 ±0.001	0.122 ±0.002	0.119 ±0.003
Oscillatoria willei	0.075 ±0.002	0.094 ±0.005	0.036 ±0.002	0.058 ±0.002	—	0.333 ±0.004	0.042 ±0.003	0.081 ±0.005	0.107 ±0.004	0.118 ±0.005	0.107 ±0.003
Synechocystis aquatilis	0.123 ±0.002	0.067 ±0.002	0.047 ±0.001	0.051 ±0.002	0.099 ±0.001	—	0.116 ±0.004	0.099 ±0.001	0.108 ±0.003	0.132 ±0.003	0.129 ±0.003
Oocystis pyriformis	0.039 ±0.002	0.048 ±0.004	0.50 ±0.003	0.042 ±0.004	0.053 ±0.003	0.052 ±0.004	—	0.063 ±0.001	0.099 ±0.002	0.097 ±0.002	0.097 ±0.003
Westiellopsis prolifica	0.063 ±0.001	0.072 ±0.002	0.104 ±0.002	0.12 ±0.003	0.08 ±0.004	0.077 ±0.002	0.092 ±0.003	—	0.055 ±0.004	0.067 ±0.003	0.073 ±0.003
Ankistrodesmus fractus	0.064 ±0.001	0.111 ±0.004	0.157 ±0.003	0.151 ±0.004	0.105 ±0.004	0.102 ±0.004	0.099 ±0.003	0.097 ±0.003	—	0.111 ±0.004	0.11 ±0.004
Merismopedia elegans	0.086 ±0.002	0.0964 ±0.003	0.107 ±0.003	0.088 ±0.001	0.092 ±0.002	0.096 ±0.004	0.094 ±0.003	0.074 ±0.003	0.088 ±0.001	—	0.100 ±0.004
Scenedesmus arcuatus	0.036 ±0.002	0.020 ±0.002	0.034 ±0.003	0.027 ±0.002	0.017 ±0.003	0.031 ±0.002	0.025 ±0.004	0.027 ±0.004	0.02 ±0.004	0.44 ±0.003	—

computed against their respective controls (Tabl 5 and 6).

Critical analysis of the aforesaid two tables depicting percentage inhibition of growth though the CFs apparently reveal that 15 days old CF in general are more potent growth inhibitor than 25 days old CF except *S. aquatilis*, *O. pyriformis* and *A. fractus*, in which cases 25 days old CF was comapratively more effective growth inhibitor. The growth inhibition over 75% was, however, recorded only with 15 days old CFs of *N. linckia*, *A. oryzae*, *M. aeruginosa*, *O. pyriformis* and with 25 days old CFs of *O. pyriformis* and *A. fractus*.

The growth inhibitory effects of CFs of ten species observed above were further examined for the production of antimicrobial compounds, if any. For this all the ten species were grown in their respective basal media in two sets - one for a period of 15 days and the other for 25 days. The cells free CFs of both ages were poured continuously in the wells on the lawn of remaining nine species in each case independently and were incubated in light. Of the ten test organisms used, only three viz., *N. linckia*, *M. aeruginosa* and *S. arcuatus* exhibited growth inhibition phenomenon on the surface of solidified basal medium. Amongst these three, *S. arcuatus* seems highly sensitive as it exhibited growth inhibitory zones in the 15 days old CFs of *N. linckia*, *A. oryzae*, *M. aeruginosa*, and *O. willei*. It was followed by *N. linckia* which showed growth inhibition phenomenon in 25 days old CFs of *S. aquatilis* and *O. pyriformis*. *M. aeruginosa* was recorded highly selective as it witnessed growth inhibitory zone in 25 days old CF of only *A. fractus*. The

Table 3. Response of test organisms to the culture filtrates of different species

	Test organisms	Age in days	GROWTH EFFECTS		
			Probiotic	Antibiotic	Neutral
A	B	C	D	E	F
1	*Nostoc linckia*	15	2,10	3,4,5,7,8,9	6
		25	–	2,3,4,5,6,7,8,9,10	–
2	*Anabaena oryzae*	15	1,3,4,5.6,7,8,9,10	–	–
		25	1,6,7,8,9	3,4,5,10	–
3	*Microcystis aerugenosa*	15	4,5,6,7,10	–	1,2,8,9
		25	2,4,5,7,9,10	6,8	1
4	*Oscillatoria willei*	15	1,2,5,6,7,8,9	3,10	–
		25	1,5,7,8,9,10	3,3,6	–
5	*Synechocystis aquatilis*	15	1,3	2,4,6,8,9,10	7
		25	9,10	1,2,3,4,6,7,8	–
6	*Oocystis pyriformis*	15	8	1,2,3,4,5,7,10	9
		25	1,2,4,5,7,8,9,10	–	3
7	*Westiellopsis prolifica*	15	1,2,3	4,6,8,10	5,9
		25	1,2,3,4,5,6,10	8	9
8	*Ankistrodesmus fractus*	15	1,2,3,4,5,6,7,9,10	–	–
		25	1,2,3,4,5,6,7,9,10	–	–
9	*Merismopedia elegans*	15	6,10	2,5,7,8	1,3,4
		25	1,2,4,5,6,10	7	3,8
10	*Scenedesmus arcuatus*	15	8	1,2,3,4,5,7,9	6
		25	9	1,3,4,5,6,7,8	2

N.B. Taxa in columns D-F indicated through their serials in column A

measurement of growth inhibitory zones exhibited by the three test organisms in the CFs of different species under reference may be seen in the Table 7. As indicated in the table under reference, the inhibition zone was measured as the clear zone around the well which varied form 9 mm to 23 mm in diameter.

A clear growth inhibition zone observed around each well was carefully scraped with a sterilized inoculation needle and added aseptically to the culture tube containing inoculum of respective test organism in 5 ml. of basal medium. The mixture was incubated in light and examined periodically for cross verification of growth inhibitory effect. The result obtained have been summarized in Table 8.

Growth inhibition and lysis of the cells recorded in the Table 7 & 8 may be presumed either due to the growth inhibitory compound of algal origin or lytic bacteria. Confirmation of the two alternative factors was made through the bioassay of growth inhibitory zone as described below.

The scraps of each of the growth inhibitory zone were inoculated in 5 ml of nutrient Broth and incubated at 28°C +_2 in dark. The turbidity appeared within 14 hrs. Microscopic examination revealed cocci and small rod-shaped bacteria of gram positive nature. On plating at suitable dilutions. isolated colonies with entire margins appeared on the agar surface.

Individual colonies were suitably grown in nutrient broth and added to the wells on the lawn of test organisms that witnessed

Table 4. Effect of culture filtrates on different test organisms

	Test organisms	Age in days	GROWTH EFFECTS		
			Probiotic	Antibiotic	Neutral
A	B	C	D	E	F
1	*Nostoc linckia*	15	2,4,5,7,8	6,10	3,9
		25	2,4,6,7,8,9	5,10	3
2	*Anabaena oryzae*	15	1,4,7,8	5,6,9,10	3
		25	3,6,7,8,9	1,4,5	10
3	*Microcystis aerugenosa*	15	2,5,7,8	1,4,6,10	9
		25	7,8	1,2,4,5,10	6,9
4	*Oscillatoria willei*	15	2,3,8	1,5,6,7,10	9
		25	3,6,7,8,9	1,2,5,10	
5	*Synechocystis aquatilis*	15	2,3,4,8	1,6,9,10	7
		25	3,4,6,7,8	1,2,10	–
6	*Oocystis pyriformis*	15	2,3,4,8,9	5,7	1,10
		25	2,7,8,9	1,3,4,5,10	–
7	*Westiellopsis prolifica*	15	2,3,4,8	1,6,9,10	5
		25	2,3,4,6,8	1,5,9,10	–
8	*Ankistrodesmus fractus*	15	2,4,6,10	1,5,7,9	3
		25	2,4,6	1,3,5,7,10	9
9	*Merismopedia elegans*	15	2,4,8	1,5,10	3,6,7
		25	2,3,4,5,6,8,10	1	7
10	*Scenedesmus arcuatus*	15	1,2,3,8,9	4,5,6,7	–
		25	3,4,5,6,7,8,9	1,2	–

N.B. Taxa in columns D-F indicated through their serials in column A

growth zones (Table 7). No lyses of test organisms were observed and hence the possibility of antimicrobial compound of bacterial origin or of bacteria being responsible for the lysis of cyanobacteria was excluded.

4. DISCUSSION

The effect of culture filtrates of ten wild strains of paddy field belonging to Chlorophyceae and Cyanophyceae have been studied on the growth of each other. The aforesaid studies revealed three categories of observations i.e. growth promoting (probiotic), growth inhibitory (antibiotic) and neutral (Table 3, 6-10).

Out of 180 sets of growth experiments conducted with two ages of CF belonging to 10 species, 91 have been proved growth promoting, 70 growth inhibitory and 19 neither of the two (Table 3). This clearly indicates that growth promoting nature of algal filtrates is comparatively common. Amongst 70 sets of growth experiments witnessing growth inhibitory phenomenon, no definite pattern could be visualized as regards the nature as well as the age of CFs. While the same filtrate proved growth promoting for one species, inhibited growth of others. These two properties varied also with age of the CF. *A. fractus* fall under the second category as its growth was never inhibited through the any age CFs of any of nine taxa at hand. Therefore, it may be regarded as the alga of high adaptability and physiologically most stable. This is in accordance with the highly pollution tolerant nature of the alga reported earlier (11-12).

Table 5. Effect of 15 days old culture filtrates of different test organisms on the growth of each other

Test organisms	Percentage increase/decrease in growth over control in 15 days old culture filtrate									
	Nostoc linckia	*Anabaena oryzae*	*Microcystis aerugenosa*	*Oscillatoria willei*	*Synechocystis aquatilis*	*Oocystis pyriformis*	*Westiellopsis prolifica*	*Ankistrodesmus fractus*	*Merismopedia eleganc*	*Scenedesmus arcuatus*
	1	2	3	4	5	6	7	8	9	10
Nostoc linckia	—	15.31	-30.63	-40.54	-38.73	0.9	-7.2	-21.62	-14.41	-5.4
Anabaena oryzae	31.25	—	6.25	55.00	62.5	32.5	18.75	8.75	31.25	20.00
Microcystis aerugenosa	-1.9	-1.9	—	26.66	9.52	16.19	9.52	1.9	-0.95	18.09
Oscillatoria willei	22.66	10.66	-10.66	—	2.66	26.66	29.33	24.00	106.66	-28.80
Synechocystis aquatilis	8.13	-30.89	32.52	-47.96	—	-47.96	2.43	-45.52	-16.66	-30.89
Oocystis pyriformis	-12.82	-33.33	-61.53	-12.82	-35.89	—	-38.46	12.82	-16.26	-3.76
Westiellopsis prolifica	12.69	68.25	112.69	-15.87	1.58	-15.87	—	-12.69	5.12	-46.03
Ankistrodesmus fractus	62.5	65.62	45.31	51.56	92.18	48.43	65.62	—	1.58	90.62
Merismopedia elegans	-1.16	-22.09	1.16	zero	-23.25	20.93	-16.27	-62.79	—	31.39
Scenedesmus arcuatus	-96.66	-88.88	-91.66	-94.44	-30.55	-8.33	-33.33	25.00	30.55	—

The remaining nine species have been found to be sensitive against the growth inhibitory effect of one or the other. The taxa falling under this category further varied with respect to their response against the age of CF. While *A. oryzae* and *M. aeruginosa* have been recorded sensitive to 25 days old CF, *O. pyriformis* has been found sensitive to 15 days old CF. Remaining six species responded irrespective of the age of culture filtrates. Out of nine sensitive taxa, only three viz., *N. linckia, S. arcuatus* and *S. aquatilis* recorded sensitive to the respective nine species. It was followed by *O. pyriformis* being sensitive to seven species. The growth of *M. aeruginosa* was inhibited through the CF of only two species and that of remaining four species through four species each. Therefore, *N. linckia, S. arcuatus* and *S. aquatilis* seem very good test organisms for any toxicity test. While studying antibiosis in cyanobacteria, Radha (13) recorded Anacystis as the best test organisms as twelve out of nineteen strains of cyanobacteria were growth inhibitory to lethal for it.

The growth inhibitory effects of the CF, however, did not vary appreciably. It was recorded minimum in 4 sets and maximum in 9 sets out of 18 sets of growth experiment as depicted in the Table 4. No definite pattern could be visualized in regards to the age of CF. However, 15 days old filtrate with maximum of 96.6% growth inhibition may be regarded more toxic than 25 days old CF with growth inhibition up to 89.52% (Table 5 and 6). It is further apparent from the

Table 6. Effect of 25 days old culture filtrate of different algae on the growth of each other

Test organisms	Percentage increase/decrease in growth over control in 15 days old culture filtrate									
	Nostoc linckia	*Anabaena oryzae*	*Microcystis aerugenosa*	*Oscillatoria willei*	*Synechocystis aquatilis*	*Oocystis pyriformis*	*Westiellopsis prolifica*	*Ankistrodesmus fractus*	*Merismopedia elegans*	*Scenedesmus arcuatus*
	1	2	3	4	5	6	7	8	9	10
Nostoc linckia	—	-7.2	-9.00	-47.74	80.18	-75.67	-27.92	-3.6	-5.4	-42.34
Anabaena oryzae	35.00	—	-7.5	-21.25	-21.25	32.5	23.75	245.00	27.5	-30.00
Microcystis aerugenosa	-3.8	28.57	—	19.04	19.04	-6.66	10.47	-89.52	16.19	13.33
Oscillatoria willei	25.33	-52.00	-22.66	—	344.0	-44.00	8.00	42.66	57.33	42.6
Synechocystis aquatilis	-45.52	61.78	-58.53	-19.51	—	-5.69	-19.51	-12.19	7.31	4.87
Oocystis pyriformis	23.07	28.2	7.69	35.89	33.33	—	61.53	153-84	148.71	148.71
Westiellopsis prolifica	14.28	65.07	90.47	26.98	22.22	46.03	—	-12.69	6.34	15.87
Ankistrodesmus fractus	73.43	145.31	135.93	64.06	59.37	54.68	51.56	—	73.43	71.87
Merismopedia elegans	12.09	24.41	2.32	6.97	11.62	9.3	-13.95	2.32	—	16.27
Scenedesmus arcuatus	-44.44	-5.55	-25.00	-52.77	-13.88	-30.55	-25.00	-44.44	22.22	—

tables under reference that the CF of all the ten species were proved growth inhibitory for variable number of test organisms but the magnitude of inhibition recorded was 75% and above only in case of *N. linckia, A. oryzae, M. aeruginosa, O. willei, S. aquatilis*, and *A. fractus*. Apparently thus, the filtrates of blue-green algae are more toxic than those of green algae. Disappearance of Chlorophycean members due to the Cyanobacterial waterblooms has been also suggested earlier by Lefevre (14).

Poisoning of water by blue green algae is virtually worldwide in occurrence. It has been frequently reported from Central North America, especially from the states of North Dacota, South Dakota, Minnesota, Iowa, Wisconsin, Illinois, Michigan and the provinces of Alberto, Saskatchewan, Manitoba and Ontario. Algal poisoning has also been reported form Russia, Morocco (15); Argentina (16), Australia (17), Israel (18), Bermuda (19), Brazil (20) and Finland (21). However, there is no information avalable on the outbreaks of algal poisoning in Indian Waters (22).

The growth inhibitory effects of the cell free CFs discussed above have been also examined on solidfied basal medium through well method. This experiment further confirmed the growth inhibitory effect of the CFs of *N. linckia, A. oryzae, M. aeruginosa* and *O. willei* on *S. arcuatus*; of *S. aquatilis, O. pyriformis* and *M. aeruginosa* on *N. linckia* and of *A. fractus* on *M. aeruginosa* by way of forming distinct growth inhibitory zone, the diameter of which ranged from 9 to 23 mm. While working on antibiosis in Cyanobacteria, Radha (13) reported the formation of

Table 7. Growth inhibition zone shown by the test organisms on the surface of solidified medium under the effect of different culture filtrates

S.N.	Test organisms	Culture filtrate	Inhibition zone (mm diam.)
1	*Nostoc linckia*	*S. aquatilis*	11-13
		O. pyriformis	10-13
2	*Microcystis aeruginosa*	*A. fractus*	9-11
		N. linckia	18-22
3	*Scenedesmus arcuatus*	*A. oryzae*	14-16
		M. aeruginosa	19-22
		O. willei	18-23

Table 8. Bioassay of Growth inhibitory factor

S.N.	Culture filtrates of alga	Age in days	Test organism	Observations
1	*Synechocystis aquatilis*	25	*Nostoc linckia*	Growth was completely checked and cells stood lysing
2	*Oocystis pyriformis*	25	*Nostoc linckia*	Symptoms observed after 3 or 4 days
3	*Ankistrodesmus fractus*	25	*Microcystis aeruginosa*	Lysing symptoms of cells appeared after 48 hrs and died within week
4	*Nostoc linckia*	15	*Scenedesmus arecuatus*	Growth was checked and the cells started lysing after 48 hrs and ultimately died within a week
5	*Anabaena oryzae*	15	–do–	–do–
6	*Microcystis aeruginosa*	15	–do–	–do–
7	*Oscilatoria willei*	15	–do–	–do–

inhibition zone through the cell free culture filtrate of seven species in the order of *Gloeotheca* > *Gloeocapsa* > *Scytonema tolypothricoides* > *Nostoc muscorum* *Anabaena oryzae* > *Tolypothrix tenuis* > *Hapalosiphon fontinales* within the range 8-26 mm in diameter. Such order of sequence amongst the seven taxa including zone of growth inhibition in the present study may be read *O. willei, M. aeruginosa N. linckia, A. oryzae, S. aquatilis, O. pyriformis, A. fractus* (Table 7).

Lyses of the population in question through the scrap of growth inhibitory zone (Table 8) confirmed that it is through the growth inhibitory compound of algal origin.

These compounds have not been identified in the present study but anatoxins (23) have been reported from *Anabaena flos-aquas*. Only one, anatoxin-a, has been chemically identified as an alkaloid-2-acetyl 9-azabicyclo (4.2.1) non 2 ene, having a molecular weight of 1959 (24) *Microcystis aeruginosa* has been reported to produce toxin of low molecular weight peptides of size ranging from 500 to 1700 daltons (25, 26). They are termed as microcystin, anatoxin c and Microcystis type c. Another source of potential environmental toxicants are the species of Schizothrix producing lipopolysaccharide endotoxins.

In general, there are not many reports on the production of toxins by marine Cyanobacteria, however, *Lyngbya majuscula* has been reported to produce two highly inflammatory agents, lingbyatoxin-A and debromoaplysiatoxin. Both compounds are potent tumor promoters and have the same activity as the well-known tumor promotor 12-O-tetradecanoyl-phorbal-13-acetate. Debromoaplysiatoxin has also been isolated from a mixture of *Schizothrix calcicola* and *Oscillatoria migroviralis* along with several related compounds, the Oscillatoxins (27-31). In the background of the reported accounts of algal toxins from different parts of the world, it is desirable to undertake further investigations on the toxic compounds contributed by the freshwater algae in the Indian habitats.

REFERENCES

1. Das HN, Prasad PK, & Verma BN, Indian Bot Soc, 73 : 1994 35.
2. Kaushik BD, Laboratory methods for blue green algae. Associated Publishing Company. New Delhi, 1987. 171.
3. Allen MB, Phycol, 14 1968 : 1.
4. Stein JR, Handbook of Phycological and Growth measurements. Cambridge university Press, 1973 pp. 448.
5. Collins CH, Lyne PM, & Grange JM. Micro biological methods. Butterworth and Co. (Publishers) Ltd. London 1989 PP. 409.
6. Flint LH, & Moreland CF. Am J Botany 33 1946 : 218.
7. Lefevre M & Jakob H, Compt Rend Acad Sc, 229 1949 : 234-236.
8. Denffer DV, Biol Zentr, 67, 1948 : 7.
9. Jorgensen G, Physiol Plantar, 9 : 1956, 712.
10. Jakob H, Rev Gen Bot, 1961 : 1.
11. Palmer CM, J Phy Col, 5 1969 : 78-82.
12. Hosmani SP & Bharati SG, Phykos, Vol. 19(1) 1980 : 23.
13. Radha S, M.Sc. Thesis, I.A.R.I., New Delhi, 1987 pp. 87.
14. Lefevre M, In "Algae and Man" Jackson DF, Plenum Press, N.Y. 1964, 337.
15. Lefevre M, Jakob H, & Nisbet M, Ann de la stat Centr d'Hydrob appl, 4 : 1952 5.
16. Muller JB, Revista de Sanidad, Asistencia social Y Trabajo, 1 1945 : 95.
17. Francis G, Nature, 18 1878 : 11.
18. Shelubsky M, Proc Intern Assoc Theoret and Appl Limnol 11, 1951 362.
19. Ingram WM, & Presscott GW, Amer Midl, Nat, 52 : 1954 75.
20. Branco SM, Algas toxical Controle das toxinas emaguas de abstic imento Revista de Departmento de Aquae Esgatos desau Paulo (Brasil) 20 : 1959, 21.
21. Hinderson R, Cornell veterinarian, 24 : 1933 1.
22. Ganpati SV, Ecology of a temple tank containing a permanent bloom of Microcystis aeruginosa. J Bombay Nat Hist Soc, 42 : 1941 65-77.
23. Carmichael WW & Gorham PR, Mitt Int Ver Limnol, 21 : 1978 285.

24. Devlin JP, Edwards OE, Gorham PR, Hunter NR, Pike RK & Stavric B, Can J Chem, 55 : 1977 1367.

25. Gorham PR & Carmichael WW. Pure Appl Chem, 52 : 1979 165.

26. Gorham PR, & Carmichael WW. Pure appl chem, 52 : 1980 165.

27. Banner AH. Hawaii Med J 19, 1959 : 35.

28. Mynderse JS, Moore RE, Kashiwage M, Norton TR. Science, 196 : 1977 538.

29. Fujiki H, Moori M, Nakayasu M, Terads M, Sugimura, T Moore, RE. Blue green algae (cited from R.E. Moore, 1981). 1980.

30. Moore RE. Carmichael, WW, ed. The Water environment. New York, Plenum Press, Algal toxins and health, 1981. pp. 15.

CHAPTER - 47

INFLUENCE OF CARBOFURAN ON THE ACTIVITY OF SOIL ENZYMES IN SUBMERGED RICE SOIL ECOSYSTEM

Z. J. Kennedy and S. Sundarav Arathan

Department of Agricultural Microbiology,
Tamil Nadu Agricultural University, Coimbatore - 641 003.

ABSTRACT

Carbofuran (2, 3 dihydro 2, 2 dimethyl, 7 benzofuranyl N-methyl carbamate) a broad spectrum, highly effective granular systemic insecticide is widely used against many insect pests of rice, sorghum, maize and several vegetable crops. It is also used for the control of nematodes in potato, banana, okra, etc. In general pesticides/ insecticides are indiscriminately used over and above the recommended dose by illiterate farmers. Hence, an attempt was made to study the effect of graded levels of carbofuran on the activity of soil enzymes which are the most accepted and valid indicators of the sum total of soil biological activity. In general, application of carbofuran at 1 and 1.5 kg a.i. ha^{-1} significantly reduced the activity of soil enzymes *viz.*, α-amylase, β-glucosidase, cellulase, urease and phosphatase upto 30 days after carbofuran application. However, application of carbofuran at the recommended level (0.5 kg a.i. ha^{-1}) had no significant effect on the activity of soil enzymes which are biologically significant as they play an important role not only in the soil chemical and biological properties but also affect the nutrient availability to plants.

1. INTRODUCTION

Insecticides are applied either through soil or foliar spray to protect the crop for sufficiently longer period. But in recent years, systemic granular insecticides are used, which are absorbed by the plants and persist for a longer peirod, so that effective control is acheived. But the persistent residues resulting from the absorption of these pesticides by the plant system and consequential residue hazards is an important factor to be considered (1,2).

When the chemicals are applied to soil some amount of residue may be left over in soil after absorption by plants and degradation by environment factors. But when insecticides are applied at higher doses, it may result in the accumulation of higher amount of residues in soil which may again affect the biological and biochemical activities in soil (3,4,5). In the present study, an attempt was made to investigate the effect of carbofuran application on the soil enzymes.

2. MATERIALS AND METHODS

The activity of six hydrolytic enzymes *viz*., amylase, invertase, cellulase, β-glucosidase, urease and phosphatase was estimated in rhizosphere soil of rice eco system at different intervals *viz*., 1, 3, 7, 15, 30 & 45 days after application of carbofuran. The carbofuran was applied at three levels *viz*., 0.5, 1.0 and 1.5 kg a.i/ha with an uninoculated control.

(a) Assay of enzymes

To a definite amount of air dried and seived soil appropriate substrate buffer and toluene were added, incubated for 24 h at room temperature (28 ± 2°C) and the product of enzyme activity was quantitatively estimated (6).

Preliminary assay of enzyme activity in the rhizophere was performed by varying the pH levels and concentrations of substrates. The most optimum pH of maximum enzyme activity and the ideal substrate concentration were used in assaying the soil enzymes.

(b) Substrates used

Sucrose, starch, carboxymethyl cellulose, arbutin, urea and disodium monophenyl phosphate were used as substrates for assaying invertase, amylase, cellulase, β-glucosidase, urease and phosphatase, respectively.

(c) Assay of amylase

The enzyme was assayed at pH 5.2 employing sodium acetate-acetic acid buffer with 8 percent soluble starch as substrate. The reducing sugars released were estimated by following Nelson's method (7).

(d) Assay of invertase

One gram of soil sample was placed in 50 ml Erlenmeyer flask and 1 ml of acetate buffer (pH 5.2), two drops of Toluene, 1 ml of 4 per cent sucrose and 3 ml of distilled water were added. The flasks were stoppered, shaken and incubated at room temperature (28 + 2°C). After 24 h, 1 ml of supernatent from each flask was withdrawn and reducing sugars were estimated (7).

(e) Assay of cellulase

Cellulase activity in rhizosphere soil was assayed by the amount of glucose released from 1 percent carboxy methyl cellulose at pH 4.8 (Phosphate buffer) by employing Nelson's Method (7).

(f) Assay of β-Glucosidase

The β- Glucosidase activity was assayed by estimating the amount of glucose liberated from one percent arbutin at pH 5.8 (*Phosphate buffer*) after 24 h of incubation (7).

(g) Assay of phosphatase

The phosphatase activity was estimated indirectly by the estimation of phenol released (8).

3. RESULTS

In the rhizosphere soil, the enzyme activity was found to decrease with increase in the levels of carbofuran application from 0.5 kg to 1.5 kg a.i./ha. The reduction in the cellulose activity was significant upto 30 days after application of carbofuran and thereafter though the activity showed a decrease, it was at par with the control. The reduction in amylase activity was noticed at all the three levels of application. The reduction in amylase activity was noticed at all the three levels of application. The reduction in the enzyme activity at 0.5 kg a.i./ha levels was only upto 15 days. Whereas, at 0.1 and 1.5 kg a.i./ha levels significant reduction in the activity was noticed upto 30 day and thereafter it was at par with control. There was no significant effect on the amylase activity at 0.5 kg level of carbofuran application (Table 1).

With the increase in the levels of carbofuran application, there was significant, decrease in the β–glucosidase activity in the rhizosphere soil of rice. The reduction was high in the initial stages after application and comparatively less at later stages of application. A significant reduction in β-glucosidase activity was observed at 1.0 and 1.5 kg a.i./ha levels. Although, there was decrease in the β-glucosidase activity at 0.5 kg a.i./ha level, it was at par with the control (Table 2). Similarly, the increase in the levels of carbofuran application reduced the invertase activity. At 1.0 kg and 1.5 kg a.i./ha levels there was significant reduction in the invertase activity upto 30 days after application. The reduction in invertase activity was noted at 0.5 kg a.i./ha level, but was not significant (Table 2).

At 0.5 kg a.i./ha level there was a slight reduction in the urease activity but at 1.0 and 1.5 kg a.i./ha levels the reduction in urease activity was markedly high. The reduction in the enzyme activity was

Table 1. Effect of different levels of carbofuran application on cellulase and amylase activity in the rhizosphere soil of rice

Levels (kg a.i./ha)	Cellulase activity (mg of glucose/g of soil) Days after application						Amylase activity (mg of glucose/g of soil) Days after application					
	1	3	7	15	30	45	1	3	7	15	30	45
0.0	0.062	0.064	0.079	0.086	0.093	0.086	0.242	0.249	0.264	0.286	0.306	0.313
0.5	0.058	0.061	0.068	0.073	0.089	0.091	0.238	0.237	0.258	0.279	0.312	0.295
1.0	0.052	0.055	0.059	0.065	0.077	0.073	0.232	0.236	0.249	0.266	0.298	0.309
1.5	0.049	0.048	0.052	0.063	0.075	0.078	0.228	0.230	0.252	0.268	0.290	0.312

	Treatments CD at 0.05%	Periods: CD at 0.05%
Cellulase	0.005	0.007
Amylase	0.007	0.009

Table 2. Effect of different levels of carbofuran application on β-glucosidase and invertase activity in the rhizosphere soil of rice

Levels (kg a.i./ha)	β-glucosidase activity (mg of glucose/g of soil) Days after application						Invertase activity (mg of glucose/g of soil) Days after application					
	1	3	7	15	30	45	1	3	7	15	30	45
0.0	0.230	0.241	0.244	0.286	0.348	0.296	0.174	0.176	0.183	0.232	0.238	0.246
0.5	0.226	0.239	0.241	0.278	0.341	0.293	0.167	0.169	0.179	0.205	0.234	0.249
1.0	0.223	0.234	0.239	0.269	0.331	0.292	0.163	0.167	0.173	0.186	0.220	0.238
1.5	0.224	0.236	0.242	0.268	0.323	0.284	0.158	0.159	0.164	0.189	0.215	0.241

	Treatments CD at 0.05%	Periods: CD at 0.05%
β-glucosidase	0.005	0.007
Invertase	0.008	0.011

Table 3. Effect of different levels of carbofuran application on urease and phosphatase activity in the rhizosphere soil of rice.

Levels (kg a.i./ ha)	Urease activity (mg of ammonia/g of soil) Days after application						Phosphatase activity (mg of Phenol/g of soil) Days after application					
	1	3	7	15	30	45	1	3	7	15	30	45
0.0	0.428	0.432	0.446	0.475	0.493	0.454	0.628	0.632	0.646	0.672	0.693	0.676
0.5	0.397	0.401	0.429	0.469	0.487	0.463	0.617	0.624	0.638	0.668	0.696	0.668
1.0	0.394	0.397	0.413	0.455	0.481	0.455	0.598	0.603	0.617	0.643	0.675	0.674
1.5	0.388	0.394	0.407	0.443	0.468	0.459	0.594	0.597	0.614	0.649	0.668	0.673

	Treatments CD at 0.05%	Periods : CD at 0.05%
Urese	0.013	0.016
Phosphatase	0.009	0.012

observed upto 30 days of application. The maximum enzyme activity was observed on 30th day at all the levels of carbofuran tested (Table 3).

In the rhizosphere soil, the phosphatase activity were found to decrease with increase in the level of carbofuran application. There was significant reduction in the phsophatase activity at 1.0 and 1.5 kg a.i./ha levels of application. The effect at 0.5 kg a.i./ha level was not significant and the enzyme activity recorded was at par with the control. The decrease in the activity was significantly pronounced upto 30 days after application. Thereafter, the decrease in the enzyme activity was not significant between the levels (Table 3).

4. DISCUSSION

The enzymatic activity in the rhizosphere region may be due to (i) microbial activity (ii) free enzymes present in soil and (iii) excreation from plant roots. Soil enzymes are often related to microbial activity and soil fertility. The information on insecticide effects on the soil enzymes is limited. In general it was found that carbofuran failed to depress the activity of amylase, invertase, urease, cellulase, β-glucosidase and phsophatase at the recommended dose. However, when excess doses of carbofuran at 1.0 at 1.5 kg a.i./ha were used, the activity of all enzymes reduced significantly. The reduction in the activity of the enzymes was noticed only at early periods after application i.e. upto 30 days. Application of carbofuran at 1.5 kg a.i./ha level, resulted in significant reduction in β-glucosidase activity (5) The activity of phosphatase was found to increase with the application of organophosphorous insecticides (9). The soil enzyme system could only be disrupted by unrealistically high doses (10). The chlorinated and organophosphorous insecticides at 5 and 10 $\mu g\ g^{-1}$ of soil inhibited the urease activity (11). Application of cartap hydrochloride inhibited the phosphatase activity (12). However most of insecticides failed to depress the amylase and invertase activity (13). Application of chloropyrifos, chlorofenvinphos, diazinon, malathion, parathion, phorate and chloradane at 5 or 10 $\mu g\ g^{-1}$ increased the amylase activity on the third day whereas application of diazion and carbofuran increased the invertase activity (14). The cellulase activity was not affected due to application of cartap hydrochloride upto 1000 ppm concentration. Under laboratory condition, organic soil incubated with carbamate at 28°C for 2 weeks revealed that carbamate insecticides failed to suppress urease activity (11).

The temporary inhibition of enzyme activity for few days may be attributed to a decline in the microbial population which recovered after 30 days of application (15). In the present study also the microbial population recovered after 30 days of carbofuran application.

REFERENCES

1. Wilcox, J & Howland, A. J Econ Ent 53 (1960) 224.
2. Var Middlem, CH & Baranowski, RM. J Econ Ent 55 (1960) 600.
3. Tewari, HK, Sidhu RS & Chahal, DS. XIIIth Annual Conf Asson Microbiologists, (1972) 4.
4. Kandasamy, D, Bhaskaran, R, Obusami, G, Krishnamoorthy KK & Subramanian, TR, Madras agric J 62 (1975) 203.
5. Purushothaman, D, Venkataraman CV & Kasirajan, C., Madras agric J 63 (1976) 515.
6. Galstyan, AS. Soviet Soil Sci 2 (1965) 170.
7. Nelson, M. J Biol Chem 153 (1944) 375.
8. Morton, RT In : Methods in Enzymology Colowick, SP eds. & Kaplan, NC eds. Academic Press inc., Publishers New York, **3** (1952) 556.
9. Prasad Reddy BV, Dhanaraj PS & Narayanan Rao VVS In : Insecticide Microbiology. Ruplal ed) Springer - Verlag, Berlin Heidelberg, New York, Tokyo (1984) 169.
10. Lethbridge Alan G, Bull & Richard G, Burns. Pestic Sci, 12 (1981) 147.
11. Tu CM. J Environ Sci Health 16 (1981) 179.
12. Endo TK, Taiki, Nobutsura T & Michihiko S. J Pestic Sci 7 (1982) 101.
13. Tu CM. Chemosphere, 11 (1982) 909.
14. Abdel Yussif RM, Zinchenko V & Gruzdev CS. Timiryazeusk Skh Akad 1 (1976) 206.
15. Purushothaman, D, Marimuthu, J & Kesavan, R. Indian J Expt Biol 12 (1974) 580.

CHAPTER - 48

MANAGEMENT AND UTILIZATION OF WASTES FROM BROCCOLI (*Brassica oleracea* var. *italica*)

Ganesh Hedawoo and R.P. Thakre

P.G. Department of Botany, Nagpur University Campus, Nagpur - 440 010, INDIA.

ABSTRACT

The present study deals with the conversion of agricultural wastes into valuable fertilizers which will be strongly beneficial to agricultural economy and environmental management. Broccoli (*Brassica oleracea* var. *italica*), a vegetable crop was used as a waste material. Seeds were harvested and whole plant (waste) was used as a substrate for composting. A simple and efficient polythene bag technique has been used for the decomposition of leafy and fleshy materials. During the process, temperature and pH of the decomposing material was noted and mesophilic/ thermophilic fungi were isolated and identified. The effect of compost on the growth of crop plants was observed and evaluated.

1. INTRODUCTION

Cole crops are the popularised and demanding major vegetable crops amongst Indians and are grown throughout the year. Harvesting of vegetables completes the life of plant and remains as agricultural/ vegetable waste. Considering the present situation, there is a prime need to convert these crop wastes into valuable high quality organic amendment or soil conditioner which strongly supports the agricultural economy and environmental management without causing any phytopathogenic risks and phytotoxicity affects.

Composting is one of the oldest solid waste treatment method for converting organic residues into the inputs for growing crops (1-2). Organic manures play a vital role in maintenance of physical and biological conditions of soil and supply of micro-macro nutrients to crops, maintenance of humic substances, improve the aeration, water holding capacity of the soil resulting in crop yields by as much as 10% (3). Crawford (4) reported that, composted farm and garden waste contain 0.4 - 3.5% nitrogen (N), 0.3-3.5% phosphorus (P_2O_5), and 0.5-2.0% potassium (K_2O). Thambirajah (5) concluded that, the most important factors i.e. substrate, micro organisms, temperature and pH affecting the composing process, are concerned with the interrelationship between the micro-organisms and the organic environment.

2. MATERIALS AND METHODS

(a) Source and characteristics of wastes

Broccoli vegetable residues used during the experiments were collected from research field, P.G. Deptt. of Botany, Nagpur University Campus. Project was planned initially for the seed production. After harvesting the threshing of seeds, remaining waste material was used for the preparation of biodung/compost.

(b) Polythene bag method

Broccoli leaves and curds collected, chopped off, placed loosely in transparent polythene bags (6-7). Two sets were prepared and kept under different conditions, one under shade and other one directly exposed to sunglight. Reduction in total solids and period required for complete degradation was noted. Details of the waste parts used are given below:

Plant parts and size (cm)	Initial weight (gm)	Polythene bag size (cm)
Leaves - 2-3 cm	500 gm	25 x 25 cm
Curds - 2-3 cm	500 gm	25 x 25 cm

(c) Pit method

Initially, the process involved the construction of pits of the size 6.5 x 4.0 x 1.5 feet and bordered by bricks. The heap was prepared by building up four inch, thick layer of plant waste + One inch thick layer of FYM + One inch, thick layer of soil. It was watered and in this way three layers were arranged in succession. Uppermost layer was supplemented with thick slurry of soil + FYM and pits were covered with thick polythene sheets to retain moisture. Broccoli crop waste was choppied off and categorised in four groups on the basis of thickness of woody (lignin) layers as under.

Pit Code	Plant & Size (inch)	Lignin Layer (mm)
A	Stem - 5-6 inch	4.55 mm ± 0.74
B	Branch/stalk - 5-6 inch	2.33 mm ± 0.47
C	Branchlets - 6-8 inch	0.61 mm ± 0.20
D	Mixed parts + entire plants	

For smooth fermentation, aeration was provided by making holes in the pits from top to bottom with the help of bamboo at 10th day. First and second unturning was done at the 20th and 40th days. We was sprinkled intermittently to maintain sufficient moisture for the process.

(d) Temperature and pH

Ambient temperature of pits was recorded after every three days at the intervals of four hours (8 AM, 1PM and 5PM). Temperature measurements was taken at a depth of 30 cm and changes. in pH was also recorded at the intervals of 7 days.

(e) Isolation and identification of Mycoflora

Samples were collected from each pit at different levels and places in three phases after 8 and 30 days and from final compost. The segment plate method described by Hutchinson and Richards (8) was employed. Czapek-Dox agar medium was used for the isolation of mesophilic fungi while YpSs Emerson agar was used for thermophiles (9). Streptomycin sulphate (25 mg/lit) was added to these media prior pouring to avoid bacterial contaminations. For the isolation of mesophilic fungi, culture plates were incubated at 25 ± 2°C and for thermophilic fungi at 45 ± 2°C. Plates were observed at the interval of two days for two weeks and the growing individual colonies were transferred to fresh sterile agar slants for further purification and identification.

(f) Utilization of biodung

Evaluation of prepared biofertilizer on two crops i.e. Barley (Monocot) and Brassica (Dicot) was studied. Effect on the growth of plants was evaluated by preparing different concentrations of biofertilizer i.e. 25%, 50%, 75% and 100%, pure soil was taken as control. 20 seeds of each crop with three replicate were sown in horticultural pots and after growth seedling height at 30th day was noted and compared with control.

3. RESULTS AND DISCUSSION

(a) Biodung formation

Degrading leaves and curds kept in direct sun and shade exhibited change in colour after 4-5 days followed by the characteristic smell of decaying linocellulosic matter. Fungal growth was prevalent after 5 days. Microbial activities was more pronounced in bags kept under shade as compared to that, exposed to sun. Gradually leaf, curd connections dissolved and it became dark brown, soft and crumpled in 10-12 days. The final product was unioformly fermented, black coloured humidified material named as biodung due to its dung like appearance (7).

Biodegradation was completed within 12 to 15 days under shade condition within 15 to 18 days under the direct sunlight. Curd degraded 2-3 days earlier than the leaves. After degradation, reduction in total solids was remarkably very high in both leaves and curds being 88-89% and 86-88% respectively. The temperature fluctuations were more pronounced in the biomass kept in direct sun as compared to those in shade. Gaur (10) viewed that, rising temperature generally accelerates the plant residue decomposition in the range of 5-30°C; but above 40-50°C its rapidity declines due to thermal kill of the organisms. In the present study also it took 3-6 days more for the process in sunlight. Curds are made up of mostly fleshy tissues with less fibres (lignin) resulted in faster degradation as compared to leaves. Reduction in total solids was reported (48-58%) in *Clerodendrum inerme* and *Leucaena leucocephala* leaves (7). Alexander (11) reported that in succulent plants the bulk of organic matter lost during decay were derived from cellulose, hemicellulose and water soluble constituents. The present study confirmed these reports as 86-89% reduction in total solids was reported in Broccoli which was attributed to its succulent nature.

(b) Compost formation

Degradation of substrates was observed first in pit C, followed by pit A and D. In pit C, it took 12-14 weeks and in pit B 14-16 weeks for complete degradation while in pits A and D 16-18 weeks, were required for the process. Gaur (10) reported that, within 8-10 weeks a good quality compost from paddy straw can be prepared with around 1.7% N. Present results were found to be directly depended on the thickness of the woody layer. In case of pit C the thickness of the layer was 0.61 mm and in case of A it was 4.55 mm thus require more days for decomposition.

(c) Temperature fluctuations

Temperature of the compost slowly increased from second day onwards and reached to its peak at the 8th to 12th day (46-49°C) and thereafter declined slowly. Temperature again raised after upturning on 20th day and subsequently declined below ambient (Fig. 1). High level of

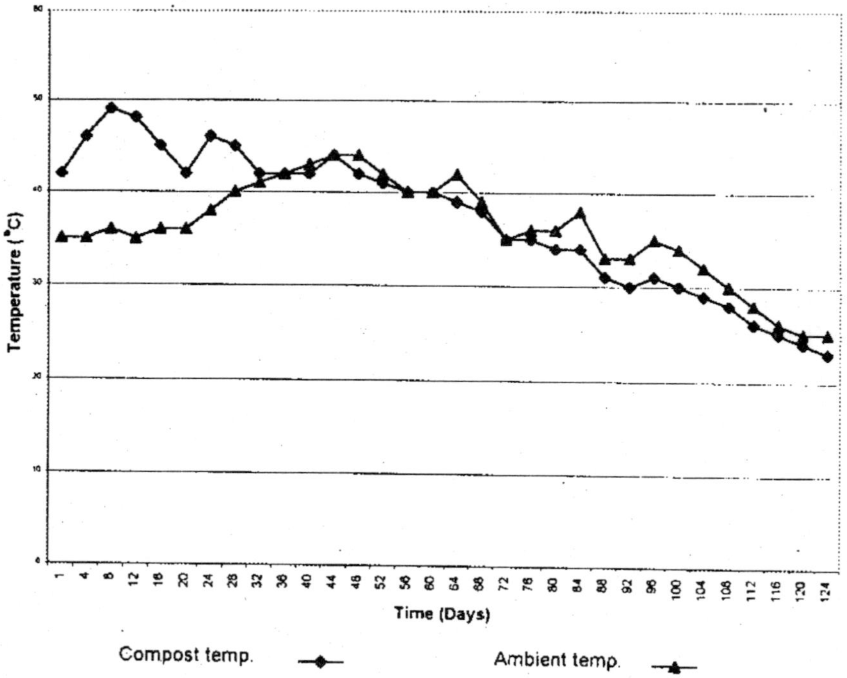

Fig. 1. Changes in temperature of composting wastes during degradation (1PM)

temperature fluctuations was also reported between 11 am to 3 pm. Fluctuation in temperature attributed to the fact that during warming up, all the very reactive compounds like sugar, starches and fats were broken down (12) and great deal of exothermic bio-oxidative reactions reported (10). The trend of change in temperature during the decomposition of biomass corroborates the observations of other workers related to the warming up and cooling down phase of composting (12-13).

(d) pH Fluctuations

Initial pH of the waste material was 5.50 and initial pH of the compost heap was generally acidic (around 5 to 6) as in the cell sap of most of the plants (14). Through the process a substantial increase in pH from 5.50 to 7.79 in the composting biomass was observed. Vallini and Antolio (15) also reported pH fluctuations in composting biomass from 4.7 to 8.6. The production of organic acids during the early stages of composting causes further acidification, but as the temperature rises pH of the compost increases to slightly alkaline in the range of 7.5 to 8.5 (12). In the present findings it appears that due to high temperature aerobic decomposition and active thermophilic microflora, there was no accumulation of acids as the intermediatory products and therefore, the pH of the biomass remained alkaline throughout the period of decomposition (Fig. 2).

(e) Mesophilic and Thermophilic mycoflora

During biodegradation 15 mesophilic fungi and 2 thermophilic fungi were isolated. Besides, white and red yellow patches of actionomycetes were also observed. Thermotolerant fungi *Algergillus fumigatus* and *A. nidulans* were most

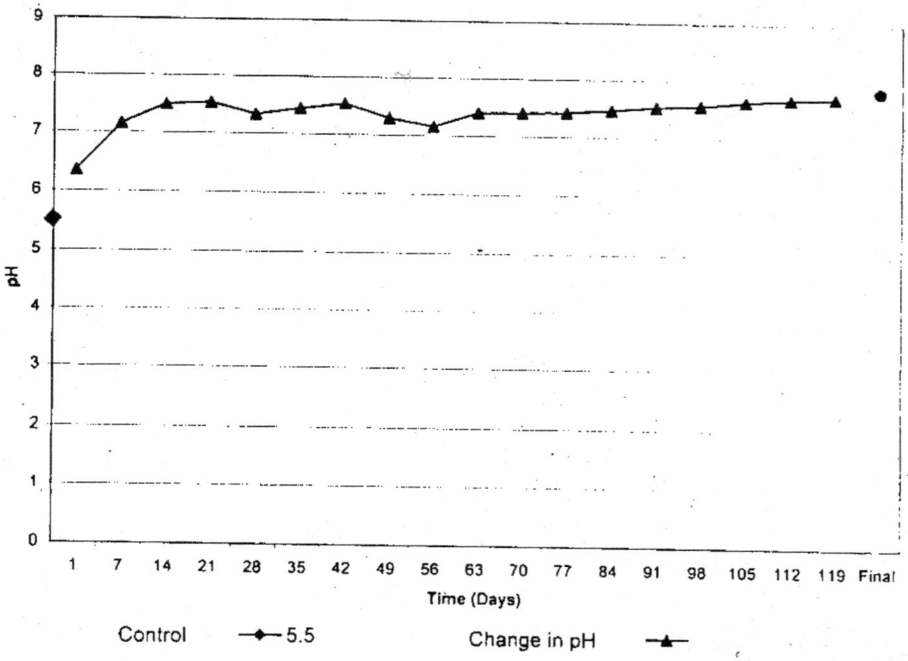

Fig. 2. Changes in pH of composting wastes during degradation (1PM)

dominant throughout the process. Some of the dominant fungi were *A. flavus, A. variecolor, Fusarium sps., Monosporium sps.*

Thermophiles observed in the samples collected after 8 days were *Torula thermophila, Mucor pusillus, Aspergillus fumigatus* and *A. nidulans*; during the period temperature range of the composting material was 46°C-49°C. Mesophiles were predominant is second and third stage (after 30 and 90 days); during the period temperature range was between 30°C-40°C. Predominent mesophiles were *Monosporium* sps., *A. niger, A. flavus, Mucorp* sps., *Fusarium* sps., and *Rhizoctonia* sps. Wood decaying fungi like *Coprinus* sps., and *Didymobotryum* sps. were also recorded.

(f) Utilization of biofertilizer

Mineralization of plant nutrients and their availability to plants is one of the major consequence of organic recycling in soil. The quantity released depends mainly on the content of a particular nutrient in organic residues undergoing decomposition. Deficiency as well as excess of any of the nutrients causes limiting or harmfull effect on the plant growth.

Effect of bulky biofertilizer was studied by taking plant height as the parameter. 100% soil was taken as control. Control soil in Barley results in 22.18 cm height. As the concentration of biofertilizer increased the seedling height also increased. 75% concentration gave the better response, resulting 30.94 cm plant hight. In Brassica, control showed 8.22 cm. height while in 75% again growth enhanced to 13.66 cm.

pH plays an important role in the vegetative growth of the plant. pH of the control showed extreme alkalinity 8.60. As the percentage of biofertilizer increased the alkalinity decreased. Higher or alkaline pH of the soil was also harmful for plant growth (yawalkar, 1981). Standard agrnomic practices recommended to be 6.5 to 7.5 pH for both the crops. In the present study better performance was reported with 75% biofertilizer.

REFERENCES

1. Biddlestone, AJ & Gray KR, The Chemical Engineer, **76** (1973) 240.
2. Gaur, AC, Neelakantan, S & Dargan, KS, ICAR, New Delhi. (1984) 159.
3. Poincolot, RP, New Haven. Bulletin (1975) 754.
4. Crawford, JH, Process Biochemistry, (1983) 14.
5. Thambirajah, JJ, Solid Waste Management and Secondary Materials. 3B : Philadelpha (1991).
6. Das, CR & Ghatnekar, SD, Indian Institute of Management, Ahemadabad (1979) 1.
7. Joshi, P & Thakre, RP, Solid Waste Management and Secondary Materials. 3B : Philadelphia. (1991).
8. Hutchinson, HB & Richards, EH, Jour Min Agri **28** (1921) 398.
9. Thakre, RP, Ph.D. Thesis, Sagar, India (1975).
10. Gaur, AC, Elsevier Science Publishers. **13** (1987) 157.
11. Alexander M, 2nd Edn. Wiley Eastern Ltd., New York. (1983) 467.
12. Dalzel, HW, (I.I.B.H.) England, (1979) 1.
13. Vimal, OP & Talashilkar, SC, Jour of Sci and Ind Res **45** (1986) 78.
14. Anonymous, Tech. Bull. 9 Sanitary Engineering, California. (1953).
15. Vallini, G & Antonio P, Biological Wastes, **29** (1989) 33.
16. Yawalkar, KS, Agrawal, JP & Bokde, S, Agri. Horticultural Publishing House. Nagpur. India (1981) 404.

CHAPTER – 49

BIOTECHNOLOGICAL IMPACT ON SULFUR METABOLIZING ENZYMES OF CYANOBACTERIUM *Anacystis nidulans*

Deepti Verma

Forensic Science Laboratory, Civil Lines, Sagar

ABSTRACT

Anacystis nidulans was able to grow with several sulfur compounds. The sulfur metabolizing enzymes, viz., ATP sulfurylase, cysteine synthase, L- and D-cysteine desulfhydrases, thiosulfate reductase and arylsulfatase were regulated by sulfur sources, particularly by sulfur amino acids and thiol containing compounds. Sulfur starvation caused reduction in ATP sulfurylase activity, while the levels of other enzymes activities were appreciated. Sulfate limitation is necessary for the development of arylsulfatase activity in the presence of its substrate p- nitro phenyl sulfate. L-cysteine repressed sulfurylase and arylsulfatase. Methionine apparently is a stronger repressor than cysteine. The data presented describe the growth of *A. nidulans* on such sulfur compounds and the influence of these compounds on enzymes of the assimilatory sulfate reduction sequence.

1. INTRODUCTION

Sulfur compounds are formed, released and degraded in nature as normal constituents of the environment in a series of reactions summarized as the sulfur cycle [1]. Besides man made sulfur compounds in the form of inorganic sulfates and sulfides the bound forms of sulfur such as catalysts, tensides organic solvents and durgs are discharged in increasing amounts to our environment [2].

Information is scanty on assimilation of these alternative sulfur compounds prevailing in nature by photosynthetic organisms. Their interactive role in regulation of enzyme systems, enabling preferential utilization of specified sulfur source is largely unknown. However, the fact is that utilizable organic and inorganic sulfur sources are numerous. Work on cyanobacterial sulfur metabolism reveals that even here wide range of sulfur sources can support growth since sulfate is the most available and assimiable form, emphasis has been given to metabolism of only this source.

Sulfate is first adenylated before being reduced and subsequently incorporated into amino acid. ATP sulfurylase (ATP : sulfate adenyl transferase; E.C. 2.77.4) the key enzyme has been purified from cyanobacteria and catalytic and regulatory properties determined [3]. The enzyme cysteine synthase (E.C. 4.2.99.4), performs the additive reactions of o-acetyl-L-serine acetate lyase and o-acetyl-L-serine sulfohydrolase. Enzyme L- cysteine desulfohydrolase is responsible for transformation of cysteine into sulfide and hence is identical to cysteine synthase though functioning in reverse direction, while enzyme D- cysteine desulfhydrase apparently has some unknown function in cyanobacteria. Sulfide, the reduced form of sulfur generated from sulfate or thiosulfate by the action of thiosulfate reductase is incorporated in cysteine. The above enzymes have been partially purified in cyanobacterium *P. uncinatum* [4]. Sulfate esters especially of phenolic and other aromatic compounds are widespread in nature occuring in plants, animals and microorganisms. Arylsulfatases are a group of enzymes catalizing the desulfonation of arylsulfates. These enzymes are found in bacteria [5] green algae [6] and higher plants [7] and characterized as Type I, Type II etc.

Although above enzymes have been detected in higher plants and green algae reports in cyanobacteria are scarce and fragmented. The planktonic cyanobacteria are often exposed to organic and anthropogenic sulfur compounds under natural habitat strikingly several planktonic algae exhibit wide choice of several compounds and take part in sulfur recycling processes. In some instances these alternative sulfur sources regulate the metabolizing enzymes and the uptake systems. Cyanobacteria exhibit remarkable diversity in accepting sulfur sources, which regulate the metabolizing enzymes to uptake, systems enabling them to take part in sulfur recycling.

In present study, we have attempted to investigate the regulatory behaviour of some sulfur compounds on various sulfur metabolizing enzymes in unicellular cyanobacterium *A. nidulans*.

2. MATERIALS AND METHODS

(a) Organism

Axenic cultures of *A. nidulans* (UTEX 625, University of Texas Culture collection) obtained from CATP- sulfurylase in Botany, Banaras Hindu University (Varanasi, India) were grown with 10 mM KNO_3 and 0.3 mM $MgSO_4$ based BG-11 medium [8]. Growth condi- tions, estimation of growth as increase in chlorophyll density and sulfur starvation/ incubation experiments were as shown previously [3].

(b) Preparation of Algal extracts

Extracts from variously grown/ incubated cells cultures were harvested by centrifugation (7,000 x g for 10 min in ultracentrifuge), washed and resuspended in 20 mM Tris-HCl buffer pH 9.0 containing 0.01 M $MgCl_2$ (breaking buffer) in a ratio of 1:5 (w:v). For breaking of cells the above suspension was sonicated. The superna- tant resulting from 12,000 rpm centrifugation was used for the assays while for ATP sulfurylase assays the extract was precipitated between 35% and 50% ammonium sulfate saturation and dialysed against 15 mM Tris/HCl pH 9.0 [4].

(c) Assay procedure for enzyme analysis

ATP-sulfurylase assay mixture contained in one ml : Tris/HCl (pH 8.5) 100 μ mol, $MgCl_2$ 10 μ mol, Na_2SO_4, 3.2 μ mol, ATP 6 μ mol, 100 μl of extract and 10 μl of

Table 1. Influence of effectors on cell free enzyme activities of sulfur metabolism in *A. nidulans*

Enzymes	Effectors	Sulfur source	Enzyme Activity (nU) at effector[1] conc. n mol/L				
			0	0.05	0.1	0.2	0.3
ATP-sulfurylase	Methionine	Sulfate	13.0	12.6	11.8	10.7	10.5
Cystein synthase	P-nitro phyenyl sulfate	Sulfate	69.0	4.5	3.9	2.5	1.8
	Methionine	Sulfate	69.0	63.4	70.1	68.7	69.2
L-Cysteine desulfhydrase	Cysteine	Sulfate	5.0	2.8	1.5	0.9	0.5
	Methionine	Sulfate	27.5	27.4	25.2	28.6	28.8
Thiosulfate reductase	Magnesium sulfate	Thiosulfate	3.3	2.2	1.7	1.2	0.6
	Cysteine	Sulfate	6.9	7.2	8.3	9.6	12.2
Arylsulfatase	Magnesium sulfate	P-nitro phyenyl sulfate	342.2	27.3	26.4	24.5	20.9
	Cysteine	Sulfate	30.6	42.6	51.2	65.6	76.8

[1]Effectors were added 10 min prior to assays.

enzyme pyrophosphatase. After 40 minutes at 37°C, reaction was terminated by adding 0.1 ml each of 1% ascorbic acid and 1% ammonium molybdate in 0.05 N H_2SO_4.

Liberation of inorganic phosphate was monitored colorimetrically at 690 nm using appropriate standard curve [4]. Production of cysteine by ninhydrin reaction in presence of DTE was measured coloimetrically for cysteine synthase activity [9]. The activities of L-and D-cysteine desulfhydrases were measured by liberation of H_2S using methylene blue method of Siegel [10]. This sulfate reductase activity was determined as sulfide production from thiosulfate with DTE as electron donor [11]. The activity of arylsulfatase was determined using P-nitro phenyl sulfate as a substrate. The assay mixture contained 50 mM Tris HCl pH 8.0, 2 mM substrate, enzyme 100 μl in a total volume of 0.5 ml. After 45 minutes at 37°C the reaction was stopped using 0.5 ml of 0.5 M NaOH. The desulfonated product was determined at 401 nm. One unit activity (1U), represents 1 mol substrates consumed/product formed min^{-1} mg^{-1} protein. The effectors were added 10 min prior to assays, in cell free extracts of modified cultures.

Chemicals used were of analytical grades from commercial sources.

3. RESULTS

A. nidulans culture grown in normal sulfate medium shows growth rate of 1 μg ml^{-1} day^{-1} chlorophyll and 2 μg ml^{-1} day^{-1} protein. In reduced glutathione isoleucine, urea and P-nitro phenyl sulfate the growth was good, while in taurine and sodium thiosulfate it was moderate. Cysteine and methionine supplemented cultures shows poor growth.

Upon starvation the initially ATP-sulfurylase activity was reduced afterwards it showed a remarkable increase, prolonged starvation causes decrease in the activity. The cysteine synthase activity increased initially with starvation but after 24 hours it showed gradual decrease. L-cysteine desulfhydrase activity increased upto 18 hours and thereafter a marginal decrease was observed. D-cysteine desulfhydrase activity showed a zig-zag pattern. Thiosulfate reductase activity increased gradually with time but prolonged starvation caused decrease in the activity. As the starvation period increased the activity of arylsulfatase increased remarkably.

The concentration of different sulfur sources were 0.3 mM and sulfate grown culture served as control. Sodium thiosulfate enhanced the activity of cysteine synthase, thiosulfate reductase and arylsulfatase but slightly affected the rest. P-nitro phenyl sulfate enhanced the activities of arylsulfatase and thiosulfate reductase. It strongly reduced the activity of cysteine synthase with marginal reduction of cysteine desulfhydrases and ATP- sulfurylase, cysteine synthase, and L-cysteine desulfhydrase and D-cysteine desulfhydrase was unaffected but the activities of thiosulfate reductase and arylsulfatase increased. Culture having methionine showed inhibition in the activities of ATP-sulfurylase and cysteine synthase but it did not show any effect on other activities. Cysteine synthase activity was not detectable in cultures grown with reduced glutathione and taurine. Lysine inhibited the activity of ATP-sulfurylase but has very high D- cysteine desulfhydrase activity. Isoleucine has highest L- cysteine desulfhydrase activity as compared to other sources. Threonine has higher activity of ATP-sulfurylase and lesser activities of thiosulfate reductase.

When sulfate, thiosulfate and P-nitro phenyl sulfate were added in combination the results showed marked changes. Sulfate and thiosulfate in combination decreased, the activities of enzymes except D-cysteine desulfhydrase. Sulfate and P-nitro phenyl sulfate combination showed drastic decrease in thiosulfate reductase and to some extent on arylsulfatase activity. While sodium thiosulfate and P-nitro phenyl sulfate combination had higher activities of ATP-sulfurylase, D-cysteine desulfhydrase and arylsulfatase.

Sodium nitrite and urea were replaced by sodium nitrate which served as nitrogen source. In both the cultures cysteine synthase was not detectable although other activities were normal. Urea served as good nitrogen source for this cyanobacterium.

Influence of effectors on cell free enzyme activities were observed by adding effectors 10 minutes prior to assays. For ATP- sulfurylase methionine was used as effector which repress the activity. In presence of effectors, P-nitro phenyl sulfate and methionine cysteine synthase activity were checked. P-nitro phenyl sulfate showed drastic reduction in activity with concentration while methionine did not show any remarkable change. L-cysteine desulfhydrase activity decreased with increased concentration of the effector cysteine. A slight change was observed in the activity of L-cysteine desulfhydrase when methionine was used as effector. Thiosulfate reductase activity decreased as magnesium sulfate concentration increased but with cysteine the activity increased with concentration. For arylsulfatase activity magnesium sulfate and cysteine served as effector and the activity decreased drastically as concentration of first one increased A slight increase in activity was observed with cysteine.

4. DISCUSSION

The experiments on sulfar starvation in *A. nidulans* apparently leads to an increase sulfate uptake signal. This uptake system is known to be controlled by the internal sulfate pool [12, 13]. Upon sulfate starvation the enhanced enzyme activities of ATP-sulfurylase, cysteine synthase and thiosulfate reductase can be explained as the consequence of limited availability of sulfur for growth, whereas prolonged sulfur

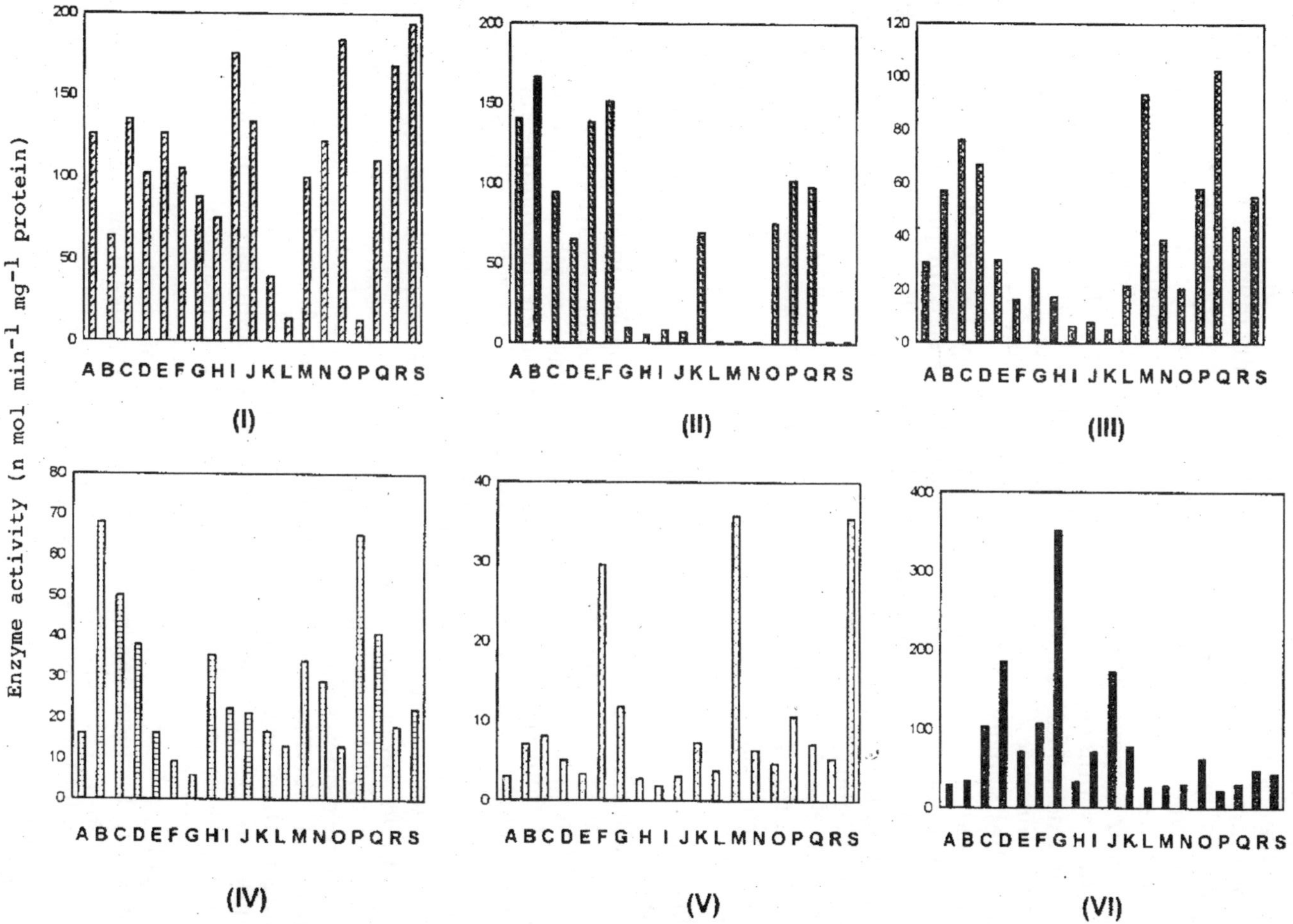

Fig. 1. Effect of sulfur compounds on enzymatic activities of sulfur metabolism in *A. nidulans*. In BG-11 medium sulfur sources were added in concentration of 0.3 m Mol/L replacing sulfate : A - sulfate free medium (0 hr), B-12 hrs, C-18hrs, D-36 hrs, E- magnesium sulfate (Control), F-sodium thiosulfate, G-P-nitro phenyl sulfate, H - Sulfate + Sodium thiosulfate, I - Sulfate + P-nitro phenyl sulfate, J - sodium thiosulfate + P-nitro phenyl sulfate, K- cysteine, L - methionine, M- reduced glutathione, N- taurine, O- threonine, P - lysine, Q - isoleucine, R - sodium nitrite, S - urea. Parameters tested were : (I) ATP - sulfurylase, (II) cysteine synthase, (III) L-cysteine desulfhydrase, (IV) D-cysteine desulfhydrase, (V) thiosulfate reductase, (VI) arylsulfatase.

starvation can be recognised by subsequent decrease in the activity of sulfate activation enzymes i.e. ATP-sulfurylase [14]. Sulfur starvation either deprivation or poor source is characterized by low to diminished growth and high specific activity of enzymes used for sulfur metabolism. The enzymes can be distinguished according to their response to sulfur starvation. Cysteine synthase, cysteine desulfhydrases and thiosulfate reductase always showed an increase in specific activity, whereas the activity of ATP-sulfurylase decreased under severe starvation. This permits differentiation between sulfur starvation or depletion of a usable substrate and limited growth due to slow metabolic conversion. Sulfate limitation is necessary for the development of arylsulfatase activity in the presence of its substrate P-nitro phenyl sulfate similar to *Chlamydomonas* [6].

ATP-sulfurylase and other key enzymes are generally partially repressed by a readily assimilable sulfur source such as sulfate. This is because cysteine and methionine are the intermediate metabolites that are produced from sulfate. Both amino acids are known to repress primarily ATP-sulfurylase and also some other enzymes viz. cysteine synthase. In given cyanobacterium these amino acids caused substantial reduction in various enzyme activities, except cysteine which appreciated thiosulfate reductase activity. Since the impact of methionine was reasonably stronger than cysteine, we predict methionine to be the putative repressor. However, cysteine to methionine conversion is uninterrupted and hence cysteine-effects could, at least partly, be due to methionine. In bacteria however cysteine is the regulator [15,16]. With cysteine but not methionine L-cysteine desulfhydrase and thiosulfate reductase activity showed concentration dependent enhancement. Since this effect is also found upon in vitro incubation, we proposed that the enzyme are activated. Therefore, cysteine emerged to be focal regulator in assimilation of reduced sulfur compound in *A. nidulans* and is mechanistically different from methionine. Thus, sulfate assimilation is subjected to positive control by nitrogen metabolism and negative control by cysteine and methionine [17]. The activities of cysteine synthase, cysteine desulfhydrases and thiosulfate reductases behave like higher plants in the presence of cysteine or methionine [18].

Release of inorganic sulfide from cysteine suggested the presence of cysteine desulfhydrases in higher plants [19]. Cyanobacteria contain a D-cysteine specific desulfhydrase which converts D-cysteine to sulfide, pyruvate and ammonia, whereas L-cysteine degradation is a side reaction of cysteine synthase activity. *A. nidulans* grew well on thiosulfate as only sulfur source. Therefore, mechanism for thiosulfate cleavage is present in this cyanobacterium. In general thiosulfate reductase activity seems to be higher under sulfur limitation and has regulation behaviour analogues to that of cysteine synthase. However with respect to cysteine/methionine regulation thiosulfate reductase differs. These effectors, on one hand represes cysteine synthase and on the other hand activates the thiosulfate reductase activity [11].

The data presented here demonstrate that arylsulfatase activity is present in cyanobacterium *A. nidulans*, however, only under specialized growth conditions. The observed data summarized in Fig. 1 allow to characterize this arylsulfatase according to Dogson and Spencer as an arylsulfatase type I, since this activity was not inhibited by sulfate and it use P-Nitro phenyl sulfate as only sulfur source. The arylsulfatase activity was induced by sulfate starvation. The presence of the substrate seems to be necessary for higher activity of the enzyme [20].

ACKNOWLEDGEMENTS

The author is grateful to the Head, Department of Biological Sciences, R.D. University for providing research facilities and to Council of Scientific and Industrial Research for financial help.

REFERENCES

1. Zehnder, AJB & Zinder S H, The Natural environment and the Biogeochemical Cycles. Berlin : Springer (1980) 105.
2. Kellog, WW, Cadle R D, Allen E R, Lazarus A L, & Martell E A. The sulfur cycle. Sciences, 175 (1972) 587.
3. Mishra, D & Schmidt A, Z. Naturforsch, 47C (1992) 95.
4. Bagchi, D & Verma D, Indian J Exp Biol, 35 (1997) 876.
5. Dogson, KS, White G F, & Fitzgerald J W, Sulfatases of Microbial Origin Vol I, CRC Press, Inc. Boca Raton, Florida (1982) 1.
6. Lien, T & Schveiner O, Biochim, Biophys Acta, 384 (1975) 168.
7. Farooqui, AA & A M M Samiullah, Planta, 133 (1977) 157.
8. Rippka, R Demalle J, Waterbury J, Herchman M, & Stainer R Y, J Gen Microbiol, 111 (1979) 1.
9. Schmidt, A, Meth Plant Biochem, 3 (1990) 349.
10. Siegel, LM, Anal Biochem, 11 (1965) 126.
11. Schmidt, A, Erdle I, & Gamon B, Planta, 162 (1984) 243.
12. Krauss, KF & Schmidt A, J Gen Microbiol, 133 (1987) 1209.
13. Diessner, W & Schmidt A, Z Pflanzenphysiol, 102 (1981) 57.
14. Smith, IK, Plant Physiol, 66 (1980) 877.
15. Hulanicka, MD, Hallquist S G, Kredich N M, & Mojica A T, J Bacteriol, 140 (1979) 141.
16. Jones, MC, Mortimer, Wheldrake J F & Pasternak C P, Biochem J 107 (1968) 51.
17. Zink, M W, Can J Bot 62 (1984) 2107.
18. Schmidt, A, Arch Microbiol, 127 (1980) 259.
19. Harrington, H M & Smith I K, Plant Cell Physiol, 65 (1980) 151.
20. Muller, S & Schmidt A, Z Naturforsh, 41C (1986) 820.

CHAPTER – 50

A STUDY OF PHYTOREMEDIATION OF SEWAGE WATER BY SEEDS OF *Strycnos potatorum* L.f. AND GUM OF *Sterculia urens* ROXB.

Ravi Upadhyay

Dept. of Botany, Govt. Degree College, WAIDHAN - 486 886 (M.P.)

ABSTRACT

Plant parts and products sometimes provide an ideal habitat for growth of different kinds of microorganisms, chiefly bacteria and fungi. These microbes are very effective in remediation of contaminated waste waters. Seeds of *Strycnos potatorum* and gum of *Sterculia urens* were used in this experiment to study their effect on sewage water treatment. Sewage water was mixed with powdered seeds and gum, and after incubating it for 24, 48, and 72 hours, the turbidity, colour, odour, pH, total dissolved solids, total suspended solids and dissolved oxygen were studied.

1. INTRODUCTION

In densly populated areas disposal of sewage and availability of drinking water are becoming major problems. Raw sewage contains 90-95% water and 5% of inorganic and organic impurities in the form of garbage, detergents, chemicals, dyes, semidigested food particles, several microbes etc. Sewage effluents from cities dotting the river banks contribute to nearly three fourths of the pollution. Mixing of sewage into rivers leads to various types of diseases and epidemics, hence treatment of sewage water, before it pours into the river is very essential. Several methods of sewage treatment are already in operation and one of them is bioremediation (cleaning by using microorganisms). In the present study, seeds of *Strycnos potatorum* L.f and gum of *Sterculia urens* Roxb, were used to study their effect on phytoremediation of sewage water.

2. METHODOLOGY

Sewage water was collected from a domestic sewage, early in the morning and kept in different jars, marked as 'CO', 'A', 'B', 'C', and 'D'. The colour, odour pH, TDS,

TSS and DO of the sewage water was noted before pouring it into the glass jars. The sewage water in the jar marked 'CO' was taken as control and hence was left as such, untreated. In Jar 'A' one gram of Strycnos seeds per litre of water was added. Similarly, in 'jar' 'B' one gram/litre of water, Sterculia gum was added and in the third jar 'C' one gram/litre of water, powdered strycnos seeds were added. Finally in the fourth jar 'D' a mixture of the powdered seeds and powdered gum in a measure of one gm/litre of the water was added.

These jars were kept open in the laboratory for four days. The change in the colour, and odour of the sewage water, its pH, Total dissolved, solids, Total suspended solids, and dissolved oxygen in it in all the jars, were recorded at different intervals of 24 hours, 48 hours, and 72 hours. After 80 hours, the coliform bacterial presence was also tested by culturing the water in McConkey's Agar medium. This was done by taking two loop full of water drops, from jars, marked 'CO' containing untreated sewage water which was dark in colour and from other marked 'D' separately and number of colonies appeared after 24 hours, were counted.

Determination of colour, odour, pH, TDS, TS, TSS, and DO was done as follows:

Colour —

Colour was visually observed.

Odour —

Odour was determined normally.

Potentia hydrogeni (pH) —

pH was recorded by using pH paper-strips and Elico pH meter.

Total Dissolved Solids (TDS) —

100 ml of the sample was fillered through Watmann filter paper No. 42. The filtrate was taken in a pre-weighed (W_1) beaker and was kept in an oven adjusted at 105°C, overnight for evaporation. On the completion of evaporation the beaker was weighed again (W_2).

$$\text{TDS in mg/l} = \frac{(W_2 \;\; W_1) \times 1000}{\text{ml sample}}$$

Total Solid (TS) —

Total solids were determined as the residue left after evaporation of unfiltered sample.

Total Suspended Solids (TSS) —

The difference between the Total solids (TS) and the Total dissolved solids (TDS) is considered as the total suspended solids (TSS).

$$\text{TSS} = \text{TS} - \text{TDS in mg/1.}$$

Dissolved Oxygen (DO) —

DO was determined by the Winkler's Azide modification method. 100 ml bottle was filled with water sample and 2ml of each Winkler's 'A' and Winkler's 'B' agents were added to it. A brown precipitate was formed, which was allowed to settle for 10 minutes. The precipitate was then dissolved in acid. The supernatent was treated against hypo with starch as the indicator and the disappearance of the blue colour was noted as the end point.

Calculation —

1 ml of Hypo = 0.1 mg of DO.

$$\text{DO in mg/1} = \frac{0.1 \text{ x ml. titrant x } 1000}{100}$$

3. RESULTS

Initial colour of sewage water was dark brown and nearly opaque. It emmitted foul odour and contained very less dissolved

Table 1. Determination of colour, odour, pH, TDS, TS, TSS, DO and Results recorded after 24, 48 and 72 hours.

Time in hours	Jar Code	Odour	Colour of water	pH	T.S.	TDS.	TSS.	DO.
0 hrs	CO	Foul	Darkbrown	6.8	864	262	602	2.0
	A	+++	Darkbrown	6.8	864	262	602	2.0
	B	+++	Darkbrown	6.8	864	262	602	2.0
	C	+++	Darkbrown	6.8	864	262	602	2.0
	D	+++	Darkbrown	6.8	864	262	602	2.0
24 hrs	CO	Foul	Darkbrown	7.2	840	250	590	2.4
	A	++	Lightbrown	7.4	482	202	280	2.6
	B	++	Lightbrown	7.4	450	195	255	2.6
	C	++	Lightbrown	7.2	460	200	260	2.8
	D	++	Lightbrown	7.5	382	142	240	4.3
48 hrs	CO	Foul	Darkbrown	7.4	820	270	550	2.8
	A	++	Lightbrown	7.4	314	182	132	2.8
	B	+	Lightbrown	7.3	297	157	140	3.2
	C	+	Lightbrown	7.3	340	136	184	3.6
	D	+	Straw	7.6	220	100	120	5.8
72 hrs	CO	Foul	Darkbrown	7.4	760	242	518	3.2
	A	+	Lightbrown	7.3	291	131	160	3.8
	B	+	Lightbrown	7.3	272	130	142	4.0
	C	+	Lightbrown	7.4	290	122	168	3.6
	D	—	Straw	7.6	186	92	94	6.2

Abbreviations

Jar CO - Control containing untreated sewage water
Jar 'A' - Containing sewage water treated with seeds.
Jar 'B' - Containing sewage water treated with gum.
Jar 'C' - Containing sewage water treated with seed powder.
Jar 'D' - Containing sewage water treated with mixture of seed powder and gum.
T.S. - Total solids, TDS - Total dissolved solids, TSS - Total Suspended solids, DO - Dissolved oxygen.
+++ - Maximum foul odour, ++ - Decreased foul odour,
— - Foul odour absent.

oxygen. There was a gradual change in water colour in all the jars from 'A' to 'D', while the water colour and its odour in jar 'CO' remained same without much change. The results are given in the Table 1.

After 80 hours of incubation, water in jar 'D' became almost clear with all its impurities settled at the bottom, while in control the colour remained dark brown emitting foul odour.

4. DISCUSSION

The goal of the biological treatment of sewage water is to employ microorganisms to completely oxidize organic components, to carbon dioxide and water. The observations indicate increased 'DO' in all the five samples but it was significantly increased in jar 'D' and also the dark colour and the foul odour of the sample in this jar was remarkably minimised. There was rapid particulate settlement during first 24 hours, in all the jars though it was maximum in jar 'D', which also showed the peculiarity of significant reduction of blue gree algae like *MIcrocystis* and *Gliocapsa* while their presence was more in all the other samples.

The presence of gum and *Strycnos potatorum* seed powder led to rapid floe and particle settlement in the D, which caused decrease in water turbidity leading to greater penetration of sun rays into it, due to which the anaerobic bacteria were killed and the turbidity further decreased along with the diffusion of oxygen in the water. Growth of some algae like *Chlorella* was observed which contributed to the increase in oxygen content in the water. The aerobic bacteria which developed and formed floccus at the bottom in the jars, might have been responsible for the degradation of the organic compounds which comprised the major portion of the suspended particulate matter of domestic sewage. This degradation was very significant in jar 'D' i.e., from 602 mg/1 to 94 mg/l.

After 80 hours, when the two samples, 'CO' and 'D' were cultured, the number of bacterial colonies in 'CO' was 36, while that in 'D' was 3, in which two were of *E. coli* and one of some other bacteria. This too is a significant development.

The above findings clearly indicate that these plant products may have greater utility if used in solar ponds and lagoons for rapid phytoremediation of sewage water before opening it into the rivers. This water may further be cleansed and filtered through charcoal filters and be used for driking purposes after chlorination. Phytoremediation, hence, may prove greatly advantageous in minimising the water pollution and lead to the increased availability of drinking water.

REFERENCES

1. Gainey, PL and TH Lord "Microbiology of Water and Sewage, Prentice Hall, Inc. Englewood Cliffs. N.J. (1952)
2. Lead better, ER, and JW Foster Bacterial, Proc. (1959) 118.
3. Foster, JW In the The Oxygenases, O. Hayaishi (ed.) Academic Prest. Inc. New York. (1962) Pp. 241-271.
4. LE Casida Jr, Industrial Microbiology, Willey Eastern Ltd, New Delhi. (1968)

CHAPTER - 51

BIODEGRADATION OF KERATIN BY KERATINOLYTIC FUNGI

Sanjana Kaul and Geeta Sumbali

Department of Biosciences, University of Jammu, Jammu - 180 006, J&K, INDIA.

ABSTRACT

Keratinolytic fungi isolated from poultry farm soils were grown on two different media containing keratin. They were found to degrade keratin releasing sulphydryl containing compounds detected as cysteine, extracellular keratinases and total proteins. Glucose supplemented medium supported maximum keratinolysis. Keratinase activity also changed the pH of the medium towards alkalinity which was more in species having strong keratinolytic activity. Different keratin substrates were used to check the preferential utilization of keratin by these fungi. Morphology of degraded human hair by some keratinolytic fungi under light microscope showed distinctly a gradual digestion of a large part of the substrate.

1. INTRODUCTION

Keratinolytic fungi are a group of highly specialized fungi which possess the ability to degrade hard keratin and utilize it as a source of protein. These fungi produce degradative enzymes by which they are capable of clearing complex keratinous substrates in nature, and, thus, are responsible for the biodegradation of keratinized structures (1). This micro-biological cycling of keratinous substrates further enhanced the interest of soil microbiologists for their detailed analysis from soils as soil provides most conducive habitat for the growth and multiplication of these fungi. Infact soil has been considered as a natural reservoir and a link that relates different ecological groups together. In general, the ecological habitats mapped out and reported to be rich in keratinophilic fungi are the soils from densely populated areas, public gardens, play grounds, poultry farms, cattle farms and other places inhabited by animals and decomposed organic matter.

2. MATERIALS AND METHODS

(a) Isolation of keratinolytic fungi from soil

One hundred soil samples were collected from ten different poultry farms of Jammu district by scraping a layer of soil not exceeding 5 cm in depth and were brought to the laboratory in presterilized polythene bags. The soil samples were screened for the keratinolytic fungi by employing "Keratin-bait-technique", of Vanbreusegham (2).

(b) Keratinolysis by poultry farm soil fungi

Two media (Medium I and Medium II) were prepared for each fungal species.

Medium I : $K_2 HPO_4$ - 1.5 g, $MgSO_4 . 7H_2 0$-0.025 g, $CaCl_2$ - 0.025 g, $FeSO_4.7H_2 0$-0.015 g, $ZnSO_4.7H_2 0$ - 0.005 g, distilled water 1000 ml, pH 7.0.

Medium II : Glucose-0.9 g, $MgSO_4.7H_2 0$-0.6 g, thiamine-0.01 g, pyridoxine-0.01 g, inositol-0.05 g, 0.028 M phosphate buffer 1000 ml, pH 7.8.

Erlenmeyer flasks containing 50 ml of sterilized basal medium supplemented with 500 mg of pre sterilized human hair as a keratin source were inoculated and incubated at 30°C for 40 days in static condition. At the end of the growth period, the culture filtrate was centrifuged at 5000 rpm for five minutes and the supernatant was assayed for change in alkalinity, release of cysteine, protein and keratinase. Estimation of cysteine was done by the method of Ramakrishna et. al. (3). Total proteins were estimated following the method of Lowry *et al.* (4) and extracellular keratinase was measured as per the method of Yu *et al.* (5).

(c) Preferential utilization of keratin

Petridishes were half filled with soil and sterilized in an oven at 180°C for 4 - 5 days. The soil was then moistened with sterile water and inoculum of the test fungus was added. Different presterilized baits were kept in these petridishes and incubated at 30°C. Preferential utilization and the degree of colonization of different baits shown by various keratinophilic fungi was recorded after an incubation period of 30 days.

3. RESULTS AND DISCUSSION

Ten different poultry farms of Jammu district yielded a total of 21 keratinophilic isolates representing 20 species belonging to 9 genera (Table 1).

All the poultry soil samples were found to be positive indicating thereby, that poultry farm soils from a favourable habitat for keratinophilic fungi. Two genera viz. *Chrysosporium* and *Aspergillus* were most frequent, being recovered from the soils of all the poultry farms and form the bulk of the isolated kerationophiles. *Malbranchea flava* has not been reported so far from India and therefore, constitute new records as well as additions to the existing Indian mycoflora.

All the test fungi could grow on human hair and degrade it, releasing sulphydryl containing compounds detected as cysteine, keratinase and total proteins (Table 2). In fact the evidence of keratinolysis lies on the ability of fungi to release soluble sulphur containing aminoacids and polypeptides into the medium in quantities significantly greater than those released by controls (6). Comparison of the data obtained from two different culture media showed that the glucose supplemented medium supported maximum release of cysteine, protein and keratinase indicating a stimulatory effect of glucose in inducing

Table 1 : Prevalence of Keratinophilic fungi in different Poultary farm solis of Jammu

No. of soil samples examined	100 (10 samples from each poultry farms)										
No. of soil samples positive	100										
Occurrence (%)	100.0										
Fungi Recorded	Number of Soil Samples Positive										
	1	2	3	4	5	6	7	8	9	10	Occurrence %
Chrysosporium kerationophilum	6	7	6	–	5	–	–	–	10	42	42
C. queenslandicum	–	–	9	–	–	8	5	–	–	7	29
C. tropicum Strain I	7	9	5	–	7	10	6	5	5	–	54
C. tropicum Strain II	–	3	7	5	–	–	–	4	–	–	19
C. pannorum	–	8	–	–	–	–	3	4	3	–	18
Malbranchea flava	–	–	6	10	7	–	–	–	10	–	33
Scopulariopsis brevicaulis	–	–	5	7	6	–	7	–	6	4	35
Microascus manganii	4	–	–	–	3	5	–	–	–	–	12
Gliocladium virens	6	–	–	5	–	7	8	–	8	4	38
Circinella muscae	5	–	2	–	4	–	–	4	–	–	15
Fusarium moniliforms	–	–	–	3	–	4	–	2	6	–	15
F. solani	–	–	–	–	3	–	5	–	–	3	11
Aspergillus candidus	6	6	–	5	7	4	–	–	5	–	33
A. deflectus	–	–	5	4	–	–	5	3	5	–	22
A. flavus	–	–	–	5	6	6	–	–	5	6	28
A. orchraceous	6	–	–	–	–	–	–	4	6	3	19
A. sydowii	–	–	–	5	–	–	5	–	4	–	14
A. versicolor	4	3	4	5	4	5	3	–	–	–	28
Penicillium aurantiogriseum	–	5	4	–	–	–	–	4	–	–	13
P. citrinum	5	5	7	3	3	–	5	–	4	6	38
P. griescofulvum	–	4	–	4	4	–	–	5	–	–	17
Fungal species recorded from each poultry farm	8	8	10	12	12	8	10	8	13	8	
Total No. of fungal species	20										

1. Machlian, 2. Prem Nagar, 3. Nihalpur Simbal, 4. Kothi Shahdaula, 5. Akhnoor, 6. Belicharana, 7. Muthi, 8. Maralia, 9. Maralia, 10. Ranbir Singh Pura.

maximum enzyme production (7-10).

Maximum cysteine was released in the glucose supplemented medium by *Chrysosporium tropicum* (28 μg/ml), followed by *C. keratinophilum* (24 μg/ml) *C. pannorum*, *C. queenslandicum* and *Malbranchea flava* (21 μg/ml). Detection of these amounts of cysteine in the culture media indicated their keratinolytic ability and also supported the activity of sulfitolysis during keratinolysis. Sulfitolysis involves the reduction of disulphide linkages of cysteine which are largely responsible for providing stability to the keratin molecule.

The degradation of hair was also monitored by determining the net protein released into the media. The test fungi released variable amounts of total protein, substantially more in keratin medium II than in medium I (Table 2). Maximum release of protein at the end of experimental period was by *Scopulariopsis brevicaulis* (65 µg/ml), followed by *Chrysosporium pannorum* (56 µg/ml). Similar values of protein released after prolonged incubation have also been reported by previous researchers (11-13). Proteolytic activity and sulphytolysis are simultaneous and perhaps complementary phenomena, allowing these fungi to transform keratin into a nutritive material.

Net keratinolytic activity was measured in keratinase units (KU). Results clearly indicate that all the test fungi secreted significant amount of keratinase (Table 2) which is known to be responsible for hydrolysis of denatured keratin. Higher keratinase activity was recorded in medium II (supplemented with glucose), where maximum release was elaborated by *Chrysosporium keratinophilum* (12.5 KU/ml)

Table 2. Keratinolytic activity of fungi (Soil isolates) on hair after 40 days of incubation.

Fungal Species	Change in pH		Cysteine µg/ml		Protein µg/ml		Keratinase KU	
	I*	II**	I	II	I	II	I	II
Chrysosporium keratinophilum	8.8	2	12.0	24.0	24.0	53	5.5	12.5
C. queenslandicum	8.1	0	15.0	21.0	14.0	45	6.2	9.5
C. tropicum Strain I	8.6	9	14.0	28.0	20.0	55	6.0	11.8
C. tropicum Strain II	8.0	6	13.0	20.0	18.0	40	7.0	10.0
C. pannorum	8.1	4	12.0	21.0	28.0	56	7.1	10.8
Malbranchea flava	8.6	0	14.0	21.0	29.0	40	5.8	12.0
Scopulariopsis brevicaulis	8.1	9	15.0	18.0	25.0	65	4.1	10.0
Microascus manganii	8.6	6	13.0	15.0	21.0	55	4.0	9.0
Gliocladium virens	7.9	1	15.0	12.0	17.0	40	6.5	7.0
Circinella muscae	7.2	9	8.0	12.0	12.0	30	2.5	5.0
Fusarium moniliforms	7.4	0	9.0	11.0	15.0	21	5.4	11.4
F. solani	7.2	3	8.0	10.0	12.0	48	6.1	10.8
Aspergillus ochraceus	7.8	5	10.0	11.0	21.0	45	2.0	8.9
A. flavus	7.9	5	11.0	10.0	20.0	55	3.0	7.9
A. sydowii	7.4	2	10.0	10.0	18.0	21	2.8	6.6
A. deflectus	7.2	4	10.0	12.0	17.0	19	3.8	5.5
A. versicolor	7.1	1	12.0	11.0	16.0	35	3.0	6.0
A. candidus	7.4	8	11.0	10.0	14.0	24	3.8	5.8
Penicillium aurantiogriseum	7.5	1	10.0	9.0	18.0	26	4.5	6.2
P. aurantiogracum	7.4	4	9.0	10.0	12.0	31	4.0	6.0
P. citrinum	7.3	2	8.0	10.0	10.0	30	2.8	9.9

Initial pH of Medium I = 7.0; II - 7.8

* Culture medium I - Contained (g/1 distilled water) :
K_2HPO_4 – 1.5g; $MgSO_4.7H_2O$ – 0.025g; $CaCl_2$ – 0.025g; $FeSO_4.7H_2O$ – 0.015g; $ZnSO_4$: $7H_2O$ – 0.005 g, distilled water 100 ml, pH 7.0.

** Culture medium II – Contained (g/1 0.028M phosphate buffer);
Glucose – 0.9g; $MgSO_4$. $7H_2O$ – 0.6g; Thiamine – 0.0lg; Pyridoxine – 0.01g; Inositol – 0.05 g; 0.028 M Phosphate buffer 1000 ml, pH 7.8.

and lowest by *Microascus manganii* (4.0 KU/ml) in medium I. Production of extracellular keratinase during the growth of keratinofers have also been established by other workers (5, 14 - 18).

It was observed that fungi which have a strong keratinolytic ability rendered the culture medium more alkaline than the less keratinolytic ones (Table 2). It is proposed that the basis of keratinolysis is the high level of deamination which renders the medium alkaline (19,20).

Five different keratin substrates (human hair, horse hair, human nails, peacock feathers and poultry bird feathers) were used to assess preferential utilization and colonisation of keratin by the keratinophilic isolates. It was noteworthy to observe that all the isolates grew on poultry feathers, peacock feathers and human hair although with variable degree of colonization, whereas, only some of them could colonize horse hair and human nails. Jain *et al.* (1985) also found nail as a poor bait in comparison to other substrates. These differences in the degree of colonization of different baits may be attributed to the soft and hard keratins that make up their composition.

Microscopic examination of hair samples attacked by keratinolytic moulds revealed extensive damage including rupturing and blistering of hair cuticle. A characteristic effect of *Scopulariopsis brevicaulis* on human hair was the lifting of the scales of the cuticle to such an extent that the hair eventually gave a frayed appearance. Fraying has also been noted by English (1965) in non-keratinolytic fungi and by Pinetii and Lostia (1966) in dermatophytes. In case of *Chrysosporium kerationophilum* and *C. tropicum*, undulation of the hair cuticle was noted in side view, which has not been observed with other fungi. Parts of the hair heavily invaded by the fungus were markedly swollen compared to untouched parts adjacent to them, possibly because the volume of the invading mycelium and spores was greater than that of the keratin it had replaced. The extent of structural damage of hair during the growth of these fungi is probably dependent on their ability to produce keratinase — a biochemical tool for keratin utilization.

Keratinase activity of keratinophiles is considered to be one of the most important traits correlated with their capacity to invade and cause pathogenic syndromes in living organisms. *In vitro* kertain digestion by the fungi including these tests is predictive of their ability to cause infection *in vivo* and thus they may pose a formidable threat.

Besides, production of keratinases by the keratinophiles is very helpful in maintaining ecological balance. As keratinous waste from various sources is dumped in our surroundings, allowing biodegradation to occur naturally without any special treatment, the excessive dumping of such organic wastes into the terrestrial and marine ecosystems can cause untoward problems unless the operation is carefully managed and monitored. In view of this, non-pathogenic keratinophilic fungi, as observed in the present investigation, can be exploited in a controlled manner for their potential to degrade keratin and thus recycle the keratinous waste.

ACKNOWLEDGEMENTS

The senior author is grateful to Council of Scientific and Industrial Research, New Delhi, for financial assistance.

REFERENCES

1. Kunert J, *J. Ind Bot Soc* 74A (1995) 89.
2. Vanbreuseghem R *Mycologia* 44 (1952) 176.

3. Ramakrishna R, Siraj P & Prakasa Sastry CS *Curr Sci* 48 (1979) 815.
4. Lowry OH, Rosebrough NJ, Farr AL & Randall *J Biol Chem* 193 (1951) 265.
5. Yu RJ, Hormon SR & Blank F *J Bact* 96 (1968) 1435.
6. Schonborn C, *Mykosen* 11 (1983) 847.
7. Jonsson, AG *Appl Microbiol* 16 (1968) 450.
8. Dug PS & Gaucher GH, *Can J Microbiol* 19 (1973) 129.
9. Singh BS & Saksena SB *J Indian Bot Soc* 67 (1988) 78.
10. Malviya HK, Rajak RC & Hasija SK *Mycopath* 120 (1992) 1.
11. Deshmukh SK & Agrawal SC *Mykosen* 25 (1982) 454.
12. Pandey A, Agrawal GP & Singh SM *Proc Nat Acad Sci India* (1988) 559.
13. Hasija SK, Malviya H & Rajak RC *Proc Nat Acad Sci* India III, (1990) 305.
14. Yu RJ, Hormon SR & Blank F *J Invest Dermatol* 53 (1969a) 166.
15. Yu RJ, Hormon SR, Watcher PE & Blank F *Arch Biochem Biophy* 135 (1969b) 363.
16. Yu RJ, Hormon SR, Grappel SF & Blank F *J Invest Dermatol* 56 (1971) 27.
17. Rajak RC, Malviya H, Deshpande H & Hasija SK *Ind J Microbiol* 31 (1991) 243.
18. Takiuchi I, Higuchi D, Takagi H & Koga M *Sabouraudia* 20 (1982) 281.
19. Chester CGC & Mathison GE *Sabouraudia* 2 (1963) 225.
20. Zielger H & Bohme H *Dermatogische Monatsschrifi* 155 (1969) 178.
21. Jain M, Shukla PK & Srivastava OP *Mykosen*, 28 (1985) 148.
22. English MP *Trans Br Mycol Soc* 48 (1965) 219.
23. Pinetti P & Lostia AL *Monograph* 13 (1966) *Rass med Sarda Caligari.*

CHAPTER - 52

THE CELLULOLYTIC ACTIVITY OF SOME BIODEGRADING FUNGI

A. A. Fulzele and R. P. Thakre

Department of Botany, Nagpur University Campus,
Nagpur - 440 010 INDIA.

ABSTRACT

Experiment was conducted with the aim to produce organic manure through decomposition of leafy biomass from agricultural waste like banana (*Musa paradisiaca* L.) and sugarcane (*Saccharum officinarum* L.) leaves, which remain as agricultural waste after harvesting. Some of the dominant isolates of some biodegrading fungi were selected to study their comparative role in decomposition of banana and sugarcane leaves by evaluating their capacity to produce C_x cellulase.

1. INTRODUCTION

Several mesophilic and thermophilic moulds have been known to degrade plant materials at faster rate. Attempts have been made to evaluate their ability to produce extra cellular enzymes. The extracellular cellulolytic system of these moulds consists of exo-β-1,4 glucanase, endo-β-1,4 glucanase and β-glucosidase which are very essential for saccharification of native cellulase (1). Reese and his co-workers, (2) first suggested a mechanism of degradation of natural cellulose to soluble sugar based on two steps sequential process.

CELLULOSE -- C_1 - - > REACTIVE CELLULOSE -- C_x --> CELLOBIOSE -- ^{Cellobiase} --> GLUCOSE

Mesophilic and thermophilic fungi were isolated from the leafy biomass during biodegradation under the present investigation to understand their role in the cellulose degradation of leaves. It was therefore, planned to study the cellulolytic activity of some mesophilic as well as thermophilic fungi with the view to understand their role in the degradation of sugarcane and banana leaves as complex cellulosic substrate and on carboxymethyl cellulose as soluble derivative of cellulose.

2. MATERIALS AND METHODS

The utilization of highly complex cellulose substrate by fungi was studied using the method of Garret, (3) and Fergus, (4). The leaves of Banana (*Musa paradisiaca* L.) and Sugarcane (*Saccharam officinarum* L.) were selected as complex source of cellulose, while carboxymethyl cellulose (CMC) as soluble cellulose derivative.

The leaves were dried in oven at 60°C till constant weight, lg of each type of dry leaves and CMC were taken in 150 ml Erlenmeyer flask to which 50 ml of medium, either Czapek-Dox ($NaNo_3$ 2g; K_2HPO_4 1g; $MgSO_4$ 0.5g; KCl 0.5g and 1000 ml distilled water) or YpSs Emerson broth (Yeast extract 4g; K_2HPO_4 1g; $MgSO_4$ 0.5g and distilled water-1000 ml) for mesophilic and thermophilic fungi respectively was added. All flasks containing cellulosic substrates and broth were sterilized in autoclave at 15 lbs psi for 20 min. Streptomycin sulphate was added (25mg/ml) to each flask before fungal inoculation to avoid bacterial contamination. Initial pH of the broth containing different cellulosic substrates was noted.

The test fungi in the present study were:

Mesophilic fungi

(1) *Coprinus comatus*,

(2) *Trichoderma viride*

Thermotolerant fungi

(3) *Aspergillus fumigatus*,

4) *Aspergillus nidulans*

Thermophilic fungi

(5) *Humicola grisea* var. *thermoidea* and

(6) *Talaromyces duponti*.

Six days old cultures grown on Czapek-Dox and YpSs agar plates at optimum temperature was inoculated and incubated at 25±1°C and 45±1°C for mesophilic and thermophilic fungi respectively while thermotolerant fungi, at 43±1°C. The experiments were performed in triplicates.

At the desired intervals three flasks of each fungus grown on single cellulosic substrate were filered through preweighted Whatman No. 1 filter paper and dried at 60°C till constant weight was obtained. The filtrates were stored at 3°C for enzyme assay. The dry wieght of the leaves at the end of incubation was compared with the dry weight of the leaves at the beginning (lg).

Enzyme assay

The extra cellular enzyme assay of the culture filtrate was determined (Cx) by the amount of reducing sugar formed due to cellulolytic activity on CMC.

The aliquot was prepared by mixing 4ml CMC solution (0.5%), 1ml sodium acetate buffer (0.2M), 5.2 pH and 2 ml of enzyme extract. The reaction mixture was kept at 45°C in a water bath for rapid enzyme activity for 60 min, amount of reducing sugar rleased during enzymatic activity was determined by Somogyi, (5) method.

Cellulolytic activity of the filtrates was expressed in terms of the amount of total reducing sugar (mg/ml) of fungal filtrate from CMC in 1 h at 45°C.

3. RESULTS AND DISCUSSIONS

Cellulase activity of test fungi

Reducing sugar analysis and loss in dry weight of different leaves by mesophilic, thermotolerant and thermophilic revealed fungi to be good decomposers of cellulose and were able to produce cellulase enzyme.

All the six test fungi produced more reducing sugar in the medium containing CMC as compared to different leafy

Table-1. Cellulolytic activity (Cx) of different biodegrading fungi

Substrates →	Reducing sugar released								
	Sugarcane leaves			Banana leaves			CMC		
Days →	7	14	21	7	14	21	7	14	21
Name of fungi									
Aspergillus-fumigatus	0.605 (0.077)	0.79 (0.080)	0.36 (0.063)	0.55 (0.030)	0.713 (0.030)	0.3 (0)	0.126 (0.028)	0.895 (0.017)	0.76 (0.014)
Aspergillus-nidulans	0.403 (0.010)	0.643 (0.011)	0.3 (0.016)	0.425 (0.17)	0.59 (0.063)	0.48 (0.053)	1.21 (0)	0.683 (0.022)	0.605 (0.038)
Humicola grisea var. *thermoidea*	0.406 (0.041)	0.75 (0.042)	0.586 (0.048)	0.525 (0.017)	0.72 (0.014)	0.54 (0.028)	1.34 (0.113)	1.08 (0.148)	0.295 (0.003)
Talaromyces-duponti	0.465 (0.010)	0.53 (0)	0.26 (0.028)	0.36 (0.007)	0.425 (0.017)	0.135 (0.017)	0.685 (0.144)	0.536 (0.046)	0.295 (0.003)
Coprinus-comatus	0.76 (0)	0.58 (0.056)	0.555 (0.074)	0.35 (0.070)	0.575 (0.102)	0.53 (0.028)	0.43 (0)	0.67 (0.007)	0.2 (0)
Tricoderma-viride	0.505 (0.053)	0.74 (0.042)	0.4 (0.056)	0.3 (0)	0.625 (0.123)	0.2 (0)	0.21 (0.007)	0.36 (0.028)	0.25 (0.035)

Note : Values in parentheses shows standard error.

Table 2. Dry weight loss of leafy substrates

Substrates →	Dry weight loss in per cent					
	Sugarcane leaves			Banana leaves		
Days →	7	14	21	7	14	21
Name of fungi						
Aspergillus-fumigatus	14.3 (0.012)	20.3 (0.013)	27.7 (0.009)	9.4 (0.011)	14.5 (0.002)	15.1 (0.010)
Aspergillus-nidulans	4.8 (0.005)	8.0 (0.003)	12.9 (0.007)	2.3 (0.017)	6.7 (0.009)	11.3 (0.033)
Humicola grisea var. *thermoidea*	13.0 (0.003)	23.5 (0.015)	26.9 (0.025)	4.5 (0)	10.3 (0.006)	11.2 (0.002)
Talaromyces-duponti	12.8 (0.003)	14.1 (0.003)	14.9 (0.001)	10.2 (0.006)	11.2 (0.006)	12.8 (0.006)
Coprinus-comatus	3.1 (0.002)	11.1 (0.005)	14.9 (0.004)	3.2 (0.007)	6.0 (0.010)	14.1 (0.007)
Tricoderma-viride	8.2 (0.009)	11.0 (0.001)	12.6 (0.009)	7.3 (0.019)	10.3 (0.008)	12.8 (0.004)

Note : Values in parentheses shows standard error.

materials as cellulosic substrates. Maximum amount of reducing sugar was liberated from CMC medium by *H. grisea* var. *thermoidea*. Similarly *Aspergillus fumigatus* on sugarcane and banana leaves showed maximum cellulase activity. All the fungi showed more cellulolytic activity on sugarcane leaves than banana leaves (may be due to waxy coating on banana leaf surface). Cellulase activity on both the substrates was increased up to 14 days and then declined. Hajny and Reese, (6) also recorded similar observation and attributed this either to the increasingly resistant cellulose residue left after degradation of susceptible portion of cellulase or to the repression in soluble hydrolysis products. Cellulase activity on CMC was found to be increased initially up to 7 days, after which it decreased, (Table 1). Thermophilic fungi inflicted more loss in weight of dry leaves (Table 2), followed by higher cellulase activity as compared to mesophiles. Satyanarayan *et al.* (7) also reported that enzymatic abilities of thermophilic fungi are superior by way of their thermostability and better production under optimum nutritional condition as compared to mesophiles.

Cellulase activity of different thermophilic and mesophilic fungi varied considerably on cellulosic substrates and different levels of enzymatic activity recorded in this study could be attributed to the differential rates and growth of fungi observed on leafy materials and on CMC.

REFERENCES

1. Johri BN & Satyanarayana T, *Ind Sci Con* held at Madurai (Abstract.) (1986).
2. Reese ET, Siu PGH & Levinson HS, *J Bacteriol* **59** (1950) 485.
3. Garret S D, *Trans Brit Mycol Soc* **45** (1962) 115.
4. Fergus CL, *Mycologia*, **61**, (1969) 120.
5. Somogyi N, *J Biol Chem* (1952) 200.
6. Hajny GJ & Reese ET, *American Chemical Society*, Washington (1969).
7. Satyanarayana T, Jain S & Johri BN, In : *Persectives In Mycology and Plant Pathology* (Agnihotri VP, Sarabhoy AK & Dineshkumar ed) X, Mehrotra Publishing House New Delhi, (1988) 24.

CHAPTER - 53

ACTIVITY OF A THERMOPHILIC FUNGUS FROM INDIA

R.V. Shukla

Department of Botany, C.M.D. College, Bilaspur, (M.P.) - 495 001, INDIA

ABSTRACT

During studies on thermophilic microorganisms a rapid growing thermophilic conidial fungus was isolated from a sample of decomposed jute bags. Initially, this was thought to be a new species of *Chrysosporium, Sporotrichum* and *Myceliophthora.* Recently, fungus is redescribed as *Corynascus thermophilus* along with its anamorph *Myceliophthora fergusii* (original isolate). The cellulolytic ability of the fungus was determined using viscometric technique, native cellulose materials and lignin like compounds, i.e. ferulic acid, p-hydroxy benzaldehyde and vanillin.

1. INTRODUCTION

During the. course of a survey a cellulolytic, thermophilic hyphomycete was isolated from a sample of decomposed jute bag collected from Banda (U.P.). In due course of intermittent observation and identification the fungus (IMI274772) was identified as *Myceliophthora fergusii* (15).

Large number of mesophilic and thermophilic fungi are known to be cellulolytic. Several investigators have enumerated that a number of micro-organisms specific growth rates increased with increasing temperature optima (1-3). In view of dual benefit by increasing hydrolysis rate and inhibiting unwanted microbial growth at high temperature two species of thermophilic group of fungi *Thielavia terrestris* and *S. Cellulophilum* have been patented (4,5). The search for thermostable cellulases has been undertaken by Durand (2) who had compared the heat stability of mesophilic and thermophilic fungi and found that none of the tested enzyme appeared to be stable at 70°C as claimed in the patent (4).

The ability to hydrolyse modified cellulose is widespread, but to degrade native cellulose is restricted (6,7). Therefore, the cellulolytic ability of present fungus was also studied at high temperature (45°C) using both, the native cellulose materials and chemical compounds like

Carboxymethyl-Cellulose, ferulic acid, p-hydroxy benzaldehyde and vanillin in various experiments (8,9). In the present studies cellulolytic ability of thermophilic fungus *M. fergusii* is being reported.

2. MATERIALS AND METHODS

Enzyme producing capability of conidial state of test fungus *M. fergusii (C. ampulliformis)* was determined using viscometric technique as described by Agrawal (8,9). The fungus was grown in two sets of flasks, one containing Carboxy-methyl cellulose (CMC) while other containing filter paper (Whatman). The basal medium used had the composition: K_2HPO_4 - 1 gm, $MgSO_4$ $7H_2O$ - 0.3 gm, Peptone - 1 gm, Microelement solution - 1 ml, $(FeNo_3)_2.9H_2O$ - 200 μgm, $CuSO_4.5H_2O$ - 38 μgm, $ZnSO_4.7H_2O$ - 225 μgm, $MnSO_4.4H_2O$ - 24 μgm, $Na_2MoO_4.2\ H_2O$ - 40 μgm, Distilled water - 1000 ml.

The pH of the medium was adjusted to 7.0 prior to sterlization. After inoculation flasks were incubated at 45° temperature for 15 days. After incubation period the content was filtered through filter papers and filterates were centrifuged for 15 min. at 10,000 r.p.m. and stored at 3°C for further use.

The reaction mixture consisting of 6.5 ml (0.55%) CMC solution in McLi vaine's citrate buffer (pH 5.5) and 1.5 ml enzyme sample was pipetted into Oswald viscometer and viscosity was measuyred at 45°C after 30 min. interval in a water bath. The percent reduction in viscosity was calculated by the following formula:

Percent reduction in viscosity

$$= \frac{Tc - Tt}{Tc - Tw} \times 100$$

where

Tc = Flow time in second for the control.

Tt = Flow time in second for the next mixture.

Tw = Flow time in second for water.

Relative enzyme activity was calculated as —

$$REA = \frac{1000}{tv75}$$

where, Tv75 represented the time in minutes required to reduce viscosity of the reaction mixture by 75% of the initial.

For the study of utilization of insoluble native cellulose materials, the method of Garrett (10) as modified by Fergus (11) was employed. For this oven-dried native materials e.g. cloth (long cloth, pure cotton) cotton fibre (Klin absorbant cotton Morol, India) and filter paper (Whatman), each approx. 4 gm. were taken in 150 ml flasks containing 35 ml of the distilled water. These are sterlized and inoculated with fungus culture disc (6 mm diam.). After incubation period of 15 days at 45°C the content was filtered through pre-weighed filter papers. The filters were then dried and weight of the native mateiral along with mycellum at the end of incubation period was compared to its initial dry weight. The percentage loss in the substrate used was calculated (12) Table 2.

To study the utilization of lignin by the test fungus the lignin like compounds, i.e., ferulic acid, p-hydroxy-benzaldehyde and vanillin were incorporated in the modified Czapek's broth (8,9,13). The percentage of each compound used in the bassal medium as carbon source is given in the Table - 3. The 150 ml flasks containing above compound and basal medium (30 ml) were inoculated at 45°C for 5 days. After incubation period the growth of test fungus was determined and the content was filtered through pre-weighed filter papers. The filter paper were then dried and mycelial weight was calculated Table 3.

3. RESULTS AND DISCUSSION

The loss in viscosity of the reaction mixture recorded over a time period of 30-90 min. Both the culture filterate i.e., filter paper and CMC of the test fungus exhibited considerable loss in viscosity over a time period of 90 minutes. This has proven that test fungus is true cellulose decomposer showing both C_1 and C_x types of enzyme (7,9). The culture filterate obtained from the flasks having filter paper as cellulose source, showed a maximum 91.4% loss in viscosity after 15 days over a time period of 90 minutes, while in the filterate of CMC it was recorded 89% (Table 1).

Table 1. Activity of cellulose

Organism	*Myceliophthora fergusii* (*C. ampulliformis*)	
CMC % loss at		
	30 min.	74.6
	60 min.	77.0
	90 min.	89.1
R E A		33.11
Filter Paper % loss at.		
	30 min.	75.2
	60 min.	84.5
	90 min.	91.4
R E A		34.01

(1) As % loss in viscosity of reaction at time 't'.
(2) Relative enzyme units.

The test fungus could degrade cotton fibres, cotton cloth and filter paper. The growth of fungus on cotton cloth and cotton fibres was found to be fair to good and showed strong cellulolytic nature. The perusal of the data shown in Table 2 showed considerable weight loss in all the above mentioned cellulosic materials.

Table 2. Degradation of different cellulosic substances by *M. fergusii* (*C. ampullliformis*)

		Weight (in mg.)
COTTON FIBRE		
	Initial wt. substrate	400
	Substrate + Fungus	332
	% Loss in wt.	17.00
	Growth	+++
COTTON CLOTH		
	Initial wt. substrate	400
	Substrate + Fungus	354.5
	% Loss in wt.	11.35
	Growth	+++
FILTER PAPER		
	Initial wt. substrate	400
	Substrate + Fungus	328.4
	% Loss in wt.	17.90
	Growth	+++
CONTROL		24

+++ = Good
Each datum given in the table is an average of three independent determinations.

Table 3. Dry Mycelial weight (in mg.) at 45°C. After 5 days of incubation.

Organism	*Myceliophthora fergusii* (*C. ampulliformis*)		
CHEMICALS			
	p-Hydroxybenzaldehyde	0.01% (w/v)	50
	Ferulic acid	0.005% (w/v)	60
	Vanillin	0.01% (w/v)	56
CONTROL – Without carbon source			28

Each datum given in the table is an average of three independent determinations.

Lignin like substances used as the sole source of carbon for fungus growth were added in low concentrations in the basal medium (8), where fungus growth was not luxuriant but was always found more than that of its control Table 3. Medium containing vanillin and ferulic acid were prefered by the fungus in comparison to p-hydroxy benzaldehyde. Although, hydrolytic activity observed with the test organism on pure chemical compounds is not evident with the natural lignocellulose. However, for industrial application of this isolate one has to consider high productivity of lignocellulose degrading enzymes from its improved strains.

The hydrolysis of CMC suggested strong C_x enzyme which was found stable at 45°C over 15 days long incubation period. The enzymes of genetically improved *S. pulverulentum* were compared to those of *Thielavia reeseil*, and observed that celluloses from *S. pulverulentum* were not suitable for saccharification process because of their low activity on crystilline cellulose at 50°C, probably related to the thermal instability of the enzyme (14). Further Durand (2) exclaimed that screening of thermophilic cellulolytic fungi should not only based on the stability of their enzymes, but also on their activity on crystilline cellulose and further more their ability to degrade natural substrates.

ACKNOWLEDGEMENT

The author is thankful to Dr. P.C. Jain, Department of Applied Microbiology & Biotechnology, Dr. H.S. Gaur University Saugar (M.P.), India, for critical examination of the manuscript.

REFERENCES

1. Brock, TD, *Science* 158, (1967) 1012.
2. Durand, H, *Enzyme Microb Technol* 6, (1984) 175.
3. Rosenburg, SL, *Mycologiz* 70, (1978) 1-2
4. Skinner, WA & Tokuyama, F, *US Pat* 408/328, (1978).
5. Komura, I. *et al.*, *US Pat.* 4106989, (1978).
6. Thakre, RP & Johri BN, *Curr Sci* 45, (1976) 271.
7. Rees, ET & Mandels, M, *Biotechnol Bioeng* 22, (1980) 323.
8. Agrawal, SC, *Philippine Jour Sci* 98, (1969) 303.
9. Shukla, RV, Ph.D. Thesis, Gour Univ Saugar (MP). (1985)
10. Garrett, SD, *Trans. Brit Mycol Soc* 45, (1962) 115.
11. Ferqus, CL, *Mycologia* 61, (1969) 120.
12. Jain, PC & Agrawal, SC, *Trans Mycol Soc* Japan 21, (1980) 513.
13. Ander, P & Erikson, KE, *Svensk papperstidn* 78, (1975) 643.
14. Durand, H & Tiraby, G, *Proc Biocon Bichem Engineering Symp* (Ghose, T.K. ed.) IIT Delhi 51 (1980).

CHAPTER – 54

DEGRADATION OF γ-HEXACHLOROCYCLOHEXANE IN SOIL AND FOOD ENVIRONMENTS BY THE BACTERIAL STRAIN BI-100 — A BIOTECHNOLOGICAL APPROACH

A.K. Maiti, J. Datta, D.P. Modak and P.K. Ray

Enviromental Sciences Section, Bose Institute,
P-1/12 C.I.T. Scheme VII-M, Calcutta 700 054.

ABSTRACT

A potent γ-HCH degrading bacterial strain designated as BI-100 isolated from agricultural field was found to grow in mineral salts medium using γ-HCH (100 mg/L) as sole source of carbon and energy when complete mineralisation occurred at 72h of growth. On the other hand, BI-100 when applied to the soil could completely mineralise γ-HCH (1 g/sq.ft.) within 20 days. Again, γ-HCH content of drinking water, milk, peanut oil etc. could be removed completely by immobilized bacteria. Due to the inherent problem of complete removal of toxic metabolic intermediates of γ-HCH from such processed food materials attempts were made to remove γ-HCH completely from soil thereby limiting its entry to the food-chain. Interestingly, γ-HCH (1 g/sq. ft.) from soil could be reduced by 23% and 6% with free and immobilized bacterial cell respectively within 5 days when the concentration of the organism was 10^8/sq.ft.

1. INTRODUCTION

γ-Hexachlorocyclohexane (γ-HCH; lindane), an organochlorine insecticide has been used worldwise since 1940s (1). Most of the countries have prohibited the use of γ-HCH due to its toxicity and recalcitrant nature. (2) Moreover, some countries are still using γ-HCH and thereby new sites are continuously being produced. From soil γ-HCH residues enter the food chain and are found in appreciable amount in vegetables, milk, egg, meat, meat product etc. and thereby causing significant biochemical and morphological changes in skin, liver, kidney, etc. and other serious health problems (3-7). However, a number of reports are available where γ-HCH was found to be present in breast milk both in technologically developed and tropical countries (8-10). Moreover, this pesticide had been found regularly in placenta, fatty tissues and fetal organs, new borns and even in dead children (11). With this background coupled to our continuous interest on removal of pesticide residues from environment, an attempt was made for removal of γ-HCH from soil and food stuffs with a potent γ-HCH degrading bacterial strain.

The bacterial strain BI-100 used in the present investigation was isolated from the agricultural field (12) and maintained on mineral agar plates supplemented with γ-HCH by regular subculturing.

2. MATERIALS AND METHODS

The bacteria was grown in nutrient broth medium for 24h and harvested by centrifugation at 10,00 X g for 10 min. The cells were then washed thrice with Trismaleate buffer (50mM, pH 6.0). Washed cells (10g) suspended in Tris maleate buffer (20 ml, 50 mM, pH 6.0) was mixed with sodium alginate solution (80 ml, 2% w/v final concentration). The resultant suspension was then dropped into a magnetically stirred solution of calcium chloride (0.1 M) through a peristaltic pump to have spherical beads of uniform size. The beads were stirred further for 1 h, finally washed and kept at-20°C.

BI-100 cells were applied to the soil at a concentration of 10^8/sq.ft. alongwith γ-HCH (1 g/sq.ft.). The same amount of cells entrapped in calcium alginate matrix were applied to the field separately under the same condition. The control field received no free or immobilized bacteria. Soil samples were collected at different time periods and assayed for residual γ-HCH.

Liquid food materials like water, milk, peanut oil etc. deliberately contaminated with γ-HCH (25 mg/1) were passed through a packed column of immobilized bacteria. The test materials were allowed to be resident over the matrix for different time periods. Residual γ-HCH content was measured by GC analysis of the eluates as described in our previous communication (12).

3. RESULTS AND DISCUSSION

The bacteria could completely mineralise γ-HCH (100 mg/1) in mineral salts medium under laboratory condition within 72 h of growth (12). The test organism when applied to the soil along with γ-HCH 28% degradation was observed at 7 day which was increased with time and ultimately complete mineralisation was observed at 3 weeks (Table 1). However, no such reports are available where assimilation of such high concentration of γ-HCH had been observed by a single microbe in the soil environment. Furthermore, γ-HCH even at a concentration of 1 ug/g soil had been shown to have an adverse effect on microbial population upto 3 months (13). For management of γ-HCH in fields where continuous contamination is occurring bioremediation was tried with immobilized bacteria with a view to get the benefit of killing of pests followed by the slow degradation of the pesticide in the agricultural fields. Interestingly, only 6% reduction of γ-HCH (1 g/sq.ft.) was observed within 7 days with Ca-alginate entrapped bacterial beads which is a quite slow degradation process compared to that with free living bacteria (23% degradation).

Table 1. Removal of γ-HCH from soil by the Bacterial Strain BI- 100

Sample	Time (Day)	Removal of γ-HCH (%)
Soil + BI-100	0	0
Soil + BI-100	7	28
Soil + BI-100	14	80
Soil + BI-100	21	100

Evidently, this approach of removal of γ-HCH from soil is quite fruitful. Nevertheless, this finding of degradation of γ-HCH in pesticide acclimated soil will not only be useful for removal of γ-HCH but also in maintaining the proper ecological balance without hampering the action of the pesticide on the target pests.

However, γ-HCH content of drinking water, milk, peanut oil etc. could be reduced by passing them through the column of immobobilized bacterial beads. This process of decontamination of liquid food-stuffs from γ-HCH is under progress.

REFERENCES

1. Lal R & Saxena DM, Microbiol Rev 36 :(1972) 85.
2. Johnson LM & Talbot HW, Experientia, 39 (1983), 1236.
3. Chandra H, Pangtey BS, Modak DP, Singh KP, Gupta BN, Bharti RS, & Srivastava SP, Bull Environ Contam Toxicol, 48 : (1992) 295.
4. Krishna G, Ram MD, Anand M, & Ray PK, Environ Ecol, 7 (1989) 571.
5. Pandey N, gundevia F. and Ray PK, Muation Res, 242 : (1990) 121.
6. Nagata K, Huang CS, Hamilton BJ, Carter and Narahashi T, Brain Res. 738 : (1996) 131.
7. Langer P, & Tajtakova M, Vintr Lek 42 : (1996) 615.
8. Mes J, Davies DJ, Turton D, & Sun WF, Food Addit Contam (England), 3(4) : (1986) 313.
9. Larsen BR, Torrio LB, Nilsson T, Iacovella N, Di-Domenico A, Montagana M, Facchetti S. Ecotoxicol Environ Saf 28: (1994) 1.
10. Schlaud M, Seidler A, Sal Je A, Behrendt W, Schwartz FW, Ende M, Knoll A, Grugel C, J Epidemiol Community Health 49 : (1995) 1995.
11. Bosse U, Bannert N, Niessen KH, Teufel M & Rose I, Zentralbl Hyg. Unweltmed 198: (1996) 331.
12. Datta J, Ghosh S. Maiti AK, Das SK, Modak DP, Bhattacharyya P, Ghosh A, Mandal NC, Chakrabarty PK & Ray PK, J Sci Ind Res (in press).
13. Tu CM, Arch Microbiol. 105 : 131.

SECTION – 4

MEDICAL BIOTECHNOLOGY

CHAPTER – 55

EXACERBATED RESPONSES OF MOUSE MACROPHAGE WITH *Mycobacterium habana*, A VACCINE CANDIDATE FOR LEPROSY

S.P. Singh[1], *V. Agarwal*[1], *R.W. Yadava*[2], *H.P. Gupta*[3], *N.B. Singh*[3] *and V.M.L. Srivastava*[1]

Divisions of [1]Biochemistry and [3]Microbiology, Central Drug Reseach Institute, Lucknow, and [2]Division of Immunology, Sanjay Gandhi Post Graduate Institute of Medical Sciences, Lucknow, INDIA.

ABSTRACT

Mycobacterium leprae and *M. tuberculosis* are known to establish and proliferate in macrophages by down regulating the effector system of infected phagocytes. *Mycobacterium habana*, a candidate mycobacterial vaccine, has earlier been reported to protect mice against infectious challenge with these pathogens. With a view to understand the mechanism involved in affording protection by the vaccine, the macrophage effector system following intradermal vaccination of mice was investigated. Peritoneal exudate cells collected from the vaccinated animals exhibited greater production of O_2^-, H_2O_2 and NO as well as higher activities of lysozyme, β-glucuronidase and acid phosphatase as compared to the normal cells. Also, these cells generated IL-1β, TNF and IL-6 at much faster rates than did the normal phagocytes. The intensity of this stimulation gradually declined with time and, on 30th day of the vaccination the changes in most of the parameters became statistically insignificant. Administration of a booster dose of the vaccine, however, markedly enhanced all the activities. *M. habana* vaccine, therefore, appears to exert its prophylactic action by potentiating both oxygen dependent and independent effector arms of the macrophage. The vaccine appears to stimulate immune apparatus also as judged by the enhancement in the cytokines production.

1. INTRODUCTION

Macrophages are associated with every stage of immune response. They provide non-sepcific defence against invadors through the release of powerful weaponds like reactive oxygen and nitrogen metabolities and a variety of acid hydrolases. They are participate in the initiation of T-cell activation for humoral and cell mediated immune (CMI) responses by processing and presenting antigen(s). In lepromatous leprosy patients macrophages have defective normal functions and therefore, act as a lawn for the growth of *Mycobacterium leprae*. Consequently, these defective phagocytic cells remain loaded with large number of the bacilli in these patients (1). There is a compelling obdy of evidence to believe that the failure of the subdued macrophages to kill the bacteria and to initiate CMI-response is mainly due to their inability to generate respiratory burst (2,3).

Vaccines and immunomodulators can play a significant role in enhancing and/or potentiating immune system of the body. Interstingly, a vaccine prepared from *Mycobacterium habana* (*M. simiae* serovar-1) protects mice against infections challenge with *Mycobacterium tuberculosis* $H_{37}Rv$ and *M. leprae* (4,5). Also, the vaccine generates delayed type hypersensitivity (DTH) in animals against the antigens of *M. leprae* and *M. tuberculosis* (6,7). It is therefore conceivable that *M. habana* vaccine may be able to provide full activation of the immunosurveillance system in the host for warding off the mycobacterial infections. As a part of comprehensive work plan to establish the mode of action of this vaccine we report here the impact of *M. habana* vaccine on the effector activities of murine macrophages.

2. MATERIALS AND METHODS

Growth and preparation of the vaccine : *M. habana* (*M. simiae* serovar - 1, TMC 5135) was obtained from TMC collection centre, Sarnac Lake, New York. The culture was maintained in Lowenstein Jensen medium and the bulk growth was obtained in Sauton's liquid medium. Vaccine from the harvested cells was prepared by irradiation at 300 K rads from a ^{60}cobalt source.

(a) Immunization of animals

Twenty female mice (AKR strain) obtained from the Institute's (CDRI, Lucknow) animal house were divided into five equal groups (A to E). The animals of groups B, C and D were treated intradermally with 63.3 μg of *M. habana* protein ($\equiv 10^8$ cells) 10, 20 and 30 days before macrophage harvest, respectively. The animals of group E were treated with two such doses 30 and 10 days before the harvest. Mice of group A, on the other hand, received saline only and served as controls. The animals were housed in plastic cages with free access to rodent pellet diet (Lipton India Ltd., Chandigarh) and water *ad libitum*.

(b) Isolation of peritoneal exudate cells (PEC)

On day 30 of the start of vaccination the animals were anaesthesized with ether and PEC isolated by the method of Mosier *et al.*, (8). In brief, the animals were injected intraperitoneally with 5 ml RPMI-1640 medium containing penicillin G (100 U/ml), streptomycin (100 μg/ml) and gentamicin (40 μg/ml) (henceforth called as RPMI) as well as heparin (5 U/ml). The cells from the lavage were collected by centrifugation and their viability was estimated by capability to exclude trypan blue. The cells from two animals of each group were used for measuring the release of O_2^-; H_2O_2, NO and cytokines while those from the other two of each group were employed for the assay of enzymes activities. Four similar experiments were performed.

(c) Generation and assay of cytokines

The peritoneal cells were suspended to a strength of 1 x 10^6 cells/ml in RPMI medium containing 10% fetal calf serum (CRPMI). 200 μl of the cell suspension were transfered to the wells of a 96-well culture plate and incubated at 37°C in a humid CO_2 incubator. After 4 h the medium was aspirated and the cells were washed twice with RPMI. 200 µl of fresh CRPMI were added to the cells and the plate was returned to the incubator. After 48 h the medium was filtered through 0.2 µm membrane filters and stored at -20°C in small aliquots.

Concentration of interleukin-6 (IL-6) and tumor necrosis factor (TNF) in the filterate was determine by biological assays using hepatoma B-9 and fibroblast L-929 cell lines showing sensitivity to the respective cytokines (9,10). The level of interleukin-1β, on the other hand, was assayed with an ELISA kit (Genzyme, USA) (11).

(d) Assay of O_2^-, H_2O_2 and NO

200 µl of the cell suspension in CRPMI were dispensed into 5 wells of one lane of a 96-well cultue plate. After 4 h the nonadherent cells were aspirated and the adherent macrophages were washed twice with RPMI. To the wells of row A was added cytochrome-c (0.2 mg) containing superoxide dismutase (50 U) whereas to the wells of rows B and C was added the solution of cytochrome-c only. After 2 h of incubation in a humid CO_2 incubator optical density was recorded at 550 nm (12). The medium obtained from rows D and E was assayed for H_2O_2 (13) and NO (14) respectively. The cells after removing the medium were washed with normal saline and dissolved in 0.1 N-NaOH. The solution was employed for protein measurement (15).

(e) Disruption of macrophages and assay of enzymes

The suspension of PEC in CRPMI was allowed to adhere to the wells of a 24-well plate and washed as described for cytokines. Subsequently 1.0 ml of KCl (1.15% w/v) was added to each well. The cells were disrupted by sonication at 20 KHz for 4 x 30 sec with successive one minute cooling intervals. The sonicate was centrifuged at 9,000 g for 30 min and the supernate was employed for assaying lysozyme (16), acid phosphatase (17) and β-glucuronidase (18). Protein estimation was done by Folin phenol reagent using bovine serum albumin as a standard (15). Statistical analysis of the data was performed by student 't' test.

3. RESULTS

Macrophages obtained from the peritoneal cavity of mice immunized with M. Habana showed greater production of O_2^-, H_2O_2 and NO as compared to those collected from the unvaccinated animals (Table 1). Enhancement in the production of these microbial effectors on day 10 was in the range of 48 to 70%. This increase gradually declined in magnitude with time and became insignificant statistically on day 30. Adminstration of a booster dose on day 20, however, activated the macrophages further and induced a sharp acceleration in the rate of production of all the three metabolites. On day 10 of the booster dose the production of O_2^-, H_2O_2 and NO exhibited an enhancement of 177, 91 and 59% respectively (Group A) as compared to 70, 50 and 48% on a similar day of the single dose of the vaccine (Group B).

The vaccination also caused marked enhancement in the activities of three enzymes studied, namely lysozymes, β-glucuronidase and acid phosphatase (Figure 1). On day 10 of the single dose all the three enzymes registered almost similar degree of activation (111-137%). The effect gradually with time and, on day 30 the stimulation in the activities of acid photophatase and lysozyme became insignifi-

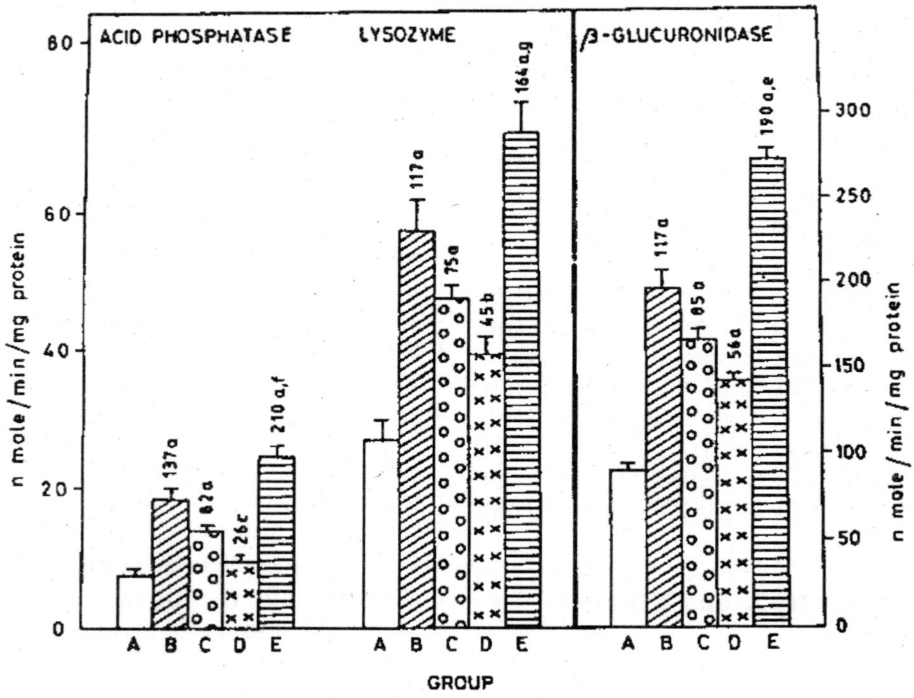

Figure 1. Activities of lysosomal enzymes in macrophages isolated from unvaccinated and *M. habana* vaccinated mice. Data are Means ± S.D. of 8 animals. Numbers above bars denote % enhancement with respect to control (group A). Details of groups (A, B, C, D and E) and p values (a to g) are given under Table 1.

cant. Administration of a booster dose showed a dramatic effect and enhanced and enzymes activities to the tune of 164 to 210%.

The release of cytokines IL-1β, IL-6 and TNF by the macrophages was also greatly enhanced upon the vaccination (Figure 2). On day 10, the stimulation in the release of these cytokines was 103, 187 and 148% respectively. There was a gradual decline in the magnitude. Thereafter and, on day 20 it became statistically insignificant in case of IL-1β. The booster dose again had an exacerbated impact on the production rates of these cytokines. Amongst the three cytokines studied, IL-6 showed most pronounced acceleration at every stage of the measurement.

4. DISCUSSION

Controversial status of BCG vaccine against tuberculosis and leprosy has aroused an all round interest for search of some other immunogenic strain(s) of mycobacteria to combat these infections. This has become all the more important due to resurgence of AIDS, increasing proportion of resistant bacilli and appearance of new and unresponsive forms of tuberculosis due to *M. avium* intracellulare scrofulaceum complex (MAIS), both in developed and developing countries (19-22).

In this endeavour we have identified one strain of a typical mycobacterium; *M. habana* which protects animals against

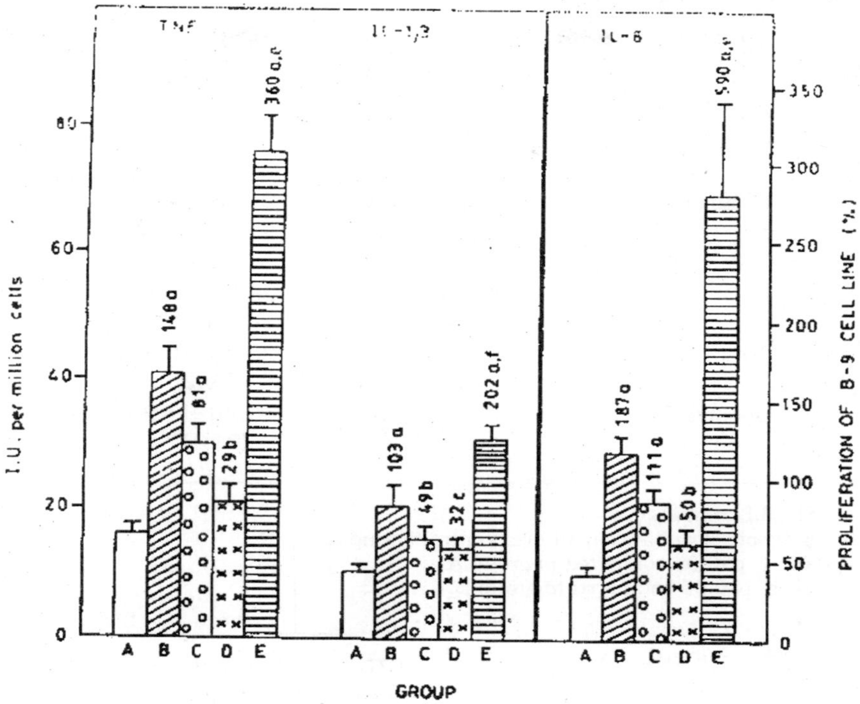

Figure 2. Impact of *M. habana* vaccination on the production of cytokines by murine macrophages. Data are means ± S.D. of 8 animals. Numbers above bars denote % change with respect to control (Group A). Details of various groups (A to E) and p values (a to g) are given under Table 1.

challenge with *M. tuberculosis* or *M. leprae* (4,5). This vaccine has been recognised as a potential one and has been granted permission by the Drugs Controller of India for conducting clinical trials. It is imperative to understand the mechanisms involved in affording protection by this vaccine. The present work is a prelude for such a study.

It has been established beyond doubt that killing of wide variety of infectious organisms by macrophages is mediated by two types of effector mechanisms; the oxygen dependent and oxygen independent (23, 24). Among these two the former due to inimical effects of its metabolic intermediates to pathogens is considered more important in overcoming pathogenic consequences. In the *in vitro* conditions *M. leprae* is highly susceptible to H_2O_2 and particularly to its halide products (6, 25). However, when *M. leprae* is ingested by phagocytic cells, it fails to stimulate the production of O_2^- and H_2O_2 (2,3). Thus like many other infectious organisms, *M. leprae* also appears to survive in the host tissues by down-regulating the respiratory burst activity. The potentiated respiratory burst activity due to *M. habana* vaccination (Table 1) therefore appears to enhance cytotoxic activity of the macrophages leading to protection against challenge organisms.

Evidences indicate that *M. leprae* resides in phagolysosomes which are known to contain a battery of hydrolytic

Table 1. Impact of *M. habana* vaccination on the production of reactive oxygen and nitrogen species by mouse macrophages

Group	Days Post treatment	n mole/h/mg protein O_2^-	H_2O_2	NO
A	Control	15.23 ± 0.92	21.38 ± 0.68	1.53 ± 0.07
B	10	25.88 ± 1.57 (69.9)[a]	32.08 ± 1.57 (50.0)[a]	2.24 ± 0.09 (47.8)[a]
C	20	21.10 ± 0.83 (38.5)[a]	28.25 ± 1.41 (32.1)[a]	2.09 ± 0.06 (37.4)[a]
D	30	19.05 ± 0.62 (25.1)[b]	25.48 ± 0.85 (19.2)[b]	1.76 ± 0.08 (16.1)[c]
E	Booster (10)	42.30 ± 1.58 (176.6)[a,e]	40.86 ± 1.20 (91.1)[a,e]	2.42 ±0.03 (59.4)[a, g]

Data are Means ± SD of 8 animals
Data in parenthesis denote % enhancement over control (group A)
a. $p<0.001$, b. $p<0.01$, c. $p>0.05$ compared to group A
e. $p<0.001$, f. $p<0.01$, g. $p>0.05$ compared to group B.

enzymes (26, 27). It therefore seems probable that *M. leprae* avoids lysosomal attack either by inactivation or down regulation of hydrolytic enzymes. Interestingly, *M. habana* vaccine activates this system of the macrophages also (Figure 1) and, thereby may induce faster digestion of infectious organisms.

As far as the pattern of proinflammatory cytokines in leprosy is concerned, the levels of IL-1, IL-6 and TNF (all produced by the macrophages) are reported to be quite low (28, 29). Interestingly, vaccination with *M. habana* markedly accelerates the release of these cytokines by peritoneal macrophages (Figure 2). These cytokines are known to stimulate the release of ACTH hormone by their direct action on hypothalamus. ACTH, in turn, induce kidney to produce corticosteroides. IL-1 and IL-6 are also known to enhance the formation of acute phase proteins by the liver. However, the mechanism by which these events induce antimycobacterial effect, is a matter of further investigation. Nonetheless it is very clear from the data that *M. habana* vaccine reverses the situation of the proinflammatory cytokines which in leprosy patients are produced in very low quanities.

It may therefore be concluded that *M. habana* vaccine activates both oxygen dependent and independent effector arms and also enhances cytokines production. This may be important in endowing host macrophages capacity to kill bacteria and hence, this may act as one of the possible mechanisms for prophylactic action of this vaccine.

ACKNOWLEDGEMENT

SPS and RWY are grateful to the Council of Scientific & Industrial Research, New Delhi for awarding them a research associateship and a senior research fellowship respectively. VA is thankful to the University Grant Commission, New Delhi for providing her a junior research fellowship.

REFERENCES

1. Thangaraj RH & Yawalkar SJ, Leprosy for Medical Practitioners and Paramedical Workers, 4th edition, Ciba Geigy Ltd, Basle (1989).
2. Holzer TJ, Nelson NE, Schauf V, Cripsen RG & Anderson BR, Infect. Immun. 51 (1986) 514.
3. Sharp AK & Banerjee DK, Clin Exptl Immunol 60 (1985) 203.
4. Gupta HP, Singh NB & Kumar A, Biol Memoirs, 13 (1987) 174.
5. Gupta HP, Singh NB, Mathur IS & Gupta SK, Ind J Expt Biol 17 (1979) 1190.
6. Singh NB, Gupta HP, Srivastava A, Kaul SM & Shukla OP, Biol Memoirs, 14 (1988) 199.
7. Singh NB, Srivastava A, Gupta HP, Kumar A & Chaturvedi VK, Ind J Leprosy, 60 (1988) 407.
8. Mosier DE, In : Methods in Enzymology (Sabato GD, Langone JJ & Vunakis HV eds), Academic Press, New York 108 (1984) 294.
9. Mossmann T, J Immunol Methods, 65 (1983) 55.
10. Saxena KC, Maheshwari RK & Dhawan BN, Cell Mediated Immunity in Tropical Diseases, Coomer Printers, Lucknow, India (1990).
11. Hogguist KA, Nett MA, Sheehan KCG, Pendleton KD, Schreiber RD & Chaplin DA, J Immunol 146 (1991) 1534.
12. Babior BM, Kipnes RS & Curnutte JT, J Clin Invest 52 (1973) 741.
13. Khan SH, Emerit T & Fringold J, Free Rad Biol Med, 8 (1990) 339.
14. Green LC, Wager DA, Glogowski J, Skipper PL, Wishnok JS & Tennenbaum SR, Anal Biochem, 126 (1982) 131.
15. Lowry OH, Rosebrough NJ, Farr AL & Randall J, J Biol Chem, 193 (1951) 265.
16. Litwack G, Proc Soc Exptl Biol Med, 89 (1955) 401.
17. Moss DW, In : Methods of Enzymatic Analysis (Bergmeyer J & Gass M eds.), Verlag Chemie, Basel, 4 (1984) 92.
18. Stahl PD & Fishman WH, In : Methods of Enzymatic Analaysis (Bergmeyer HU ed), Verlag Chemie, Weinheim (1984) 246.
19. Rieder HI, Cauthen GM, Kellery GD, Block AB & Snider DE, J Am Med. Assoc. 262 (1989) 385.
20. Quinn TC, Rev Infect Dis, 11 (1989) S 379.
21. Smith DW, Protective effect of BCG in Experimental Tuberculosis, S. Karger, Basel (1985).
22. Andersen P, Infect Immun, 62 (1994) 2536.
23. Babior BM, N Engl J Med, 298 (1978) 659.
24. James SL, Exptl Parasitol, 73 (1991) 223.
25. Klebanoff SJ & Shepard CC, Infect Immun, 52 (1984) 534.
26. Steinhoff U, Golecki JR, Kazda J & Kaufmann SHE, Infect immun, 57 (1987) 1008.
27. Marolia J & Mahadevan PR, Ind J Leprosy, 56 (1984) 776.

28. Jayaraman T, stimulation of antigen reactive T cells in lapromatous laprosy : Role of lymphocyte growth factor, Ph.D. thesis submitted to All India Institute of Medical Sciences, New Delhi (India) (1988).

29. Yamamura M, Ulemura K, Deani RJ, Weinberg TH, Rea TH, Bloom BR & Modlin RL, Science, 254 (1991) 277.